International Symposium on Dynamic Deformation and Fracture of Advanced Materials

(D2FAM 2013)

Journal of Physics: Conference Series Volume 451

Loughborough, United Kingdom
9-11 September 2013

ISBN: 978-1-62993-075-6
ISSN: 1742-6588

Printed from e-media with permission by:

Curran Associates, Inc.
57 Morehouse Lane
Red Hook, NY 12571

Some format issues inherent in the e-media version may also appear in this print version.

Printed by Curran Associates, Inc. (2013)

Institute of Physics
Dirac House, Temple Back
Bristol BS1 6BE UK

Phone: 44 1 17 929 7481
Fax: 44 1 17 920 0979

techtracking@iop.org

Additional copies of this publication are available from:

Curran Associates, Inc.
57 Morehouse Lane
Red Hook, NY 12571 USA
Phone: 845-758-0400
Fax: 845-758-2634
Email: curran@proceedings.com
Web: www.proceedings.com

TABLE OF CONTENTS

PAPERS

EXPERIMENTAL STUDIES

CRACKS IN MATERIALS AND STRUCTURES

NUMERICAL STUDIES

SPECIAL SESSION ON PARAMETRIC RESONANCE, VIBRO-IMPACT AND RELATED PHENOMENA

BIOMEDICAL PROBLEMS

INDUSTRIAL PROCESSES

Journal of Physics: Conference Series **451** (2013) 011001 doi:10.1088/1742-6596/451/1/011001

Editorial

V Silberschmidt
Wolfson School of Mechanical and Manufacturing Engineering,
Loughborough University, Loughborough, Leicestershire, LE11 3TU, UK
E-mail: V.Silberschmidt@lboro.ac.uk

Intensification of manufacturing processes and expansion of usability envelopes of modern components and structures in many cases result in dynamic loading regimes that cannot be presented adequately employing quasi-static formulations of respective problems of solid mechanics. Specific features of dynamic deformation, damage and fracture processes are linked to various factors, most important among them being: a transient character of load application; complex scenarios of propagation, attenuation and reflection of stress waves in real materials, components and structures; strain-rate sensitivity of materials properties; various thermo-mechanical regimes. All these factors make both experimental characterisation and theoretical (analytical and numerical) analysis of dynamic deformation and fracture rather challenging; for instance, besides dealing with a spatial realisation of these processes, their evolution with time should be also accounted for.

To meet these challenges, an International Symposium on Dynamic Deformation and Fracture of Advanced Materials D2FAM 2013 was held on 9–11 September 2013 in Loughborough, UK. Its aim was to bring together specialists in mechanics of materials, applied mathematics, physics, continuum mechanics, materials science as well as various areas of engineering to discuss advances in experimental and theoretical analysis, and numerical simulations of dynamic mechanical phenomena. Some 50 papers presented at the Symposium by researchers from 12 countries covered various topics including: high-strain-rate loading and deformation; dynamic fracture; impact and blast loading; high-speed penetration; impact fatigue; damping properties of advanced materials; thermomechanics of dynamic loading; stress waves in micro-structured materials; simulation of failure mechanisms and damage accumulation; processes in materials under dynamic loading; a response of components and structures to harsh environment. The materials discussed at D2FAM 2013 ranged from traditional ones such as metals, alloys, polymers and composites to advanced and emerging materials, such as foams, cellular materials and metallic glasses, as well as bio-materials.

Within the framework of the Symposium, a Special Session "Parametric Resonance, Vibro-impact and Related Phenomena" was organised by partners of the FP7 IAPP project PARM-2: "Vibro-impact machines based on parametric resonance: Concepts, mathematical modelling, experimental verification and implementation." The Session focused on the topics, directly related to the project: excitation, stabilization, control and applications of parametric resonance (PR); multiple degrees of freedom of PR-excited systems; basic principles of PR-based macro and micro tools; design and technological aspects of PR-based machines; vibro-assisted machining; fatigue under high-amplitude vibro-impact conditions and corresponding optimal design; localisation near defects in dynamic response of elastic lattices and structures; dispersive waves and dynamic fracture in non-uniform lattice systems; thermally induced surface-breaking cracks, etc.

This issue presents a selection of research papers presented at the International Symposium on Dynamic Deformation and Fracture of Advanced Materials D2FAM 2013. The Symposium Organisers would like to acknowledge its sponsors: Institute of Physics, International Centre of Vibro-Impact Systems and Marie Curie Action: Industry-Academia Partnerships and Pathways of the Seventh Framework Programme (FP7) of the European Commission (PARM-2 consortium). The PARM-2 consortium sponsored twenty scholarships for early-stage researchers to participate in this Symposium.

Journal of Physics: Conference Series **451** (2013) 011002 doi:10.1088/1742-6596/451/1/011002

Peer review statement

All papers published in this volume of *Journal of Physics: Conference Series* have been peer reviewed through processes administered by the proceedings Editors. Reviews were conducted by expert referees to the professional and scientific standards expected of a proceedings journal published by IOP Publishing.

Characterization of sandwich panels for indentation and impact

M Shazly, Y Bahei-El-Din[1] and S Salem
Centre for Advanced Materials
The British University in Egypt
El Sherouk City, Misr-Ismailia Road;
Cairo, Egypt, Postal code: 11837, P.O. box 43.

E-mail: ybahie@bue.edu.eg

Abstract. The integrity of sandwich structures which are susceptible to impact may deteriorate significantly due to collapse of the core material and delamination of the face sheets. The integration of a thin polyurethane interlayer between the composite face sheet and foam core is known to protect the core material and substantially improve the resistance to impact. The objective of the present work is to characterize the response of sandwich panels, as well as that of the constituents to impact. In particular, the response of polyurethane and foam samples under a range of quasi-static and dynamic loading rates is determined experimentally. Furthermore, the response of sandwich panels to quasi-static indentation and low velocity impact is examined to quantify the extent of damage and how it is affected by the integration of polyurethane interlayers in their construction. This information is useful in the modelling of high velocity impact of sandwich panels; an effort which is currently underway. The results illustrate the benefit of using polyurethane interlayers within the construction of sandwich panels in enhancing their performance under quasi-static indentation and impact loads.

1. Introduction

Motivated by the desire to develop damage-tolerant composites and sandwich structures for wind turbine blades, the present paper discusses an experimental program for quantification of the performance of these structures and their constituents. Of particular interest is the behaviour of sandwich panels since they are used extensively in the designs of modern wind turbine blades. It is well known that standard sandwich designs, which consist of laminated face sheets bonded to structural foam core often exhibit collapse of the core under both quasi-static and dynamic compressive loads. Collapse of the foam is usually accompanied by delamination of the face sheet when bending moments are dominant [1].

Under lateral impact and blast loads, induced compression waves propagate through thickness of the sandwich structure and cause permanent compression of the foam core [2]. Recent work by Bahei-El-Din et al. [3-5] showed that permanent compression of the foam core can be partially suppressed if a thin membrane of a flexible material is inserted between the face sheet and the core on the loading side. Various types of materials were investigated for the proposed membrane including polyurethane, polyurea, elastomeric foam and their combinations. Massab`o [6] studied the interaction

[1] To whom correspondence should be addressed.

of multiple damage mechanisms in multilayered and sandwich structures subjected to static and dynamic loading conditions and showed that these loadings on laminated and multilayered systems affect crack growth characteristics and energy absorption. Abrate [7] introduced some models capable of predicting the ballistic limit and the extent of damage in composites and sandwich structures. Gupta and Shukla [8] studied the performance of sandwich structures subjected to high velocity impact accompanied with varying service temperatures. Their results showed a significant decrease in flow stress with increasing the service temperature. Gardner and Shukla [9] studied the effect of implementing a polyurea interlayer behind the foam core and in front of the back face sheet using a shock tube; the results obtained emphasized that implementing a polyurea sheet will reduce the back face deflection and in-plane strain, which lead to enhancing the overall blast performance.

Review of the literature on this subject indicates that the current state of the art informs that implementing a ductile interlayer in composite sandwich structures enhances their structural performance. While several trials have shown that introducing polyurethane in the design of sandwich panels leads to significant improvements, there is a lack of sufficient experimental data for sandwich panel to quantify it. Consequently, the current work aims at investigating the properties of a modified sandwich structure, where a flexible polyurethane (PUR) layer is introduced.

The paper examines the behaviour of sandwich panels and their constituents under quasi-static and dynamic loading conditions, with focus on how effective the inclusion of PUR interlayers is in enhancing the resistance to indentation and impact. The paper first describes in Section 2 the materials involved in the present work. Section 3 discusses the experimental methods and tools involved. Section 4 presents results obtained along with discussions, followed by conclusions in Section 5.

2. Materials

Conventionally designed sandwich plates consist of laminated fibrous composite face sheets bonded to closed cell structural foam core, figure 1a. The modified sandwich design, on the other hand, inserts a polyurethane (PUR) interlayer between the face sheet and the foam core on one side where the load is expected, figure 1b.

E-glass/epoxy composites were used for fabrication of the composite face sheets of the sandwich panels. The glass fibres were obtained from Selcom Multiaxial Technology, Italy as nonwoven fabric, and the epoxy resin was provided by Huntsman, Switzerland. The core material used in the sandwich panels is Divinycell P, a closed cell structural foam provided by DIAB. Axson Technologies of France provided the polyurethane (PUR) used in modified sandwich designs, figure 1b. Table 1 lists the material data received from respective supplier.

3. Experimental Work

3.1. Uniaxial quasi-static tests

To quantify the mechanical properties, samples of the above materials were prepared and tested under uniaxial tensile load. Composite specimens are fabricated using both hand layup and vacuum infusion methods. Tensile tests performed on unidirectionally reinforced specimens provided a fibre volume fraction of 0.542 when the measured longitudinal Young's modulus was compared to that estimated from the Mori-Tanaka estimate [10]. This was verified by comparing the measured and calculated overall tensile strength in the longitudinal direction and the overall density. Table 2 shows measured and calculated elastic properties of E-glass epoxy unidirectional composites produced by the vacuum infusion technique.

Samples of polyurethane (PUR) are prepared from two parts, which are mixed and cast in moulds prepared to provide the net shape of the samples. The latter are cylindrical samples with a diameter of 38 mm and height of 6 mm. The compression tests are performed at various strain rates for two sets of boundary conditions, unconfined and laterally confined. In addition, the friction between the loading platens and the PUR samples is reduced in one group of samples by means of a lubricant. Foam samples having a diameter of 50 mm and thickness of 20 mm are cored from Divinycell P150 structural foam sheets and tested at different strain rates under the quasi-static regime.

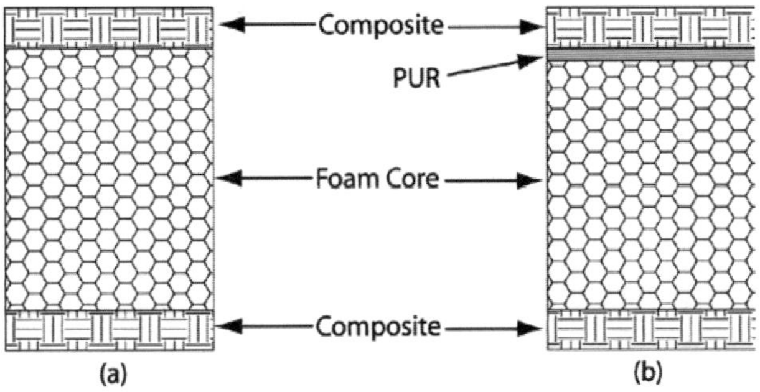

Figure 1. Sandwich construction, (a) Conventional design, (b) Modified design.

Table 1. Material data

Nonwoven E-Glass			
Designation	UNIE1050M50	ETXL 1200 – SBIL	ETXL 900
Fabric Arrangement (areal density g/m^2)	0° (1005) 90° (50) Random (50)	0° (595) +45° (303) -45° (303)	+45° (224) 90° (452) -45° (224)

Divinycell P Structural Foam		
Density (kg/m^3)	Compressive Strength (MPa)	Compressive Modulus (MPa)
150	9.4	152

Polyurethane Casting Elastomer UR5801 /UR5850		
Hardness (Shore A1)	Tensile Strength (MPa)	Elongation at Break (%)
50	5	1100

Table 2. Elastic properties of E-glass/epoxy unidirectional composites

ρ (kg/m^3)	E_1 (GPa)	$E_2 = E_3$ (GPa)	$G_{12} = G_{13}$ (GPa)	G_{23} (GPa)	$\nu_{12} = \nu_{13}$	ν_{23}
1848*	42*	10.33	4.31	3.77	0.238	0.37

*** measured**

3.2. Split Hopkinson Pressure Bar Tests

The Split Hopkinson Pressure Bar (SHPB) apparatus used in the present work consists of an operated gas gun, a striker, an incident bar and a transmitted bar made of aluminium alloy. A data acquisition system consisting of a 120 ohm Wheatstone bridge connected to an amplifier and a PC-Based oscilloscope is employed in the present work. However, testing of soft materials presents a challenge to the conventional SHPB. Chen et al. [11] have investigated techniques for dynamic testing of soft materials and suggested modifications to be made to the conventional split Hopkinson pressure bar in order to test low-strength, low-impedance materials. A hollow transmission bar made of aluminum alloy together with a pulse shaper are used to measure the dynamic mechanical responses of soft materials such as PUR and foam tested in the present work. In addition, one dimensional wave analysis of signals recorded, i.e. incident, reflected and transmitted, is modified to account for the new arrangement. Foam cylindrical specimens, which are 10 mm long and 18 mm in diameter are tested at different levels of strain rates. Samples of PUR having a diameter of 5 mm and 1 mm or 2 mm thickness are cored from sheets of the same thickness. All tests are carried out using copper pulse shaper and grease between the surfaces of the specimen and the ends of the bars.

3.3 Indentation Tests

Indentation tests are performed according to ASTM D6264–98 on sandwich panels. The purpose of the experiments is to measure the resistance to damage under the effect of a concentrated load applied at the centre of the specimens by means of an indenter with a hemispherical tip. The rate of loading is slow enough to qualify for quasi-static loading conditions.

Square specimens with a side length of 152 mm are cut from larger panels which were fabricated by both hand layup and vacuum infusion. One set of specimens contained sandwich construction with PUR interlayers on the indentation side, figure 1b. In this case and to insure flow of the resin through the specimen thickness during vacuum infusion, a network of small holes was introduced in the PUR interlayer. Experiments are conducted for both simply supported and fully clamped boundary conditions. A universal testing machine was used to apply the load in displacement control mode at 2 mm/min. The diameter of the hemispherical tip of the indenter is 20 mm.

Low velocity impact is performed using a drop tower facility in which the indenter resembles that used in the quasi-static indentation tests. Three test specimens of each sandwich construction are impacted with equivalent impact energy of 67 Joule induced by a mass of 3.4 Kg dropped from a height of two meters. Specimens are securely clamped between two steel plates.

4. Results and discussion

Figure 2 shows the stress-strain behaviour of PUR under quasi-static compression. The loading branch of the stress-strain curves is slightly nonlinear. On unloading, the stress-strain response of the specimens with no lubrication on the surfaces which support the load is nonlinear with a large hysteresis forming before the specimens are fully unloaded, figure 2a. This is attributed to friction between the compression platen and the specimen, which tend to retard the lateral deformation and hence affect the axial displacement. The hysteresis however disappears for the specimens where lubrication was applied at the loaded surfaces, figure 2b. The material exhibits moderate increase in strain hardening as the strain rate increases by one order of magnitude. As expected, lateral confinement of the PUR specimens enhances its

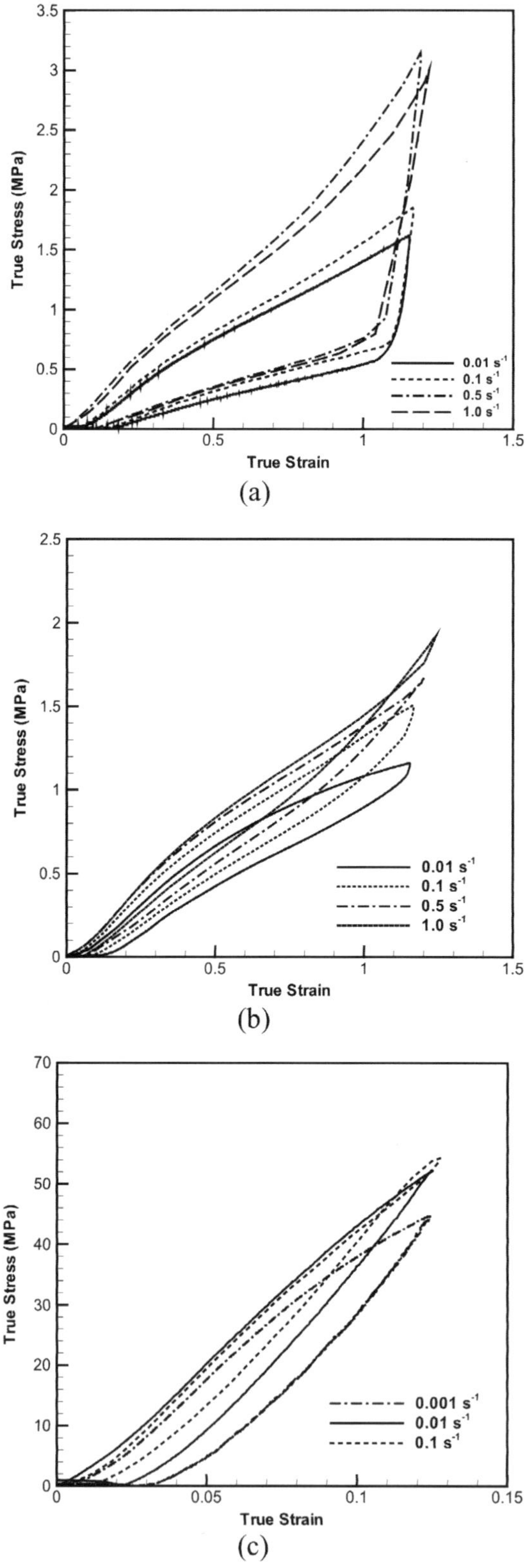

Figure 2. Measured response of PUR under uniaxial compression, (a) Unconfined and unlubricated, (b) Unconfined and lubricated, and (c) Confined.

load carrying capacity, figure 2c. The dynamic response of PUR on the other hand shows extremely high dependency on the strain rate, figure 3. For example, the flow stress at the highest strain rate of 8800 s^{-1} is almost 35 times that found at a quasi-static rate of 1.0 s^{-1} and a strain of 0.25.

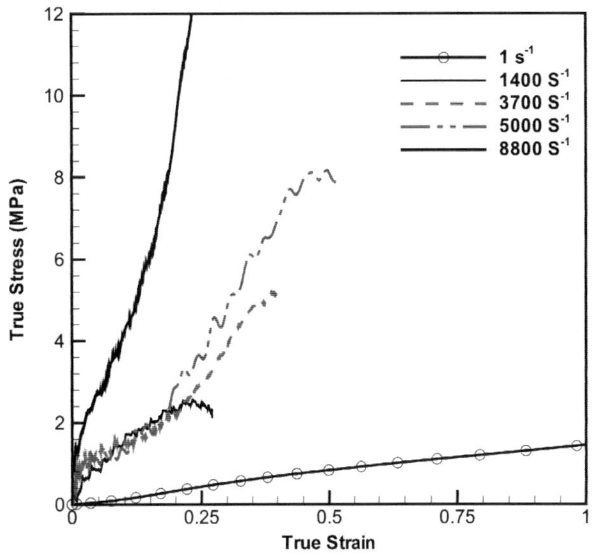

Figure 3. Stress-strain response of PUR.

Samples of structural foam are tested under compression. The measured quasi-static behaviour is given in figure 4. The response shows a linearly elastic behaviour with Young's modulus of 150 MPa up to a yield stress of 2.0 MPa. This is followed by perfectly plastic response up to about 50% strain, which is caused by collapse of the foam interior cells. Increasing the strain beyond this magnitude initiates locking of the foam closed cells, which causes a sharp increase in the compression load. Full densification of the foam is reached at strain of about 85%.

The experiments reveal that behaviour of the structural foam is independent of the rate of loading at the quasi-static regime. Under dynamic loading, however, there is a remarkable dependency on the strain rate as shown in figure 5. At low and intermediate strain rates up to 2800 s^{-1}, the foam samples show a weaker behaviour as compared to the quasi-static response. At higher strain rates on the other hand, the foam exhibits much larger stiffness under the applied load in comparison with loading at quasi-static conditions. This discrepancy in behaviour could be attributed to the rate and sequence of collapse of the foam cells.

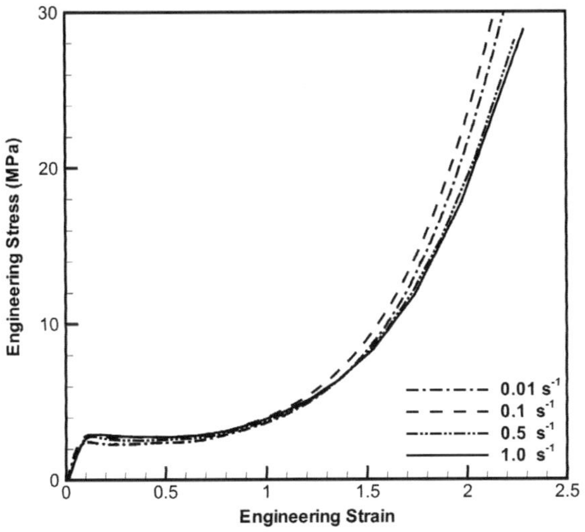

Figure 4. Response of cellular foam under uniaxial, quasi-static compression

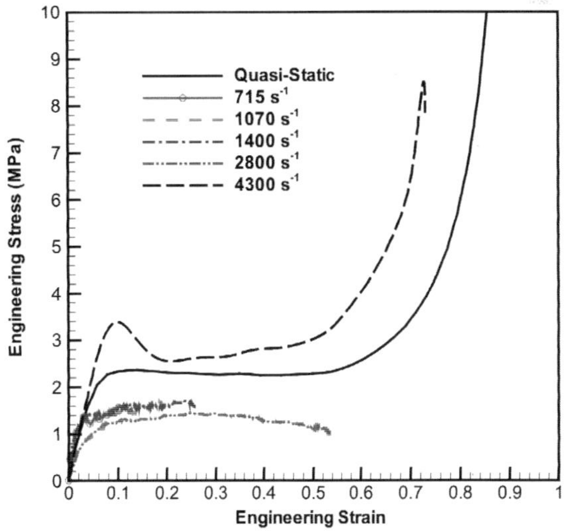

Figure 5. Dynamic response of cellular foam under uniaxial compression

Energy absorption during quasi-static indentation up to complete perforation for different layups and manufacturing techniques is shown in figure 6. It is seen that energy absorption in the specimens fabricated by vacuum infusion is slightly higher than that found in the specimens fabricated by hand layup. This is attributed to the fully consolidated specimens produced by vacuum infusion. Moreover, the addition of PUR interlayer in the sandwich panels led to enhancement of the performance under penetration. It should be noted that, although the clamped specimens are expected to give higher energy absorption as compared to the simply supported specimens, the load-stroke curves for both experiments are very close as shown in figure 7. The short span and the fact that the deformation process is dominated by face sheet perforation lead to almost identical behaviour.

In low velocity impact experiments, the effect of introducing PUR interlayers to the sandwich panels can be quantified by measuring the depth of indentation and observing failure. The tests are conducted on different and multiple specimens and the depth of indentations are recorded and averaged. Sample results for the drop tower tests are given in figure 8. It is seen that inserting thin PUR layer between the face sheet and the foam core improves the resistance to impact substantially. For a 1 mm thick PUR layer, the impact resistance increased by 30% for specimens made by vacuum infusion and by 20% for specimens made by hand layup. The impact resistance was increased favourably by 60% for both hand layup and vacuum infusion specimens when a 2mm thick PUR layer is incorporated in the sandwich construction.

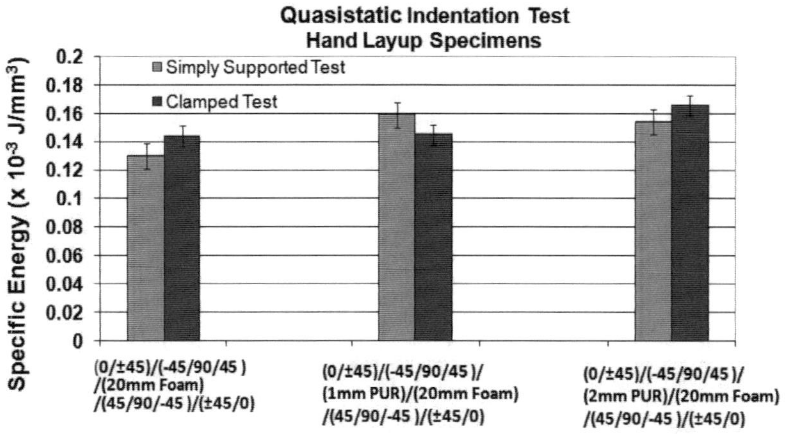

(a)

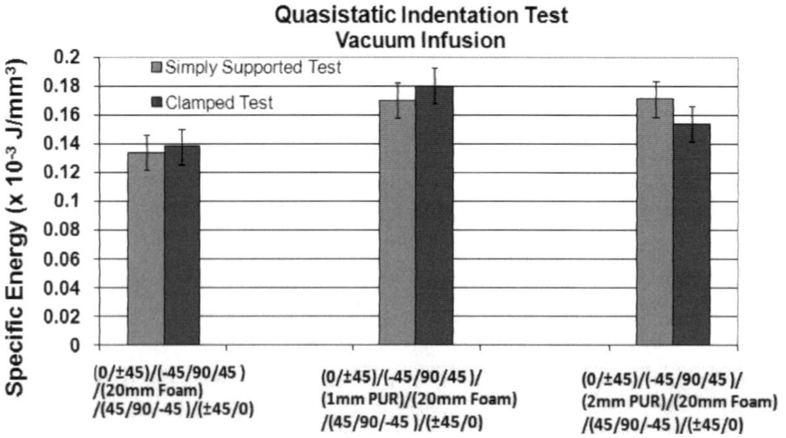

(b)

Figure 6. Comparison of measured specific energy for simply supported and fully clamped sandwich specimens subjected to quasi-static indentation, (a) Hand layup, (b) Vacuum infusion.

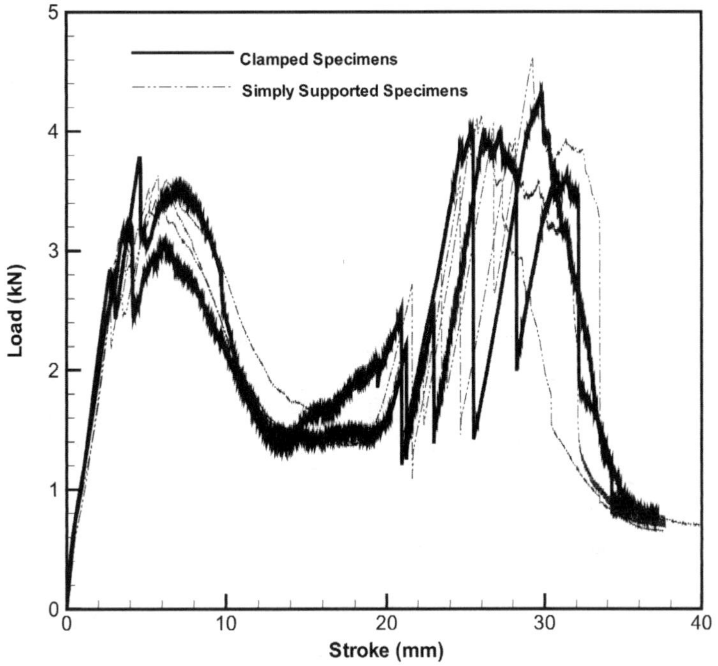

Figure 7. Representative load-stroke curves for simply supported and fully clamped sandwich specimens subjected to quasi-static indentation for [(0/±45)/(-45/90/45)/(20mm Foam)/(45/90/-45)/(±45/0)] specimens produced by vacuum infusion.

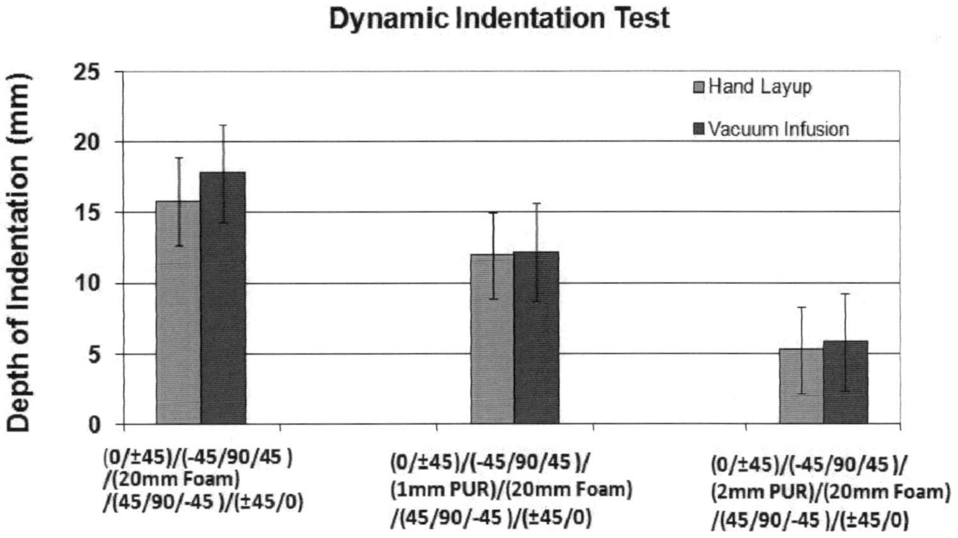

Figure 8. Depth of indentation caused in sandwich plates by low velocity impact at 67 Joule.

5. Conclusions

The present research measured experimentally the response of the constituents of composite sandwich panels under quasi-static and dynamic loading ranges. It also investigated the effect of modifying conventional sandwich designs by adding polyurethane (PUR) interlayers on resistance to indentation.

The following salient results summarize our findings:

1. PUR shows significant dependency on test conditions and loading rate, including quasi-static and dynamic rates, with the maximum resistance to deformation exhibited at high strain rates.
2. Cellular foam samples show distinct behaviours under quasi-static and dynamic loading rates.
3. Specimens fabricated by vacuum infusion exhibit enhanced resistance to indentation over specimens fabricated by hand layup.
4. The energy dissipated due to progressive damage and perforation of sandwich panels show an increase of about 20% for specimens produced by vacuum infusion over those produced by hand layup.
5. The energy dissipated by progressive damage and perforation under indentation is enhanced by the addition of PUR interlayers to sandwich designs. This increase in energy dissipation amounts to about 10-15% for thin PUR sheets on the order of 1-2 mm thick.
6. Under velocity impact, a 1 mm thick PUR layer, increases the impact resistance by 30% for specimens made by vacuum infusion and by 20% for specimens made by hand layup. The impact resistance is increased favourably by 60% for both hand layup and vacuum infusion specimens when a 2 mm thick PUR layer is incorporated in the sandwich construction

Acknowledgements
This work was performed under research grant number MED/2008/172-215(C1/S1/116) from the European Commission through EU-Egypt Innovation Fund. Research, Development and Innovation Program of the Ministry of Scientific Research, Egypt served as project monitor and Yehia Bahei-El-Din served as the Principal Investigator.

References

[1] Fleck NA 2004 *Int. J. Mech. Sci.* **46** 585
[2] Bahei-El-Din YA and Dvorak GJ 2007 *Mech. Adv. Mater. Struc.* **14** 465
[3] Bahei-El-Din YA *et al* 2006 *Int. J. Solids Struct.* **43** 7644
[4] Bahei-El-Din YA and Dvorak GJ 2007 *J. Sandw. Struct. Mate.r* **9** 261
[5] Bahei-El-Din YA and Dvorak GJ 2008 *Composites Part B* **39** 120
[6] Massabo R 2010 *Dynamic Interaction of Multiple Damage Mechanisms in Composite Structures* in Major Accomplishments in Composite Materials and Sandwich Structures pp 133-168 Springer
[7] Abrate S 2010 *Dynamic Interaction of Multiple Damage Mechanisms in Composite Structures* in Major Accomplishments in Composite Materials and Sandwich Structures pp 465-501 Springer
[8] Gupta S and Shukla A 1999 *Effect of Temperature on the dynamic performance of the core material, face-sheets and the sandwich composite* Proceedings of the IMPLAST 2010 Conference October 12-14 2010 Providence, Rhode Island USA, Society for Experimental Mechanics
[9] Gardner N and Shukla A 2011 *The Blast Response of Sandwich Composites With a Functionally Graded Core and Polyurea Interlayer*, Dynamic Behavior of Materials, Volume 1 Conference Proceedings of the Society for Experimental Mechanics Series pp 215-223
[10] Mori T and Tanaka K 1973 *Acta Metall* **21** 571
[11] Chen W *et al* 1999 *Exp Mech* **39** 81

Study of factors influencing the mechanical properties of polyurethane foams under dynamic compression

E Linul[1], L Marsavina[1], T Voiconi[1] and T Sadowski[2]

[1] Mechanics and Strength of Materials Department, "Politehnica" University of Timişoara, 1 Mihai Viteazu Avenue, 300222 Timişoara, Romania
[2] Faculty of Building and Architecture, Lublin University of Technology, Nadbystrzyka 33 str., 20-618, Lublin, Poland

E-mail: linul_emanoil@yahoo.com

Abstract. Effect of density, loading rate, material orientation and temperature on dynamic compression behavior of rigid polyurethane foams are investigated in this paper. These parameters have a very important role, taking into account that foams are used as packing materials or dampers which require high energy impact absorption. The experimental study was carried out on closed-cell rigid polyurethane (PUR) foam specimens of different densities (100, 160 respectively 300 kg/m^3), having a cubic shape. The specimens were subjected to uniaxial dynamic compression with loading rate in range of 1.37–3.25 m/s, using four different temperatures (20, 60, 90, 110°C) and two loading planes (direction (3) – rise direction and direction (2) - in plane). Experimental results show that Young's modulus, yield stress and plateau stress values increases with increasing density. One of the most significant effects of mechanical properties in dynamic compression of rigid PUR foams is the density, but also the loading speed, material orientation and temperature influences the behavior in compression

1. Introduction

Cellular materials are being increasingly used in many applications because of their crashworthiness characteristics. Research on their behaviour under quasistatic and high strain rates is of value for engineering applications such as those related to impact absorption (sandwich panel, packaging of electronic products) [1]. Foams can be compressed to a relatively high strain under an approximately constant load.

Characterization of cellular materials under compressive loading conditions covering elastic properties, failure modes, constitutive modeling and other related issues associated with foams have been investigated by several researchers [1-6]. The behavior of a family of polymeric structural foams with various initial bulk densities (porosity levels) was experimentally investigated under quasistatic and high strain rate loading conditions by Subash et al. [1]. In this context, Gibson and Ashby [2] performed an extensive study of the research results on cellular solids with applications from packaging cushions to lightweight structures. The impact response of rigid PUR foams has been studied by Linul et al. [3] using different densities (40, 80, 120 and 200 kg/m^3).

The dynamic compressive stress – strain behavior of a rigid polyurethane foam with four values of density (78, 154, 299, and 445 kg/m^3) has been determined by Chen et al. [4] in the strain-rate range of 1000-5000 s^{-1}. Uniaxial strain impact experiments have been performed by Chahabildas [5] to obtain shock compression and release response of a 0.22 g/cm^3 polyurethane foam in a configuration where the foam impacts a thin target witness plate. Compressive mechanical behavior of three common polymeric foams (expanded polystyrene, high-density polyethylene, and polyurethane) under quasi-static, medium and high strain rate conditions (ranging from 0.0087 to 2500/s), have been studied by Oullet et al. [6].

This study was motivated by a need to understand foam properties for energy – absorbing applications in different loading directions (direction (2) – in plane and direction (3) – rise direction) at high loading rates (up to 3.25 m/s), respectively high temperature (up to 110°C). The rise direction of foams is given by the used manufacturing process.

2. Mechanical testing in compression of rigid PUR Foams

Experimental tests were performed in the Strength of Materials Laboratory at Lublin University of Technology, Poland. Compressions tests were carried on with a 40 KN Instron – Dynatup impact testing machine shown in figure 1.

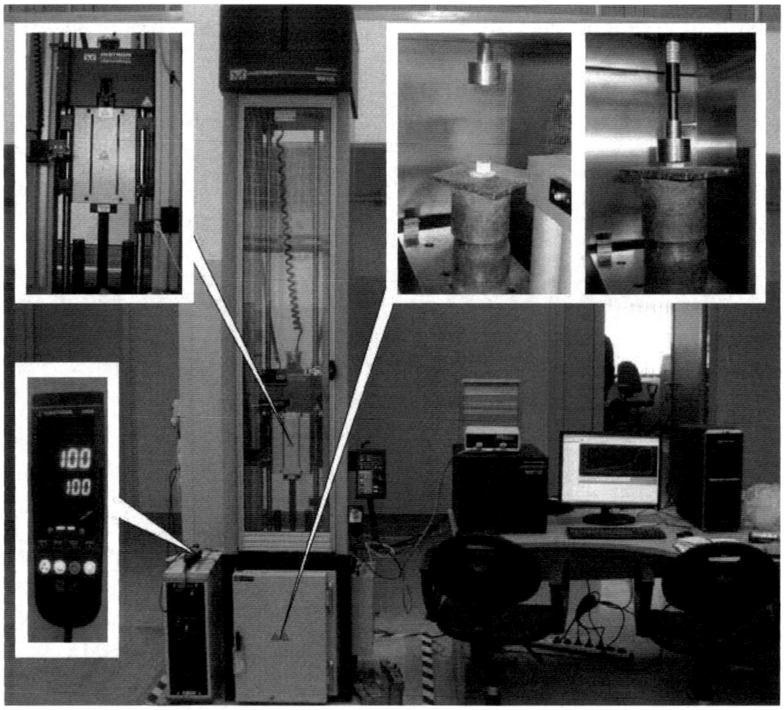

Figure 1. 40 KN Dynamic Instron – Dynatup testing machine

The specimens were subjected to uniaxial dynamic compression with maximum loading rate up to 3.25 m/s. An environmental chamber was used for tests at high temperature (up to 110°C). For tests performed at higher temperatures than room temperature, the samples were placed in the environmental chamber for 30 min before the test, to ensure thermal equilibrium [7].

The tests were performed following the ASTM D1621-00 standard [8] using cubic specimens (15 mm x 15 mm x 15 mm) sown in figure 2a. Also in the same figure are shown fully densified state for the same 3.09 m/s loading rate (for each density). Different densities of PUR foams (100, 160 and 300 kg/m^3) were used in this investigation for the characterization of mechanical behavior on dynamic compression. The relative densities were 0.085, 0.137 and 0.256, respectively. Samples were cut from a semi-finished plate using a pendulum saw, and then cut to the desired length using a Proxxon DS 230/E cut machine to avoid deformation of the specimen during cutting. Special care was taken when cutting samples to ensure that the specimen's end surfaces were parallel [6].

For each testing case (density, temperature, speed and forming plane), five samples were tested.

Figure 2. a) Photographs of the as-received PUR foams specimens; (b) Micrographs of the interior pore structure of the specimens

Figure 2b shows the corresponding pore structures in the interior sections of the specimens at the 1000x magnification. An examination of the microstructure (figure 2b) indicates that the foam has a typical closed-cell structure [4, 9]. It is seen that the low-density (100 and 160 kg/m^3), foams exhibit a wide variation in pore size and shape (wall thickness in range 3-18 μm), where as the high-density (300 kg/m^3), foam exhibit smaller uniform sized pores separated by large amount of solid polymer (wall thickness in range 8-48 μm). The measured porosity for each family of foams is also indicated in figure 2a [1].

3. Results and discussions

Conventional force–displacement data was obtained from the impact tests and converted to stress–strain data using the sample dimensions. A comparison of typical stress-strain curves for three different densities of rigid PUR foams (100, 160 and 300 kg/m^3), subjected to dynamic compression are shown in figure 3. This data was obtained with 3.09 m/s loading rate from direction (3) – rise direction at room-temperature (20°C).

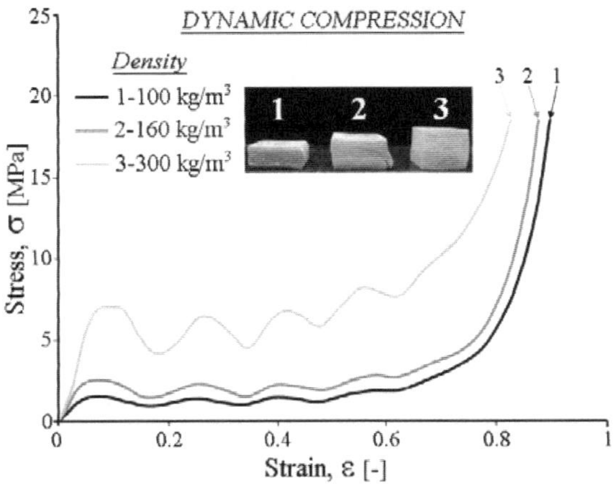

Figure 3. Stress-strain curves. Influence of density

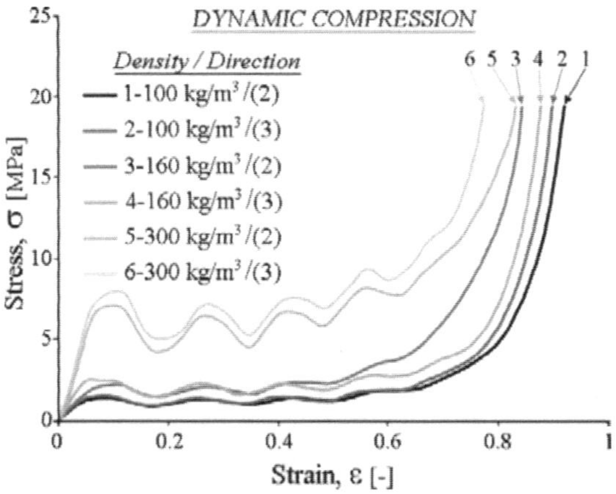

Figure 4. Stress-strain curves. Influence of forming plane

In figure 3 it can be seen easily an increase of mechanical properties with density, which means the density has a major influence in dynamic compression of cellular materials.

Figure 4 presents the influence of loading direction on forming plane, at room temperature and 3.09 m/s loading rate. For the tested compression specimens, the rise direction of the foam was notated as direction (3) and in-plane directions as direction (2).

From stress-strain curves, the basic properties of foams in compression were determined such as Young's Modulus, Yield Stress, Plateau Stress and Densification. Mean values of these properties for both loading directions: direction (2) and direction (3), depending on density at 3.09 m/s loading rate are presented in table 1.

Table 1. Mean values of mechanical properties depending on density and loading direction

Density [kg/m^3]	Loading direction	Young's Modulus [MPa]	Yield Stress [MPa]	Plateau Stress [MPa]	Densification [%]
100	(2)	26.06	1.52	1.18	54.15
(Foam A)	(3)	32.69	1.59	1.17	53.98
160	(2)	39.13	2.29	1.91	49.58
(Foam B)	(3)	58.85	2.58	1.91	56.20
300	(2)	139.26	7.21	5.55	61.83
(Foam C)	(3)	152.05	7.80	6.02	59.57

The results obtained with a load in the rise direction of the foam presents a small increase compared with those obtained in – plane loading direction.

For the same density, loading direction (rise direction) and temperature (20 °C) yield stress, plateau stress and densification remain approximately constant while a decrease in Young's Modulus with the increase in loading rate was observed (except for the first two small loading rates where results are not conclusive). Mean values of mechanical properties on dynamic compression for rise direction of foam depending on density and loading rate are listed in table 2.

Table 2. Mean values of mechanical properties depending on density and loading rate

Density [kg/m^3]	Loading rate [m/s]	Young's Modulus [MPa]	Yield Stress [MPa]	Plateau Stress [MPa]	Densification [%]
	1.37	39.91	1.58	1.16	53.95
	1.94	30.08	1.51	1.15	54.88
100	2.39	37.55	1.58	1.15	54.31
(Foam A)	2.77	35.95	1.42	1.15	54.44
	3.09	32.69	1.59	1.17	53.98
	3.25	30.16	1.59	1.17	57.65
	1.37	61.84	2.51	1.97	55.90
	1.94	66.57	2.61	1.92	57.97
160	2.39	65.22	2.57	1.74	55.10
(Foam B)	2.77	63.13	2.55	1.93	55.49
	3.09	58.85	2.58	1.91	56.20
	3.25	56.17	2.48	1.90	54.96
	1.37	166.54	7.52	6.29	58.32
	1.94	190.38	8.13	7.14	57.98
300	2.39	163.85	7.55	6.46	59.18
(Foam C)	2.77	157.33	7.50	6.55	57.14
	3.09	152.05	7.80	6.02	59.57
	3.25	136.61	7.04	5.58	60.59

The effects of temperature (20, 60, 90 and 110°C) on stress-strain curves for 300 kg/m^3 foam density at 2.77 m/s loading rate are shown in figure 5. Load was applied in the rise direction of foam.

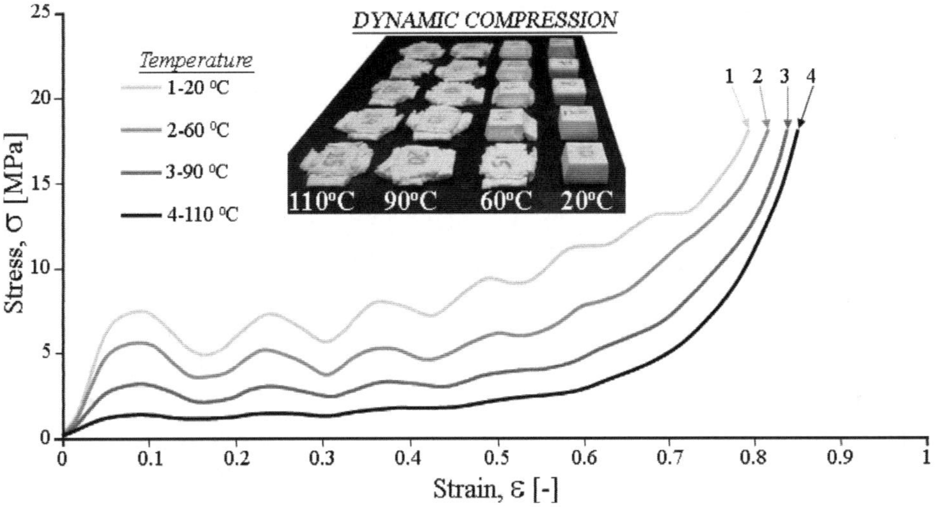

Figure 5. Stress-strain curves. Influence of temperature.

In table 3 are shown mean values of mechanical properties on dynamic compression depending on temperature for a load applied according to the rise direction of foam

Table 3. Mean values of mechanical properties depending on density and temperature

Density [kg/m^3]	Temperature [°C]	Young's Modulus [MPa]	Yield Stress [MPa]	Plateau Stress [MPa]	Densification [%]
	20	35.95	1.42	1.15	54.44
100	60	18.48	0.85	0.73	53.88
(Foam A)	90	8.50	0.68	0.60	53.00
	110	5.36	0.54	0.44	52.16
	20	63.13	2.55	1.93	55.49
160	60	49.11	2.23	1.69	54.20
(Foam B)	90	32.55	1.69	1.28	52.89
	110	24.82	1.28	1.06	52.00
	20	157.33	7.50	6.55	57.14
300	60	108.86	5.48	4.40	54.86
(Foam C)	90	56.35	3.11	2.67	53.99
	110	38.97	1.62	1.52	52.01

According to the results presented in figure 5 and table 3 we observed that the mechanical properties significantly decrease with the increase of temperature. After compression tests the foam shows a total destruction of cells, which increases the stress delivered to an almost constant strain known as densification. In the moment of densification, due to the filling of the gaps in the foam, this one acts almost like a solid material [3].

4. Conclusions

This paper presents an investigation of few parameters like effect of density, loading rate, material orientation and temperature on dynamic compression behaviour of rigid polyurethane foams. Compression tests were made for three different densities of closed-cell PUR foams (100, 160 and 300 kg/m^3). After dynamic compression tests the following conclusions can be drawn:

- From typical stress-strain curves the basic properties of foam were determined such as Young's Modulus, Yield Stress, Plateau Stress and Densification. These parameters have a very important role, taking into account that foams are used as packing materials or dampers which require high energy impact absorption.
- Also, from the analysis of results presented in figure 3 and table 1 it can be seen that with increasing of density we obtain a significant increase of the mechanical properties of foam. For example Young's Modulus is approximately 5 times higher for 'Foam C' than for 'Foam A' for the same loading rate, material orientation and temperature conditions.
- It was also found that for the same density, loading direction (rise direction) and temperature (20 °C) the Young's Modulus decrease with increasing of loading rate while Yield Stress, Plateau Stress and Densification remain approximately constant, (see table 2).
- The results obtained with a load in the rise direction of the foam show a small increase compared with those obtained in-plane loading direction, (see figure 4 and table 1).
- According to the results presented in figure 5 and table 3 we observed that the mechanical properties significantly decrease with increases of temperature. For example Young's Modulus for 'Foam C' is approximately 7 times lower at 110°C than 20°C for the same material orientation and loading rate conditions.

As the final conclusion we can say that one of the most significant parameter on mechanical properties in dynamic compression of rigid PUR foams is the foam density, but also we must consider the temperature, loading rate and material orientation.

Acknowledgement
This work was supported by the grant PN-II-ID-PCE-2011-3-0456. The experiments were performed in the facilities of the *Center of Excellence for Modern Composites Applied in Aerospace and Surface Transport Infrastructure* (European Union Seventh Framework Programme (FP7/2007 – 2013) supported by FP7-REGPOT–2009–1, under grant agreement No: 245479 and Polish Ministry of Science and Higher Education - Grant No 1471-1/7.PR UE/2010/7) from Lublin University of Technology, Poland.

References

[1] Subash G, Liu Q and Gao X-L 2006 Quasistatic and high strain rate uniaxial compressive response of polymeric structural foams *Int. J. of Imp. Eng.* **32** 1113-1126
[2] Gibson LJ and Ashby MF 1997 *Cellular solids, Structure and properties, Second edition.* Published by the Press Syndicate of the University of Cambridge
[3] Linul E, Marsavina L, Apostol DA, Constantinescu D and Sadowski T 2012 Effect of density, loading rate, material orientation and temperature on dynamic compresion behavior of rigid polyurethane foams *CD ROM Proc of 10th Int Conf on Foam Mat & Technology*, Barcelona
[4] Chen W, Lu F and Winfree N 2002 High-strain-rate compressive behavior of a rigid polyurethane foam with various densities *Experimental Mechanics* **42** 65-73
[5] Chhabildas LC, Kipp ME, Reinhart WD and Wong MK 1999 Polyurethane Foam Impact Experiments and Simulations, *Topical Conference on Shock Compression of Condensed Matter*, Snowbird, UT (US)
[6] Ouellet S, Cronin C and Worswick M 2006 Compressive response of polymeric foams under quasi-static, medium and high strain rate conditions *Polymer Testing* **25** 731–743
[7] Benderly D and Putter S 2004 Characterization of the shear/compression failure envelope of Rohacell foam *Polymer Testing* **23** 51-57
[8] ASTM D1621 Standard Test Method for Compressisive Properties of Rigid Cellular Plastics.
[9] Linul E, Marsavina L, Sadowski T and Kneć M 2012 Size Effect on Fracture Toughness of Rigid Polyurethane Foams *Solid State Phenomena* **188** 205-210

Indentation tests of aluminium honeycombs

A Ashab[1], Y C Wong[1], G Lu[2] and D Ruan[1, 3]

[1]Faculty of Engineering and Industrial Sciences, Swinburne University of Technology, Hawthorn, VIC 3122, Australia
[2]School of Mechanical and Aerospace Engineering, Nanyang Technological University, Singapore 639798

Email: druan@swin.edu.au

Abstract. Aluminium honeycomb is a type of cellular material which has high strength to weigh ratio and is a good energy absorber. They are used as structural components in various engineering applications. Comprehensive study has been conducted on the compressive behavior of aluminium honeycombs. However, the research of aluminium honeycombs subjected to other type of loading, such as indentation, is still limited. In this paper, quasi-static and dynamic indentation tests were conducted to study the deformation and energy absorption of three types of HEXCELL® aluminium honeycombs with different cell sizes and cell wall thicknesses. Quasi-static tests were conducted by using a universal MTS machine at velocities of 0.05 mm/s, 0.5 mm/s and 5 mm/s, respectively. Dynamic tests were conducted by using a high speed INSTRON machine at a velocity of 5 m/s. Force-displacement curves were plotted in which the total energy absorbed was calculated. The deformation of aluminium honeycombs in indentation tests includes the compression of honeycomb cells under the indenter and tearing of honeycomb cell walls along the indenter edges. The energy dissipated in compression and tearing were calculated and discussed. The effects of cell size, cell wall thickness and loading velocity or strain rate on the plateau stress and energy absorption were analyzed.

1. Introduction

Aluminium honeycombs are lightweight and have good energy absorption capabilities. They are widely used as core materials in various fields of engineering such as aerospace and automotive because of their extensive engineering properties. Some of the applications of aluminium honeycombs are landing gear door and flaps of aircraft. For example crushable aluminium honeycombs had been used in the Apollo 11 lading module. The large plastic deformation in hexagonal honeycombs is the main energy absorption region. It has been demonstrated that the plastic deformation and energy absorption characteristics of aluminium honeycombs depends on both the geometrical configuration and mechanical properties of cell wall materials [1, 2]. Cell size, cell wall thickness and strain rate are important parameters that affect plateau stress as well as impact response of aluminium honeycombs. Gibson and Ashby [1] pointed out that the deformation pattern and compressive strength varied with the direction of loads, for example, aluminium honeycombs strength is higher in the out-of plane direction compared with the two in-plane directions. A schematic diagram of aluminum honeycomb structure is shown in Figure 1. The cell wall thickness, t and $2t$ are known as single wall and double wall respectively. The cell edge length is denoted as l while h is the thickness of honeycomb.

[3] To whom any correspondence should be addressed.

Buckling, folding, stretching, de-bonding and tearing are the major failure mechanisms of aluminium honeycombs. Many researchers [3-6] studied aluminium honeycombs experimentally, numerically and theoretically, to characterize their crushing behaviour and failure mechanism under both quasi-static and dynamic compression. Zhou and Mayer [7] conducted compressive and indentation punch tests on two types of aluminium honeycombs with different cell sizes (6.4 mm and 19.1 mm respectively). In compression, the honeycomb with larger cell size was found to be strain rate dependent, while the 3honeycomb with smaller cell size did not show evident strain rate sensitivity. Quasi-static tearing strength of honeycomb was defined, measured and calculated from the indentation punch tests. Two types of honeycombs deformed differently. However, since no dynamic indentation punch test was conducted, the strain rate sensitivity is still unknown.

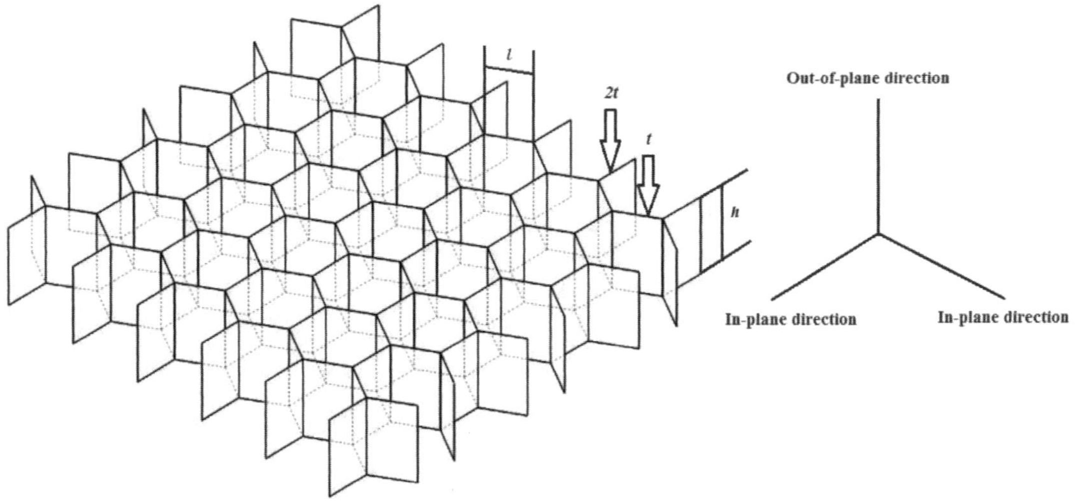

Figure 1. Schematic diagram of an aluminium honeycomb structure.

In the present paper, out-of-plane quasi-static and dynamic indentations, and compressive tests have been conducted using MTS and high speed INSTRON machines to investigate tearing and compressive strengths. Three types of aluminium honeycombs having different cell sizes and wall thicknesses have been used in the experiment. For quasi-static tests, velocities of 0.05 mm/s, 0.5 mm/s and 5 mm/s were used and for dynamic tests a velocity of 5 m/s was used. The tearing and compressive strengths were calculated from the experimental results and the dependency of the dissipated energy on the loading speed was also discussed.

2. Experiments

2.1. Specification of materials and specimens

HEXCELL® aluminium honeycombs with different cell sizes and wall thicknesses were used in the experiments. The specific material used is aluminium alloy 5052 with an H39 temper. The properties of three types of honeycombs are listed in Table 1. Each of the hexagonal honeycomb cell contain four single walls and two adjacent double walls of which all the cells are inter-connected to each other. The thickness of the double walls is twice of the single wall thickness. In compressive tests, honeycomb specimens used were 90 mm × 90 mm. In indentation tests, honeycomb specimens used were 180 mm × 180 mm. All honeycomb specimens were 50 mm in height.

The dimension of the indenter used in the indentation tests was a 90 mm × 90 mm which covers at least 9 × 9 cells of all types of honeycombs studied. This cross-section of the indenter used was to ensure almost uniform compression of honeycomb cells. A circular plate with holes was used as fixed platen where the honeycomb specimen was placed. In order to avoid specimen dis-connected with the platen rubber bands double sided tape was used to fasten the specimen in the platen. The holes in the circular plate are for the entrapped air to escape during testing. Therefore in this experiment the effect of entrapped air could be ignored.

Table 1. HEXCELL® 5052 aluminium hexagonal honeycomb.

Name	Material	Cell size D (mm)	Cell wall thickness t (mm)	Nominal Density ρ (kg/m^3)	Modulus (GPa)	Crush Strength (MPa)	No. of cells covered by Indenter
A	3.1-3/16-5052-.001N	4.763	0.0254	49.66	0.52	0.90	19×19
B	4.2-3/8-5052-.003N	9.525	0.0762	67.28	0.93	1.52	9×9
C	4.5-1/8-5052-.001N	3.175	0.0254	72.09	1.03	1.79	28×28

2.2. *Quasi-static tests*

Figure 2. Out-of-plane quasi-static indentation tests conducted on the 50 kN MTS machine.

Out-of-plane quasi-static tests were conducted on an MTS machine (Model LPS.504) which had a load capacity of 50 kN that could reach a velocity of 10 mm/s. In the quasi-static tests, the specimens were placed on the lower fixed circular platen. The fixed platen consisted of holes around the centre to allow the entrapped air to escape during testing. Figure 2 shows the experimental setup of the MTS machine for quasi-static tests. The indenter was connected to the upper load cell which moved downwards to crush the specimens. Fixed position of the specimens on the circular platen was maintained for all the tests. Both indentation and compressive tests were conducted using the MTS machine at velocities of 0.05 mm/s, 0.5 mm/s and 5 mm/s respectively. The corresponding nominal strain rates were 10^{-3}, 10^{-2} and 10^{-1} s^{-1} respectively for the specimen with a thickness of 50 mm. Compressive force, F_C, was measured in the compressive tests. The total force measured in the indentation tests, F_T, was the sum of the compressive force, F_c, and tearing force, F_t. Thus the tearing force could be calculated as, $F_t = F_T - F_c$.

2.3. Dynamic Tests

Figure 3. Out-of-plane dynamic indentation tests conducted on the high speed INSTRON machine.

Out-of-plane dynamic tests were conducted on a high speed INSTRON (VHS8800) machine which has capacity to generate a maximum velocity of 10 m/s. Both indentation and compressive tests were conducted at a velocity of 5 m/s. Figure 3 shows the experimental setup of the INSTRON machine for dynamic indentation tests. In the experiment the circular fixed platen was connected to the upper load cell. The specimens were attached to the fixed upper platen by using rubber bands for the indentation tests and specimens were fasten to the indenter by double-sided tape in compressive tests. The indenter was connected to the lower piston of the machine which moved upwards to apply the impact force to the specimens.

3. Results and Discussions

The force-displacement curves of three types of aluminium honeycombs subjected to indentation and compressive loads at velocities of 0.05 mm/s, 0.5 mm/s 5 mm/s and 5 m/s are shown in Figures 4-7 respectively.

It can be seen from Figures 4-7 that the plateau forces in the quasi-static tests (Figures 4-6) for all types of honeycomb specimens are generally constant. While in the dynamic tests (5 m/s, Figure 7) the transient forces fluctuate significantly in the plateau region.

The average plateau force was calculated by taking the average force within a displacement range of 3 - 38 mm. The mean plateau stress was calculated as the ratio of average force to the cross-sectional area of the honeycomb specimen. The mean plateau stresses of three different honeycomb materials at different velocities, 0.05 mm/s, 0.5 mm/s, 5mm/s and 5m/s are listed in Table 2. It has been observed that the mean plateau stress in both indentation and compressive tests increases with the increase in velocity. However, the tearing strength, $F_t = F_T - F_c$, did not show any trend with the change of the velocity.

The area under the force-displacement curve is the total energy absorbed by honeycombs. The energy absorbed by honeycombs in compression, E_c, is the area under the force-displacement curves recorded in compressive tests.

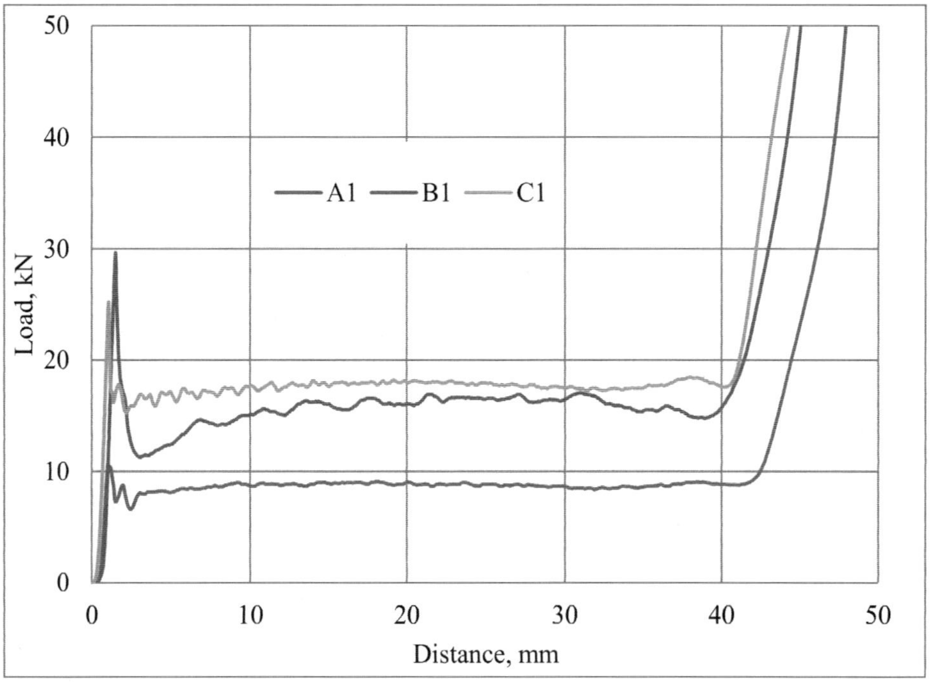

(a)

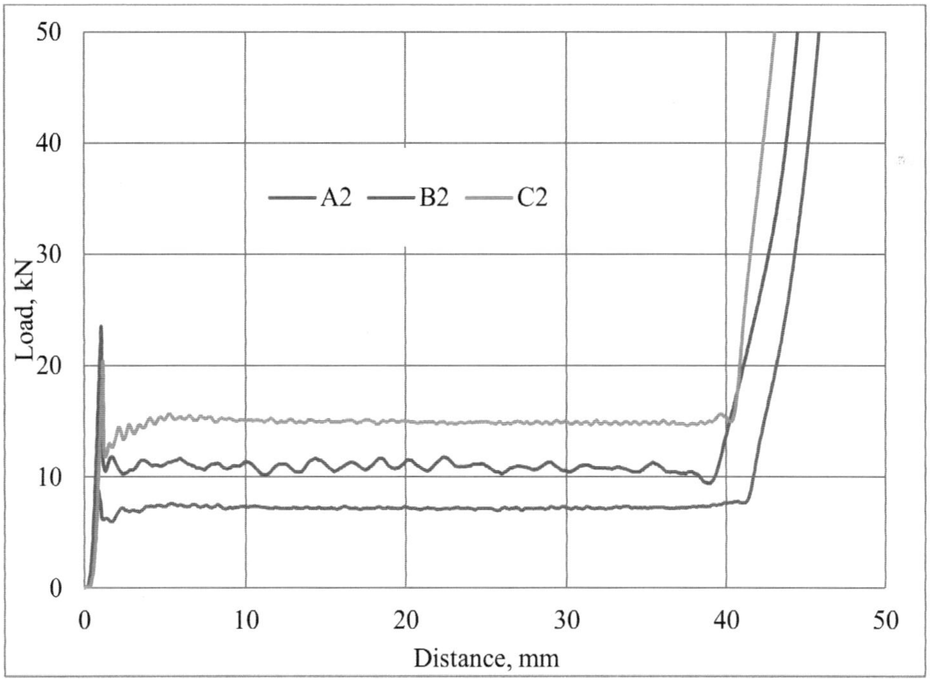

(b)

Figure 4. Quasi-static force-displacement curves of three types of aluminium honeycombs at a loading velocity of 0.05 mm/s in: (a) indentation tests; (b) compressive tests.

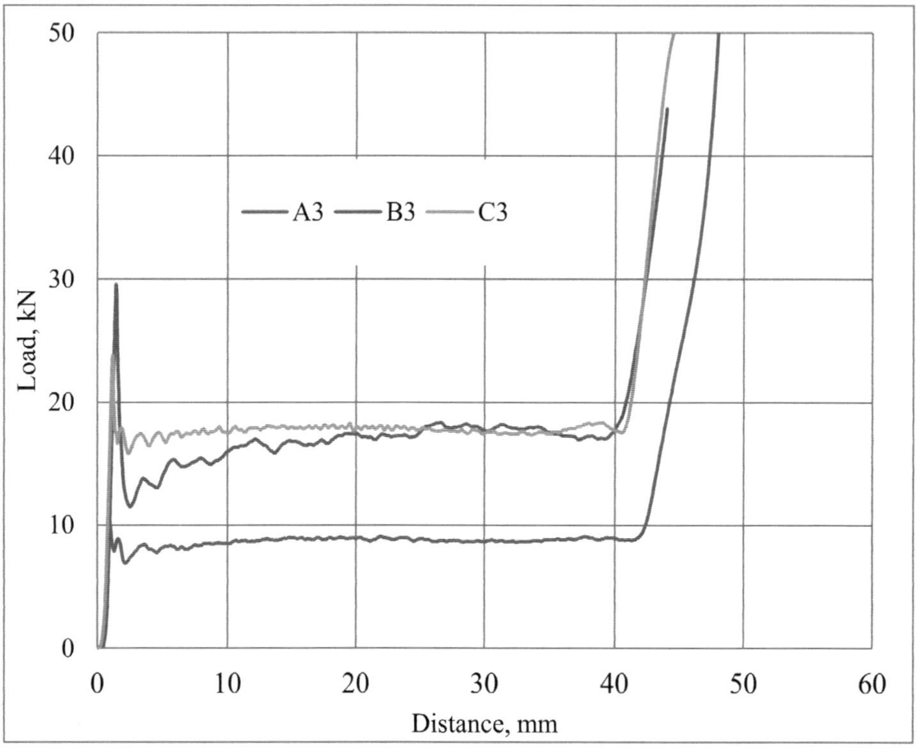

(a)

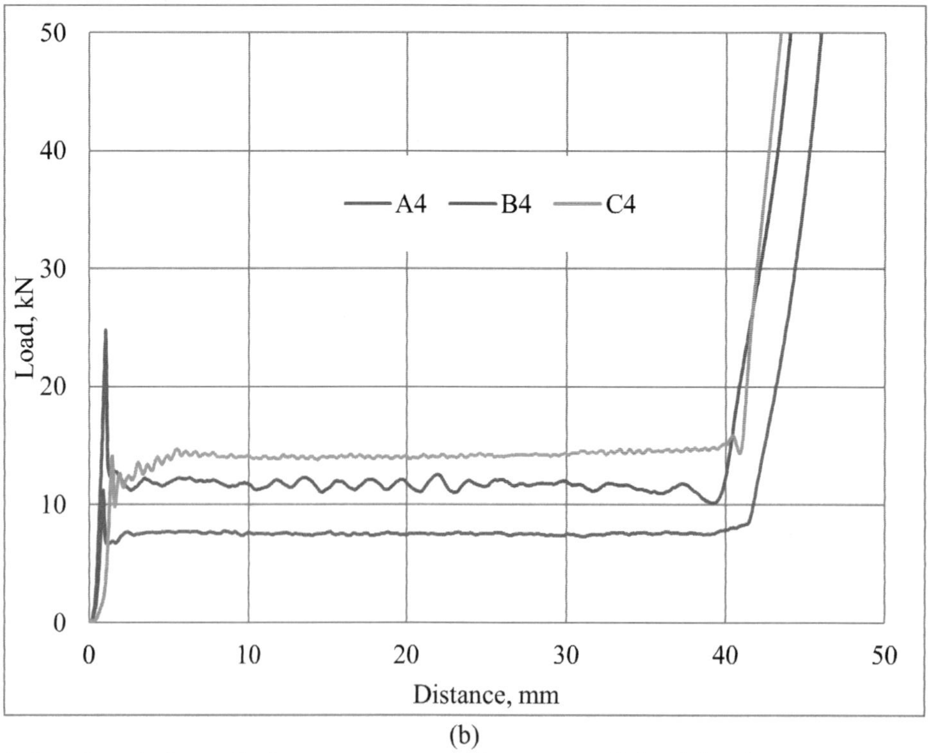

(b)

Figure 5. Quasi-static force-displacement curves of three types of aluminium honeycombs at a loading velocity of 0.5 mm/s in: (a) indentation tests; (b) compressive tests

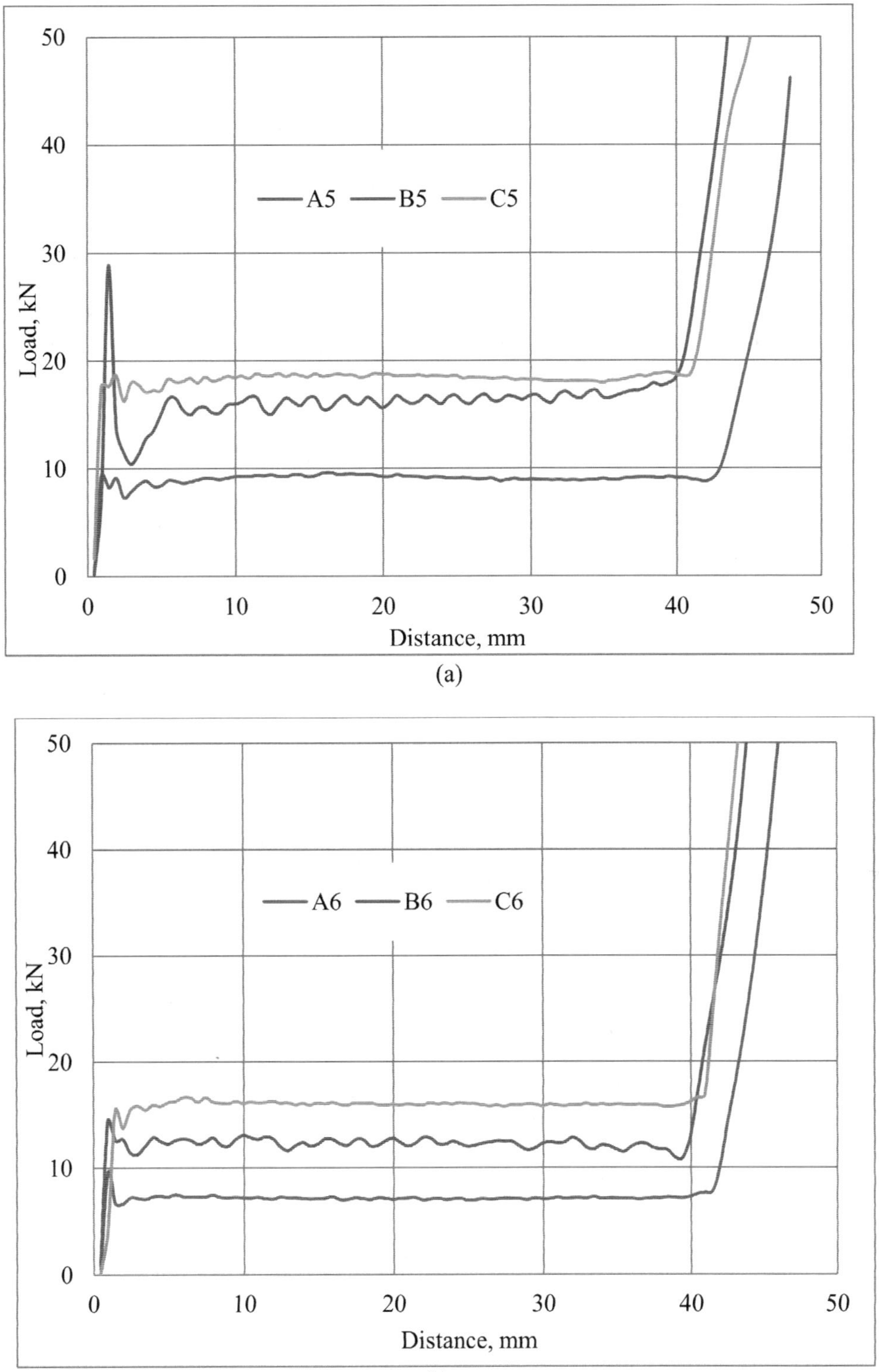

Figure 6. Quasi-static force-displacement curves of three types of aluminium honeycombs at a loading velocity of 5 mm/s in: (a) indentation tests; (b) compressive tests

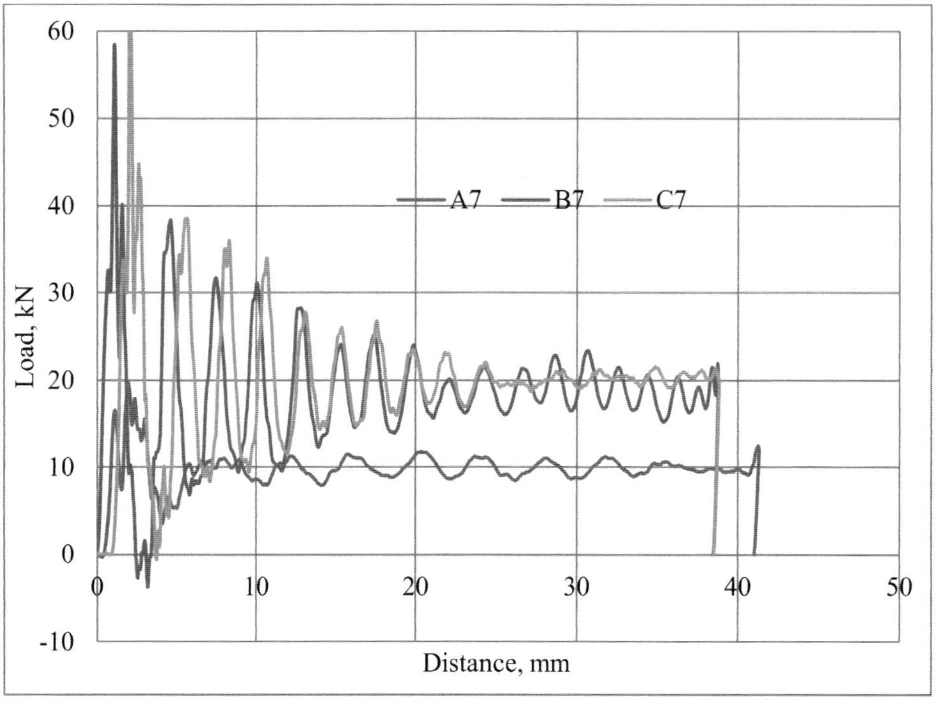

(a)

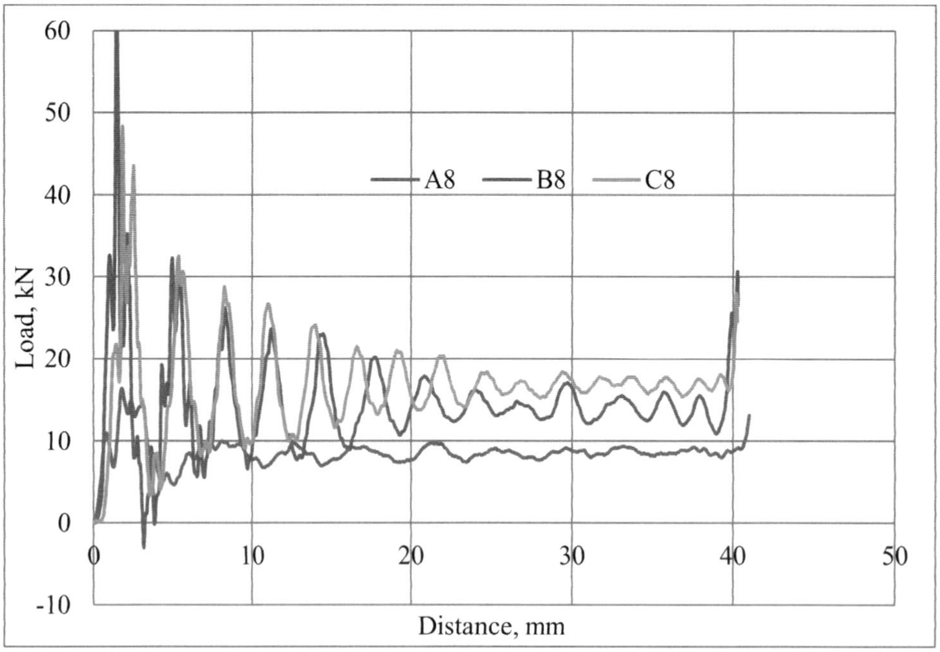

(b)

Figure 7. Dynamic force-displacement curves of three types of aluminium honeycombs at a loading velocity of 5 m/s in: (a) indentation tests; (b) compressive tests

The total energy absorbed by honeycombs in indentation, E_T, is the area under the force-displacement curves recorded in indentation tests and listed in Table 3. Since both compression of honeycomb cells and tearing along the 4 edges of honeycomb specimen occurred simultaneously in the indentation of honeycombs, the total energy dissipated in tearing, E_t, is calculated by the difference of E_T and E_c.

The percentages of compressive energy, E_c, and tearing energy, $E_{t,}$ at different strain rates are listed in the Table 3. For honeycombs material 3.1-3/16-5052-.001N (A1, A3, A5, A7) the percentage of compression and tearing energies varied in the range of 83% - 87% and 13% - 17% respectively. For the honeycomb material 4.5-1/8-5052-.001N (B1, B3, B5, B7) the percentages for compression energy and tearing energies were similar to those for honeycombs material 3.1-3/16-5052-.001N. However, for honeycombs material 4.2-3/8-5052-.003N the percentages of compression and tearing energies varied in the range of 70% - 74% and 26% - 30% respectively, i.e. the tearing energy was higher than that of the other two types honeycomb materials. This is mainly due to the thicker cell walls of honeycomb material (4.2-3/8-5052-.003N) compared with the other two types of honeycomb materials. However, the tearing energy, $E_t = E_T - E_c$, did not change significantly with loading velocity.

Table 2. Summary of experimental results

Test no.	Material	Test type	t/l	Velocity (m/s)	Mean plateau stress σ^*_{pl} (MPa)	Total Dissipated energy (J)
A1	3.1-3/16-5052-.001N	Indentation	0.00924	5×10^{-5}	1.07	304
A2	3.1-3/16-5052-.001N	Compression			0.89	252
A3	3.1-3/16-5052-.001N	Indentation	0.00924	5×10^{-4}	1.08	305
A4	3.1-3/16-5052-.001N	Compression			0.93	264
A5	3.1-3/16-5052-.001N	Indentation	0.00924	5×10^{-3}	1.12	323
A6	3.1-3/16-5052-.001N	Compression			0.89	268
A7	3.1-3/16-5052-.001N	Indentation	0.00924	5	1.2	338
A8	3.1-3/16-5052-.001N	Compression			1.04	295
B1	4.2-3/8-5052-.003N	Indentation	0.0139	5×10^{-5}	1.95	546
B2	4.2-3/8-5052-.003N	Compression			1.35	382
B3	4.2-3/8-5052-.003N	Indentation	0.0139	5×10^{-4}	2.09	586
B4	4.2-3/8-5052-.003N	Compression			1.44	409
B5	4.2-3/8-5052-.003N	Indentation	0.0139	5×10^{-3}	2.01	598
B6	4.2-3/8-5052-.003N	Compression			1.53	444
B7	4.2-3/8-5052-.003N	Indentation	0.0139	5	2.27	643
B8	4.2-3/8-5052-.003N	Compression			1.74	460
C1	4.5-1/8-5052-.001N	Indentation	0.0139	5×10^{-5}	2.18	616
C2	4.5-1/8-5052-.001N	Compression			1.84	522
C3	4.5-1/8-5052-.001N	Indentation	0.0139	5×10^{-4}	2.19	619

C4	4.5-1/8-5052-.001N	Compression			1.76	526
C5	4.5-1/8-5052-.001N	Indentation	0.0139	5×10^{-3}	2.28	661
C6	4.5-1/8-5052-.001N	Compression			1.98	553
C7	4.5-1/8-5052-.001N	Indentation	0.0139	5	2.42	683
C8	4.5-1/8-5052-.001N	Compression			2.03	576

Table 3. Effect of strain rate on the dissipated energy

Test no.	**Strain rate (s^{-1})**	**Compression energy %**	**Tearing energy %**
A1	10^{-3}	83	17
A3	10^{-2}	86	14
A5	10^{-1}	83	17
A7	100	87	13
B1	10^{-3}	70	30
B3	10^{-2}	70	30
B5	10^{-1}	74	26
B7	100	71	29
C1	10^{-3}	84	16
C3	10^{-2}	84	16
C5	10^{-1}	83	17
C7	100	84	16

4. Conclusions

The out-of-plane indentation and compression behaviours of three types of aluminium honeycombs were investigated experimentally by using an MTS machine and a high speed INSTRON machine. In both quasi-static and dynamic tests, constant velocity was achieved in all tests. Force-displacement curves were recorded in all tests and presented. Mean plateau stress and energy absorbed in compression and indentation of honeycombs were calculated. Results indicated that both mean plateau stress and total energy absorbed in compression and indentation increased with the density of honeycombs and loading velocity or strain rate. The tearing strength and tearing energies increased with the cell wall thickness of honeycomb, no trend was observed with the change of loading velocity or strain rate in this study. Further detailed study will be carried out in the near future with a focus on the strain rate effect on tearing strength and energy.

Acknowledgements

The authors are grateful to Swinburne University of Technology for the financial support through a postgraduate scholarship, the Australia Research Council for the financial support through a Discovery grant and Dr Shanqing Xu for his help in the experiments.

References

[1] Gibson L J and Ashby M F 1997 *Cellular solids: structure and properties* 2nd ed. (Cambridge: Cambridge University Press)
[2] Yamashita M and Gotoh M 2005 *Int. J. of Impact Eng.* **32** 618
[3] Zhang J and Ashby M F 1992 *Int. J. of Mech. Sci.* **34** 475
[4] Xu S, Beynon J H, Ruan D and Lu G 2012 *Compos. Struct.* **94** 2326
[5] Masters I G and Evans K E 1996 *Compos. Struct.* **35** 403
[6] Wierzbicki T 1983 *Int. J. of Impact Eng.* **1** 157
[7] Zhou Q and Mayer R R 2002 *J. Eng. Mater.-T ASME* **124** 412

Journal of Physics: Conference Series **451** (2013) 012004 doi:10.1088/1742-6596/451/1/012004

Role of specimen size upon the measured toughness of cellular solids

I Christodoulou and P J Tan

Department of Mechanical Engineering, University College London,
Torrington Place, London WC1E 7JE, U.K.

E-mail: pj.tan@ucl.ac.uk

Abstract. It is well known that the mechanical properties of cellular solids depend critically upon the specimen size and that a 'sufficiently' large test specimen is needed to obtain representative bulk values. Notwithstanding, the fracture toughness of cellular solids is still measured experimentally based on standards, such as the ASTM E399 and E813, developed for solid materials that do not possess an intermediate, 'cell-level' length scale. Experimental data in the literature appears to show that the toughness of stochastic 3D foams is, also, size-dependent. This paper presents the results of a detailed finite element (FE) study that will quantify, and identify the physical origin of, the size-dependent effect. Three-point bending of a single-edge notched (or SEN(B)) specimen, with a 2D Voronoi micro-architecture, is modelled numerically to obtain estimates of fracture toughness which are compared to those obtained with a 'boundary-layer' analysis.

1. Introduction

An important consideration in the mechanical testing of cellular materials is the effect of specimen size upon their measured properties. It is well known that the elastic modulus and yield strength of stochastic 3D foams increase with specimen size until their bulk values are reached; by contrast, their shear modulus and shear strength decreases. The former is due to a reduction, and the latter an increase, in the area fraction of weak boundary layer that develops next to the stress-free edges with increasing specimen size. Under bending, the weak boundary layers that develop at the top and bottom surface have an enhanced contribution to the mechanical response compared to the bulk material, leading to greater sensitivity to specimen size than, say, in compression [1–3].

Size effects were also reportedly observed in toughness measurements of stochastic 3D foams [4–6]; however, data scatter did not allow conclusive inference to be made. In this paper, the results of a detailed FE study are presented to quantify, and identify the physical origin of, the size dependent effect. The single edge notch specimen, with a 2D Voronoi micro-architecture, is modelled numerically using FE to obtain estimates of its fracture toughness which are compared to those obtained with a 'boundary-layer' analysis described in [7, 8].

2. Methodology and model set-up

Voronoi tessellations are employed to generate random planar lattices. A global cell regularity parameter, viz. $\Lambda = d/d_0 \in (0,1]$, controls the minimum distance d between the nuclei of any two adjacent cells and d_0 is the corresponding distance in a regular hexagonal lattice, see Christodoulou and Tan [7] for details. $\Lambda = 1$ and $\Lambda \to 0^+$ corresponds to a regular hexagonal and a completely random Γ-lattice, respectively. It was shown in [7] that the lattice ligaments deform by bending regardless of Λ; all specimens modelled have identical cell regularity of $\Lambda = 0.5$. Since the effects of relative density $\overline{\rho}$ upon toughness are already well known, see [8], only results for $\overline{\rho} = 0.1$ are shown here.

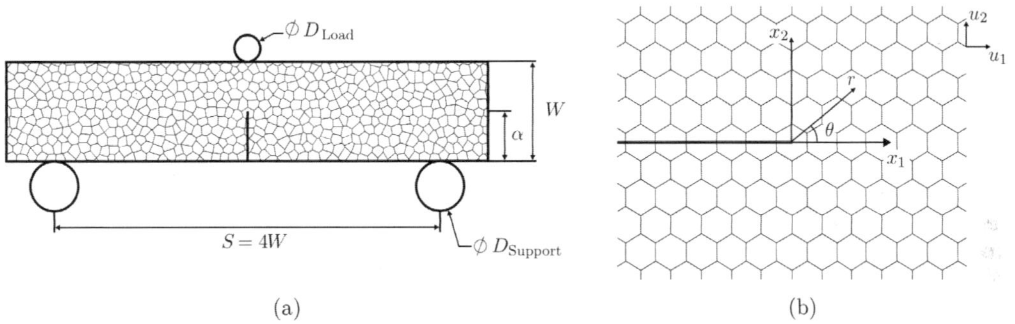

Figure 1. (a) SEN(B) specimen geometry and (b) Schematic of model set-up in the 'boundary-layer' analysis

Twenty Voronoi tessellations with $\Lambda = 0.5$ were generated. Each tessellation is trimmed to size giving seven SEN(B) specimens of different width, viz. $W/d_0 = 10, 15, 20, 25, 30, 50$ and 100 where $d_0 = \sqrt{3}l$ for a regular hexagonal lattice with cell wall length l. The crack-tip in each specimen, generated from a given tessellation, is always centred at the same location. This eliminates variability of the calculated toughness caused by different crack-tip cell geometry [7]. A schematic of the SEN(B) specimen is shown in Figure 1a. The crack length $\alpha(= 0.5W)$ and span $S(= 4W)$ of the specimen were chosen according to [9]. The minimum diameter of the loading and support pins satisfy $D_{\text{Load}}/W > 0.25$ and $0.5 < D_{\text{Support}}/W < 1$, respectively [9]. Fracture toughness of the SEN(B) specimen is calculated using [9]

$$K_{\text{IC}} = 3\frac{P_Q}{\sqrt{W}}\frac{S}{W}\left[\sqrt{\frac{\alpha}{W}}\frac{1.99 - \left(\frac{\alpha}{W}\right)\left(1-\frac{\alpha}{W}\right)\left[2.15 - 3.93\frac{\alpha}{W} + 2.7\left(\frac{\alpha}{W}\right)^2\right]}{2\left(1+2\frac{\alpha}{W}\right)\left(1-\frac{\alpha}{W}\right)^{3/2}}\right] \tag{1}$$

where P_Q is the peak load during testing at the onset of fracture.

Additionally, the toughness of each tessellation is also calculated using a 'boundary-layer' analysis, see details given in [7,8], where the displacement and rotational fields corresponding to the asymptotic K-field are applied to the nodes along the boundary of the mesh shown in Figure 1b. The size of the mesh used has been chosen to avoid size effects.

Both the SEN(B) and 'boundary-layer' specimens were implemented in ABAQUS/standard® where each ligament is modelled using four Timoshenko beam (B21) elements. Surface-solid body interaction was used to model contact between the pins and specimen, with the former as solid non-deformable bodies. Load is applied by prescribing a finite displacement to the rigid loading pin in the negative x_2 direction and constraining all its other degrees of freedom. This approach is preferred to a prescribed displacement BC which would have resulted in a more compliant response. An elastic-brittle failure criterion is adopted where ligament failure is assumed to occur if the predicted tensile stress reaches the fracture strength of the ligament σ_f.

3. Results

3.1. Effect of pin size

The toughness is found to be sensitive to the diameter of the pins and the manner in which they come into contact with the specimen. To illustrate these effects, a regular hexagonal lattice is used. Given a SEN(B) specimen with even number of cells along its width (W), the loading pin comes into contact with a closed cell whilst the support pins come into contact with open cells. The opposite occurs with odd number of cells along its width.

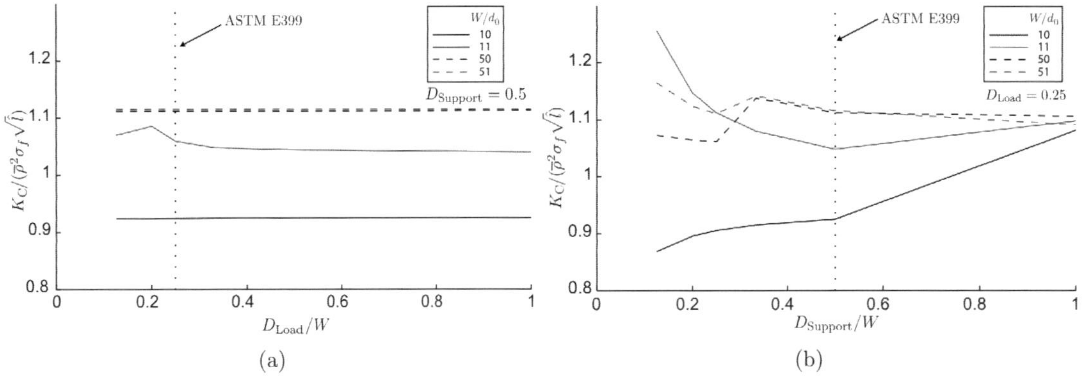

Figure 2. Variation of non-dimensional fracture toughness with (a) loading and (b) support pin diameters for specimen with $W/d_0 = 10, 11, 50$ and 51 cells. Black and red lines indicates a specimen with even and odd number of cells along its width, respectively. The vertical dotted line indicates the minimum pin diameter recommended by the ASTM E399 [9].

Figure 2a plots the variation of the non-dimensional fracture toughness $K_{\text{IC}}/(\sigma_f\bar{\rho}^2\sqrt{l})$ against loading pin diameter D_{Load}/W for a fixed $D_{Support}/W = 0.5$. Results from both small ($W/d_0 = 10, 11$) and large ($W/d_0 = 50, 51$) sized specimens with odd and even number of cells along its width are shown. The diameter of the loading pin has a negligible influence upon the toughness of the specimen. The recommended value of $D_{\text{Load}}/W = 0.25$ by the ASTM E399 [9] gives converged results for all values of W/d_0.

By contrast, the diameter of the support pins has a considerable influence upon the specimen toughness; this is clearly evident in Figure 2b. For the smaller-sized specimens ($W/d_0 = 10, 11$), there is a significant difference between their predicted toughness for the entire range of D_{Support}/W. Even at $D_{\text{Support}} = W$, the results do not appear to have converged in a satisfactory manner. The larger specimens ($W/d_0 = 50, 51$), on the other hand, give converged results between $0.5 < D_{\text{Support}}/W < 1$. The numerical data suggest that smaller-sized specimens are likely to be the more sensitive to local cell topological variations along the specimen boundary where the pins come into contact. As it wouldn't be possible to control the local morphological variations of the cells in stochastic lattices where the support pins come into contact, one could expect a significant increase in the variations of the measured toughness if smaller-sized specimens were used.

3.2. Specimen size

SEN(B) specimens of various size $W/d_0 = 10, 15, 20, 25, 30, 50$ and 100 were generated from each of the twenty Voronoi tessellations created. Figure 3 plots the non-dimensional toughness for three, out of the twenty, tessellations. The predicted toughness by 'boundary-layer' analysis is also shown for comparison. The diameter of the pins used were that recommended by the ASTM E399 [9].

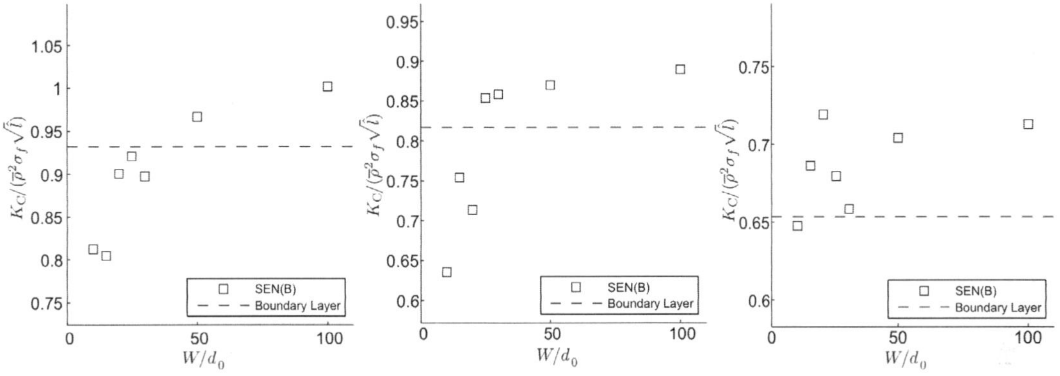

Figure 3. Effect of specimen size upon the non-dimensional fracture toughness of three different Voronoi tessellations. The corresponding toughness predicted by a 'boundary-layer' analysis is also shown.

In general, an increase in toughness is observed with increasing specimen width W for a given d_0. Each Voronoi tessellation gives slightly different dependence upon specimen size. For instance, the increase in toughness with W/d_0 is more dramatic for the tessellation shown in Figure 3b than in Figure 3a. Figure 3c, on the other hand, appears to show converged results although this was only observed in two, of the twenty, tessellations studied. More importantly, the 'boundary-layer' analysis consistently predicts lower toughness compared to that estimated by the largest SEN(B) specimen, for all tessellations. Note that the toughness predicted by the 'boundary-layer' analysis is representative of the lattice material since it is unaffected by size effects.

Figure 4 plots the average toughness of twenty tessellations obtained from the SEN(B) specimens and 'boundary-layer' analyses. The error-bars and shaded region correspond to ± one standard deviation. An average reduction in toughness, by up to ≈24%, is noted as specimen size decreases. The results show that a minimum specimen size of approximately 50 cells is needed to obtain representative lattice toughness if it is to be unaffected by size effects. The lattice toughness predicted by the 'boundary-layer' analysis is considerably lower, by approximately 10%, compared to one obtained using a SEN(B) specimen.

4. Discussions

To better appreciate the physical origin of the size effects as seen in Figure 4, continuous strain maps are needed to explain the cell-level deformation in the specimens. Strain maps are generated by triangulating each cell in the lattice, using a technique known as Delaunay triangulation, and assuming that each of the triangulated regions have a constant strain, i.e. a constant strain triangle; the details are described in [7]. Figure 5 plots the normalised equivalent strain, $\bar{\varepsilon}_{eq} = (\sigma_f\sqrt{3/2\varepsilon_{ij}\varepsilon_{ij}})/(\bar{\rho}E_s)$, for three SEN(B) specimens of different size, and the those obtained by a corresponding 'boundary-layer' analysis, at the onset of first ligament fracture. The strain maps shown are the average of twenty tessellations.

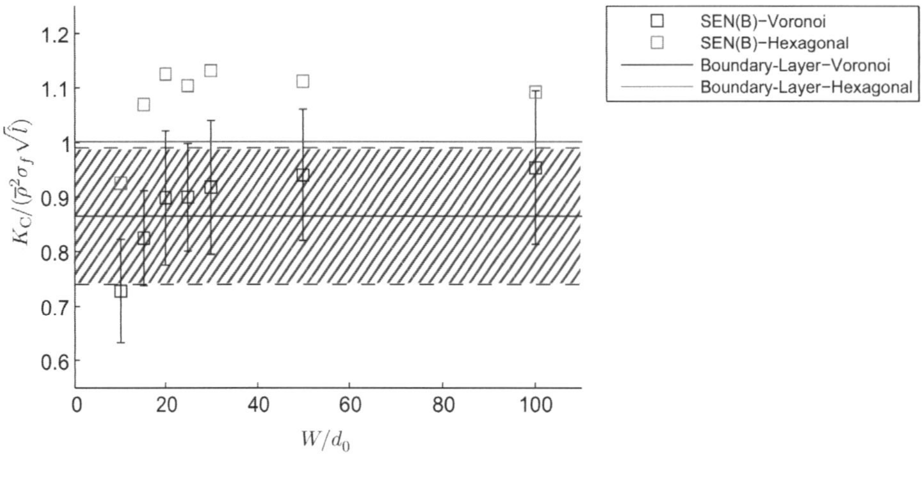

Figure 4. Variation of non-dimensional toughness with specimen size for Voronoi lattices (average of twenty tessellations are given) and regular hexagonal lattices. Results from the 'boundary-layer' analysis are shown for comparison and the shaded region corresponds to ± one standard deviation.

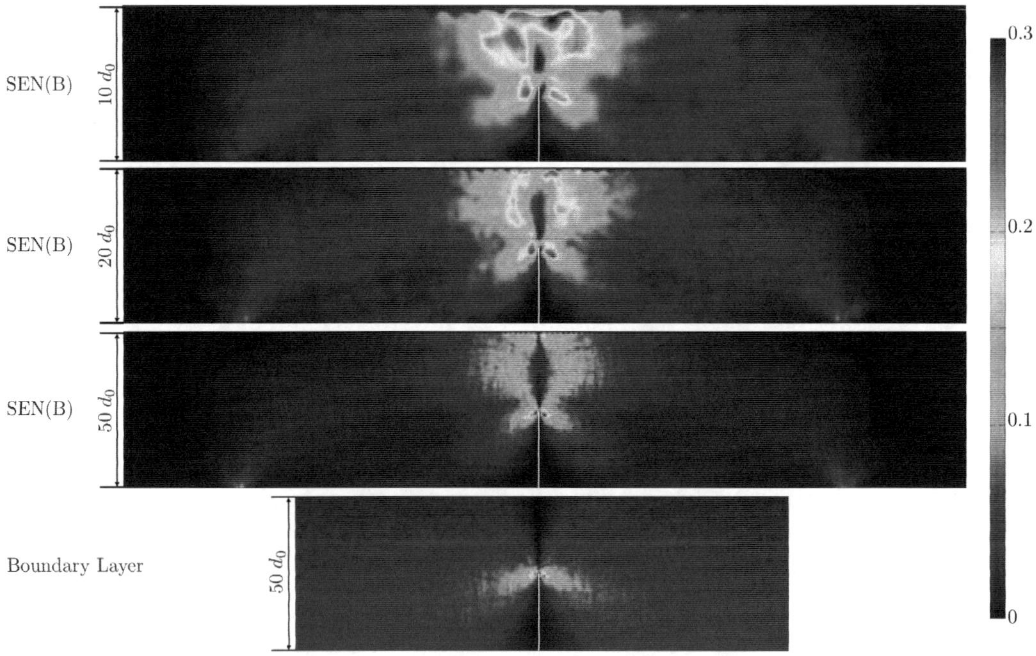

Figure 5. Strain maps showing the normalised equivalent strain $\bar{\varepsilon}_{eq}$ for three SEN(B) specimens of different size $W/d_0 = 10, 20, 50$. The corresponding strain map from the 'boundary-layer' analysis is also shown. The strain maps shown are the average of twenty Voronoi lattices.

For the smaller SEN(B) specimen ($10d_0$ and $20d_0$), the strong interaction between the crack tip fracture process zone and the specimen boundary (upper), in the vicinity where the loading pin makes contact with the specimen, is clearly evident. A region of elevated equivalent strain appears in the un-cracked region ahead of the crack-tip; interestingly, the highest equivalent strain occurs near the specimen boundary rather than the crack tip. This is contrary to using a pre-cracked specimen where the purpose is to generate a region of K-dominance around the crack tip. As W increases, the interaction between the fracture process zone and the specimen boundary becomes somewhat weaker; at $W = 50d_0$, the highest equivalent strain now occurs in the vicinity of the crack-tip and is comparable to that observed in a 'boundary-layer' analysis. The strain maps appear to show that the physical origin of size effect in lattice materials are due to interactions between the crack-tip fracture process zone with the stress field generated by local indentation due to the loading pin. This disrupts the K-field in the vicinity of the crack-tip and renders LEFM inapplicable.

Since fracture toughness is a material property, it is not expected to be sensitive to specimen geometry and the measurement technique employed. However, the difference in predicted toughness by the two methods is significant, approximately 10%, between specimens thought to be unaffected by size-effects, see Fig. 4. Figure 5 shows that, unlike the SEN(B) specimen of width $W = 50d_0$, hardly any deformation is observed ahead of the crack tip in the 'boundary-layer' analysis. This suggests that there exist an additional component of the deformation field in the SEN(B) specimen which might not be present in a 'boundary-layer' analysis. A likely explanation is the presence of positive T-stress ahead of the crack-tip in the SEN(B) specimen. The biaxiality ratio of the SEN(B) specimen modelled here is estimated to be $B = 0.2$ and would typically result in a 200% increase in fracture toughness for hexagonal and random lattices [7, 10]. This is part of an on-going investigation to be reported elsewhere.

5. Conclusions

The predicted fracture toughness of random Voronoi 2D lattices is compared for single-edge notch specimen and one obtained by a 'boundary-layer' analysis. The SEN(B) specimen was found to be sensitive to size effects and at least 50 cells along the specimen width are needed to avoid size effects. It was found that size effects in lattice materials are caused by interactions between the crack-tip fracture process zone with the stress field generated by local indentation due to the loading pin, which disrupts the K-field in the vicinity of the crack-tip. Additionally, a significant difference between the predicted toughness of the SEN(B) specimen and the 'boundary-layer' analysis was found to exist between specimens unaffected by size effect and this is attributed to the presence of higher order terms in the stress-field around the crack-tip.

References

[1] Tekoglu C, Gibson L, Pardoen T and Onck P 2011 *Prog Mater Sci* **56** 109–38
[2] Onck P, Andrews E W and Gibson L J 2001 *Int J Mech Sci* **43** 681–99
[3] Andrews E W, Gioux G, Onck P and Gibson L J 2001 *Int J Mech Sci* **43** 701–13
[4] Noury P, Shenoi R and Sinclair I 1998 *Int J Fra* **92** 131–51
[5] Motz C and Pippan R 2002 *Acta Materialia* **50** 2013–33
[6] Kabir M E, Saha M C and Jeelani S 2006 *Materials Science and Engineering* A **429** 225–35
[7] Christodoulou I and Tan P J 2013 *Engng Fract Mech* **104** 140 – 161
[8] Fleck N A and Qiu X 2007 *J Mech Phys Solids* **55** 562–88
[9] E399-09 2009 *Standard Test Method for Plane-Strain Fracture Toughness K_{IC} of Metallic Materials* (American Society for Testing and Materials)
[10] Leevers P S and Radon J C 1982 *Int J Fracture* **19** 311–25

Behaviour of advanced materials impacted by high energy particle beams

A Bertarelli[1,4], F Carra[1], F Cerutti[1], A Dallocchio[1], M Garlasché[1], M Guinchard[1], N Mariani[1], S D Marques dos Santos[1], L Peroni[2], M Scapin[2] and V Boccone[1,3]

[1] European Organization for Nuclear Research CERN, CH-1211 Geneva 23, Switzerland
[2] Politecnico di Torino, Corso Duca degli Abruzzi, 24 - 10129 Turin, Italy
[3] University of Geneva, Rue du Général- Dufour 24, CH-1211 Genève 4, Switzerland

Email: alessandro.bertarelli@cern.ch.

Abstract. Beam Intercepting Devices (BID) are designed to operate in a harsh radioactive environment and are highly loaded from a thermo-structural point of view. Moreover, modern particle accelerators, storing unprecedented energy, may be exposed to severe accidental events triggered by direct beam impacts. In this context, impulse has been given to the development of novel materials for advanced thermal management with high thermal shock resistance like metal-diamond and metal-graphite composites on top of refractory metals such as molybdenum, tungsten and copper alloys. This paper presents the results of a first-of-its-kind experiment which exploited 440 GeV proton beams at different intensities to impact samples of the aforementioned materials. Effects of thermally induced shockwaves were acquired via high speed acquisition system including strain gauges, laser Doppler vibrometer and high speed camera. Preliminary information of beam induced damages on materials were also collected. State-of-the-art hydrodynamic codes (like Autodyn®), relying on complex material models including equation of state (EOS), strength and failure models, have been used for the simulation of the experiment. Preliminary results confirm the effectiveness and reliability of these numerical methods when material constitutive models are completely available (W and Cu alloys). For novel composite materials a reverse engineering approach will be used to build appropriate constitutive models, thus allowing a realistic representation of these complex phenomena. These results are of paramount importance for understanding and predicting the response of novel advanced composites to beam impacts in modern particle accelerators.

1. Introduction

The introduction in recent years of new, extremely energetic particle accelerators such as the Large Hadron Collider (LHC) [1] has required the development of advanced methods to predict the behavior of beam intercepting devices in case of direct beam impact. Such events can be simulated resorting to complex wave propagation codes like Autodyn or LS-Dyna [2]. These numerical tools, however, require accurate material models which scarce in scientific literature, especially for non-conventional alloys or composites. In order to probe and evaluate such models at the extreme conditions induced by particle beam impacts, a first-of-its-kind experiment was recently carried out at CERN HiRadMat facility [3]. The performed tests entailed the controlled impact of intense proton pulses on specimens

[4] To whom any correspondence should be adressed

made of six different materials, including advanced metallic alloys and newly developed composites: the present work focuses on modeling and results of Inermet® 180 (tungsten heavy alloy W-Cu-Ni), while the studies on the other materials are still ongoing. For a comprehensive characterization, experimental data were acquired relying on extensive embedded instrumentation (strain gauges, temperature and vacuum sensors) and on remote-acquisition devices (laser Doppler vibrometer and high-speed camera).

Two different specimen shapes were chosen for each material (Figure 1): cylindrical disks (*type 1*) for medium intensity tests, to measure axially symmetric shockwaves; cylinders with a half-moon cross section (*type 2*) for high intensity tests, allowing extreme surface phenomena (melting, material explosion, debris projections etc.) to be visualized and optically acquired. Specimens were arranged into a sample holder maintained under vacuum; a total of six Inermet® 180 specimens have been tested: three at medium intensity (*type 1*) and three at high intensity (*type 2*).

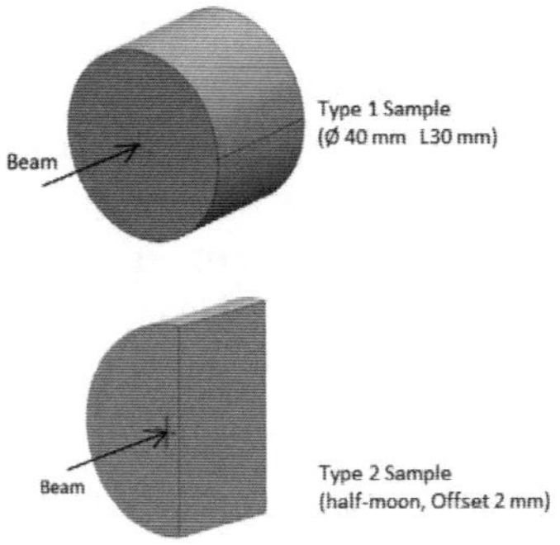

Figure 1. Material specimen shapes for tests at medium and high intensity

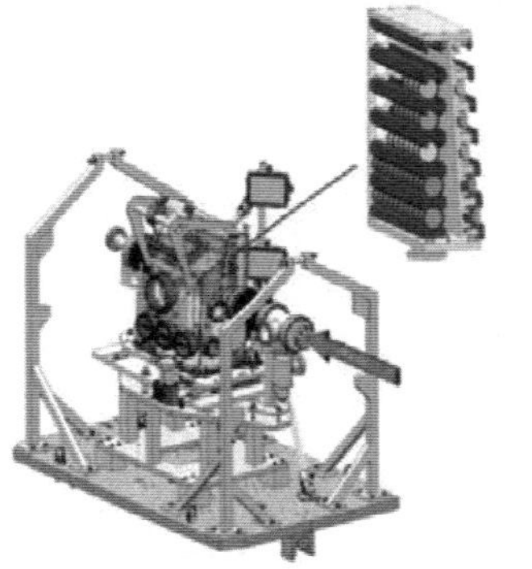

Figure 2. General assembly of the HRMT-14 test-bench. Red arrow represents the beam

2. Numerical methods

When solids are impacted by highly energetic particle beams, phenomena such as significant change of density, changes of phase and explosions can occur. Wave propagation codes used to study these complex cases rely on material constitutive models encompassing a very large range of densities and temperatures [4]. The input for numerical simulations is the map of the thermal load generated by the beam impact on the component. The deposited energy can be calculated through a Monte-Carlo particle transport code and is proportional to the density of the impacted material. Simulations described in this paper are performed with Autodyn [5], while energy deposition 3D maps are calculated with FLUKA [6],[7].

2.1. Lagrangian method

Impacted components can often be modeled with a lagrangian mesh: even when high deformations or material spallation occur, it is possible to overcome the limitations of the lagrangian method by, for example, implementing the erosion of elements at given thresholds of stress, temperature or deformation. This approach was therefore adopted to simulate the response of *type 1* samples impacted at medium energy during the test in the HiRadMat facility. When importing the thermal load in Autodyn, a subroutine interpolates the energy deposition map calculated by FLUKA and assigns the corresponding energy density to the Autodyn elements: ideally, no interpolation artifacts are generated throughout the process if the two meshes are identical. On the other hand, in order to decrease the simulation time, the mesh in Autodyn is often coarser and is refined only around the impact point, i.e. in the volume containing the most loaded elements.

The impinging proton pulse is constituted by trains of bunches spaced by 25 ns, while the impact duration of each bunch is 1 ns. The energy distribution deposited by one bunch, calculated with FLUKA, is therefore uploaded in Autodyn every 25 ns: this means that the time step of the simulation should be lower than 1 ns in order to correctly reproduce the impact phenomenon. The time step can be increased after the end of the beam impact up to 10 ns for numerical stability.

2.2. Smoothed-particle hydrodynamics (SPH)

The abovementioned lagrangian method is no longer adequate when the main goal of the study is to model the material fragments ejected after high-energy impacts. In such case, one has to adopt a different approach, such as the smoothed-particle hydrodynamics (SPH) technique. In this computational method, the material is modeled by discrete elements (particles) with a spatial distance of interaction (smoothing length) over which their properties are weighted by a kernel function. For example, material density in a given position x is:

$$\rho(x) = \sum_{J=1}^{n} m_j W\left(\left|x - x_j\right|, h\right) \tag{1}$$

where m_j is the mass of particle j, W is the kernel function and h is the smoothing length.

This method was used for the simulation of high energy deposition on HiRadMat *type 2* samples; since the SPH algorithm requires high computational time, only the inner volume of the specimen was modeled with SPH particles, while all the other regions were meshed with lagrangian elements.

3. Inermet180 material model

3.1. Equation of state

An equation of state (EOS) is a constitutive relation between the state variables of a material; usually it expresses the pressure (or the internal energy) as a function of two independent variables such as density and temperature. Most widely adopted EOS for wave propagation in solids are Shock, Tillotson and Mie-Grüneisen: these formulations involve complex theoretical models based on the band theory and combine them with shock compression experimental data. However, analytical modeling can describe only a single region of the EOS [8], excluding the description of phase transitions. Tabular EOS overcome this problem, integrating different analytical formulations to describe the behavior of different phases and the related phase transitions [9].

The EOS used in this work is the tabular SESAME N.3550 [10]; it has been devised for pure tungsten, for which experimental Hugoniot data are close to those of W-Ni-Cu heavy alloys [11].

3.2. Strength model

The strength model simulates the deviatoric behavior of the material and has to take into account the phenomena occuring in the matter during the beam impact, such as the melting of the impacted volume (loss of material strength), the variations in elastic and plastic behavior in the termally-altered zone and the hardening due to high strain rates in order to evaluate the propagation speed, the intensity and the damping during the travel inside the sample and obtain the wave intensity on the free surface.

A widely adopted model that considers the abovementioned effects is the Johnson-Cook (JC) model [12], whose parameters can be obtained through a set of experimental tests, which include Hopkinson Bars, Taylor cylinders, tensile and compression tests at different temperatures. JC model expresses the flow stress as:

$$\sigma_y = \left(A + B\varepsilon_{pl}^{n}\right)\left(1 + C\, ln\frac{\dot{\varepsilon}_{pl}}{\dot{\varepsilon}_0}\right)\left(1 - \left(\frac{T - T_m}{T_m - T_r}\right)^{m}\right) \tag{2}$$

where A is the elastic limit, B and n are the work hardening parameters, C and $\dot{\varepsilon}_0$ express the strain-rate sensitivity and m describes the thermal softening. Parameters for the JC model used in the present work are reported in Table 1; such values are based on the experimental results on W-Fe-Ni alloys and were slightly modified to fit Inermet® 180 mechanical properties.

3.3. Failure model

This model contains the set of conditions which lead to the bulk failure of the material; its choice heavily depends on the physical fracture mechanism. In case of particle beams impacting close to a free surface, the compressive shockwave immediately reflects back and turns into a tensile wave in the adjacent volume, causing the bulk failure of external crust and the projection of melted material.[13].

This mechanism could be reproduced in Autodyn through the Minimum Hydrostatic Pressure model (P_{min}); the model also implements the energy necessary for crack formation, calculated on the basis of the material fracture toughness. Reference values have been obtained experimentally with laser-induced spallation tests on polycrystalline tungsten [14].

Table 1. Inermet® 180 material model.

Equation of state	**Strength Model Johnson Cook**		**Failure Model Hydrostatic Minimum Pressure**	
SESAME 3550	Shear Modulus	140 GPa	P_{min}	2.7 GPa
	A	715 MPa	Fracture Energy	515 J/m^2
	B	177 MPa		
	n	0.12		
	C	0.016		
	m	1.00		
	T_{melt}	1616 K		
	Ref. Strain rate	1 s^{-1}		

4. Numerical simulations and experimental results

Table 2 reports the characteristic values of the impacting beam during tests on Inermet® 180 in the HiRadMat facility. Numerical simulations adopted the same parameters, except for the beam transverse dimension which was set to 2.5 x 2.5 mm^2.

Table 2. Beam parameters for tests performed on Inermet® 180 during HRMT-14 experiment

	Medium intensity test	**High intensity test**
Proton energy	440 GeV	440 GeV
Number of bunches	24	72
Pulse intensity	2.7e12 protons	9.05e12 protons
Energy deposited on the most loaded specimen	8.35 kJ	25.1 kJ
Theoretical impact point	Centre of *type 1* specimen	2 mm from *type 2* specimen flat surface
Beam transverse dimension	1.4 x 2 mm^2	1.9 x 1.9 mm^2

4.1. Medium intensity impacts

Strain gauges measured axial and hoop strains on the external surface of *type 1* samples, while the laser Doppler vibrometer (LDV) acquired the radial velocity. Acquired raw data were then compared to the results of numerical simulations (Figure 3,-Figure 4). A strong electromagnetic noise induced by the particle beam perturbed the strain gauge measurements during the first few microseconds after the impact, concealing the first deformation peak. However, this effect was limited to the beam impact duration, allowing to capture the remainder of the phenomenon. Measured and simulated signals are in good accordance during the first three reflections of the shockwave; afterwards, the modifications in the material bulk induced by the shock wave lead to an overestimation of the wave speed. Random spikes in the signal of gauges and LDV will be treated during more accurate signal processing.

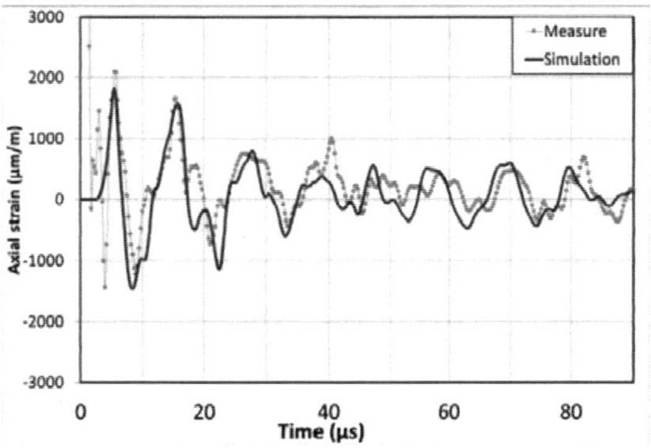

Figure 3. Axial strain measured with strain gauge at r = 20 mm, L = 15 mm and numerically simulated.

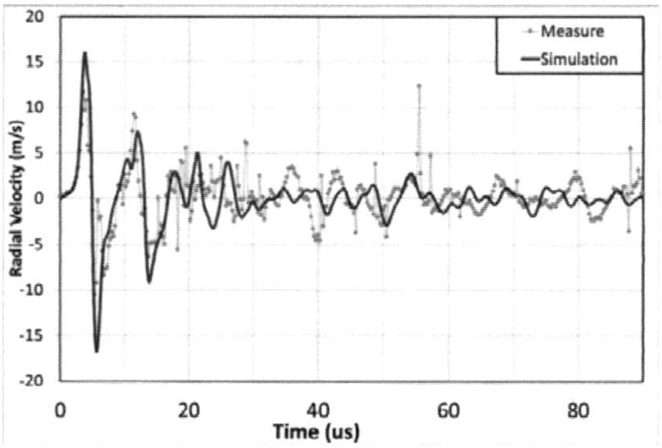

Figure 4. Radial velocity measured with LDV at r = 20 mm, L = 15 mm and numerically simulated.

4.2. High intensity impacts

The developed high-speed camera system allowed for the first time, to the best of authors' knowledge, to record images of the impact of a proton beam on solid targets and of the effects induced.

As shown in Figure 5, a large quantity of hot material was ejected at high velocity from the two most loaded Inermet® 180 *type 2* samples, generating the large cavities shown in Figure 7.

Concerning the SPH simulation results, ejected material front shape and calculated absolute velocity (ABS VEL in Figure 6) are both consistent with the high-speed camera measurements,even considering the differences in beam size between real and simulated scenarios. The velocity of the fragment front has been estimated by measuring the displacement between two successive frames and is about 275 m/s, well matching the calculated velocity of 316 m/s (difference is about 15%).

5. Conclusions

A complex and comprehensive experiment has recently been carried out at CERN, aiming at the characterization, mostly in real time, of six different materials impacted by 440 GeV/c intense proton pulses.

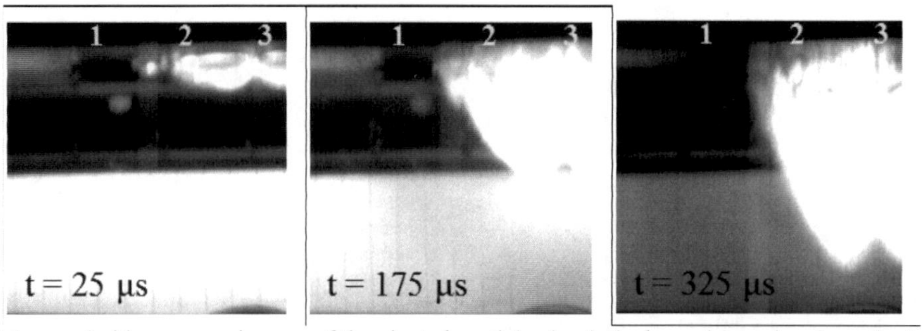

Figure 5. High-speed video camera images of the ejected particle cloud: the beam is coming from the left while the cloud is moving downwards; three Inermet® 180 samples are partially visible (numbered 1 to 3)

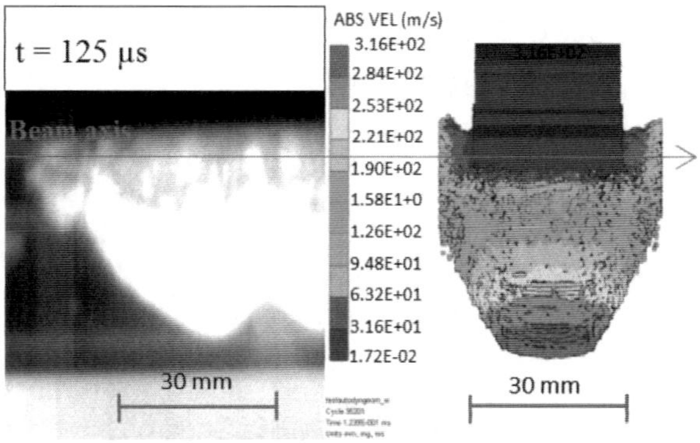

Figure 6. Comparison between simulation (SPH) and acquired image.

Figure 7 Error! No bookmark name given.. Post-mortem observations.

Preliminary measurements on Inermet® 180 specimens well match results of advanced computations, providing promising indications on the validity of the constitutive models used for this material. A large amount of data is still to be treated and will hopefully help deriving constitutive models for the less known composite materials. An extensive post-irradiation campaign, implying direct observations, non-destructive and destructive testing, is to be launched in future months to complete the experiment and provide additional valuable information.

References

[1] The LHC Design Report, http://ab-div.web.cern.ch/ab-div/Publications/LHC-DesignReport.html, July 2004

[2] Bertarelli A et al Limits for Beam Induced Damage: Reckless or Too Cautious? *Proc. Chamonix 2011 Workshop on LHC Performance*, January 2011

[3] Bertarelli A et al., An experiment to test advanced materials impacted by intense proton pulses at CERN HiRadMat facility A. Bertarelli et al., CERN-ATS-Note-2013-005 TECH

[4] Scapin M, Peroni L and Dallocchio A Effects induced by LHC high energy beam in copper structures 2012 *J. Nuclear Mater.* **420**

[5] ANSYS Autodyn User Manual, Release 13.0, ANSYS Inc., 2010

[6] Battistoni G et al., The FLUKA code: Description and benchmarking 2006 *Proc. of the Hadronic Shower Simulation Workshop* 2007 *Fermilab AIP Conference Proc.*

[7] Fassò A, Ferrari A, Ranft J and Sala P R FLUKA: a multi-particle transport code CERN-2005-10 (2005), INFN/TC_05/11, SLAC-R-773

[8] Fortov V E and Lomonosov I V Equations of state of matter at high energy densities, 2010 *Open Plasm. Phys. J.* **3** 122-130, 186-5343/10

[9] Kerley G I Equations of state for Be, Ni, W and Au, Sandia Report Sand 2003-3784, 2003

[10] *SESAME: the Los Alamos National Laboratory Equations of State Database*, ed: S P Lyons and J.D. Johnson Group T-1 LA-UR-92-3407

[11] Compendium of Shock Wave Data *Lawrence Livermore Laboratory report* UCRL-50108, Rev. 1, June 1977, ed M van Thiel

[12] Johnson J R and Cook W H A constitutive model and data for metals subjected to large strains, high strain rates and high temperatures 1983 *7th Int. Symp. on Ballistic* pp 541–547

[13] Grady D E The spall strength of condensed matter 1988 *J. Mech. Phys. Solids* **36**, 535-384

[14] Hu L, Miller P and Wang J, High strain-rate spallation and fracture of tungsten by laser-induced stress waves 2009 *Mater. Scien. Engin.* A **504** 73-80

Impact tensile properties and strength development mechanism of glass for reinforcement fiber

T Kim[1], K Oshima[1] and H Kawada[2]

[1] Department of Applied Mechanics and Aerospace Engineering, Graduate School of Waseda University, 3-4-1 Okubo, Shinjuku, Tokyo 169-8555, Japan
[2] Department of Applied Mechanics and Aerospace Engineering, Waseda University, 3-4-1 Okubo, Shinjuku, Tokyo 169-8555, Japan

E-mail: [1] sayoitu1126@akane.waseda.jp, [2] kawada@waseda.jp

Abstract. In this study, impact tensile properties of E-glass were investigated by fiber bundle testing under a high strain rate. The impact tests were performed employing two types of experiments. One is the tension-type split Hopkinson pressure bar system, and the other is the universal high-speed tensile-testing machine. As the results, it was found that not only the tensile strength but also the fracture strain of E-glass fiber improved with the strain rate. The absorbed strain energy of this material significantly increased. It was also found that the degree of the strain rate dependency of E-glass fibers on the tensile strength was varied according to fiber diameter. As for the strain rate dependency of the glass fiber under tensile loading condition, change of the small crack-propagation behaviour was considered to clarify the development of the fiber strength. The tensile fiber strength was estimated by employing the numerical simulation based on the slow crack-growth model (SCG). Through the parametric study against the coefficient of the crack propagation rate, the numerical estimation value was obtained for the various testing conditions. It was concluded that the slow crack-growth behaviour in the glass fiber was an essential for the increase in the strength of this material.

1. Introduction

In recent years, an application of the fiber-reinforced plastics (FRP) is growing, and the introduction rate in the fields of the aerospace and the automobile is higher than in other industries. Designing a structural member in these industries, the crashworthiness subjected to an impact load is an important design factor. Therefore, when applying FRP to a structural member, it is important to fully understand the characteristics under impact loading conditions. It has been reported that FRP has different strain-rate dependencies according to the reinforcements [1-3].

In previous studies, Taniguchi et al. investigated the influence of the strain rate on the tensile properties of FRP based on the thermoplastic epoxy resin. They found that strain rate has very small effect on tensile properties in case that the carbon fiber is employed as reinforcement, on the other hand, glass fiber composite was obviously affected by the strain rate [4].They concluded the fiberglass contributed more significantly to the strain rate dependency of FRP than the matrix. Xia et al. studied about impact properties of glass fiber [5-6]. From the result of the impact tensile test, they established a statistical model for the strain rate dependency and determined the constitutive equation of the glass fibers. Although the proposed model are in good agreement with the experimental results, it is not enough shows experimental data necessary for the material design such as the influence of fiber diameter and fiber glass composition.

In this study, impact tensile properties of glass fibers were investigated in order to obtain useful knowledge for material and structural design. First, the impact tensile test was performed about the fiber bundle of E-glass, and the strain rate dependencies on the mechanical properties were investigated. Then, impact tensile tests were performed on of the glass fibers having different diameters, and the effect of fiber diameter on the impact properties was investigated. Furthermore, the tensile fiber strength was estimated by employing the numerical simulation based on the slow crack-growth model (SCG).

2. Experiment

Figure 1 shows the geometry of the specimen used for impact tensile tests. E-glass fiber bundle (the fiber diameter is 17μm, the number of fibers is 1000) was used as the specimen, and the aluminum tabs were bonded at both ends. The impact tests were performed employing two types of experiments. One is the tension-type split Hopkinson pressure bar (SHPB) system, and the other is the universal high-speed tensile-testing machine (Shimadzu Hydro Shot, HITS-T10). Figure 2 shows schematics of the SHPB apparatus. It consists of an input bar with a flange, an output bar, and a cylindrical striker.

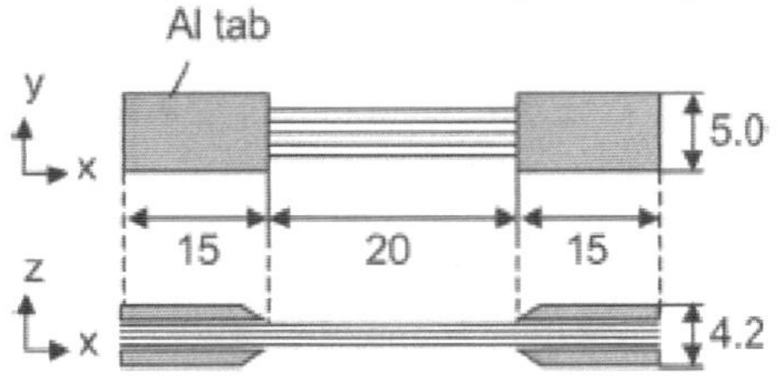

Figure 1. Geometry of the specimen used for impact tensile test.

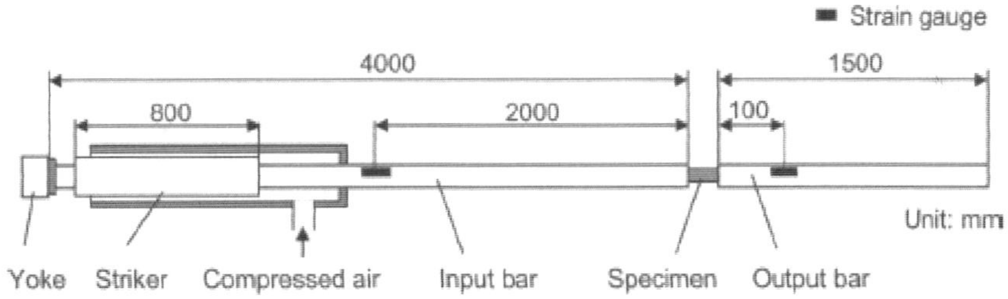

Figure 2. Tension-type SHPB apparatus

It is observed that a tensile incident wave is produced after the striker hits the yoke. In the SHPB system, the stress, strain and strain rate of a specimen can be calculated using the strain gauge outputs from the input and output bars on the basis of the one-dimensional wave propagation theory.

3. Experimental result and Discussion

3.1. *Effect of strain rate on mechanical properties*

Figure 3 shows the comparison of stress–strain curves of E-glass under quasi-static loading condition and strain rate of $250s^{-1}$. In Figure 3, the initial slope of the stress strain curve is not changed. In other words, there is no change in Young's modulus. On the other hand, with respect to the maximum stress and maximum strain, it can be seen that increasing has occurred in strain rate of $250s^{-1}$ compared to under quasi-static loading.

Figure 4 (a) - (c) shows the relationships between Young's modulus, tensile strength, fracture strain and strain rate, respectively. In addition, since the universal high-speed tensile-testing machine only measures the load history, strain rate dependency of the initial Young's modulus and fracture strain are examined only in the results of Hopkinson bar method. Figure 4 (a) shows the strain rate have very small effect on the initial Young's modulus. It is estimated that this is because E-glass fiber exhibits little viscoelasticity. Figure 4 (b) also shows the tensile strength of E-glass is improved with the strain rate. The tensile strength increases up to the strain rate of $50s^{-1}$, and after that the rate of increase becomes gradual. As a result, it increases by 60% compared to the quasi static condition in the strain rate of $250s^{-1}$. This is the same trend with the study by Xia et al [6].

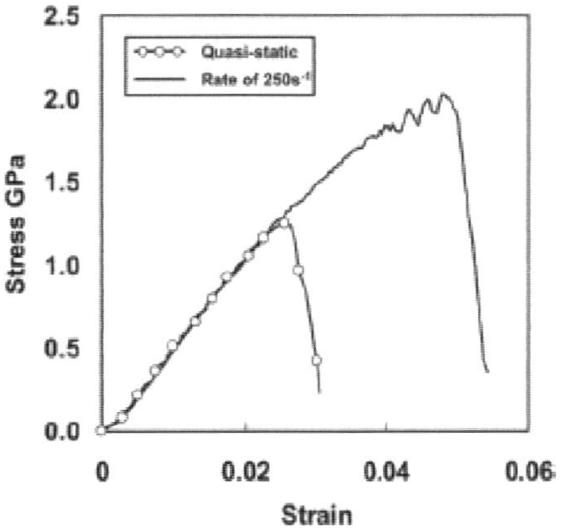

Figure 3. Comparison of tensile stress –strain curves of E-glass under quasi -static and strain rate of $250s^{-1}$.

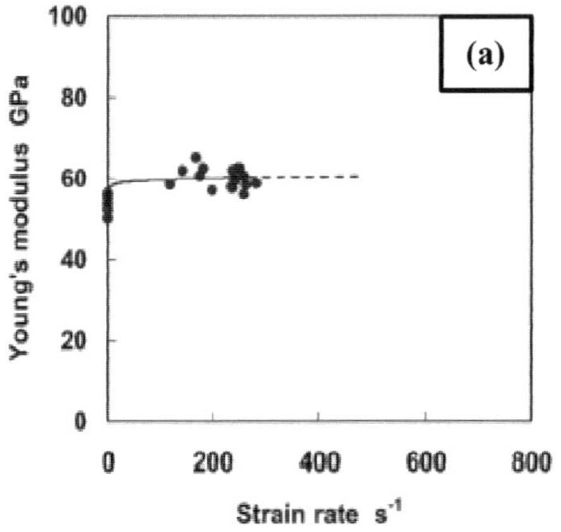

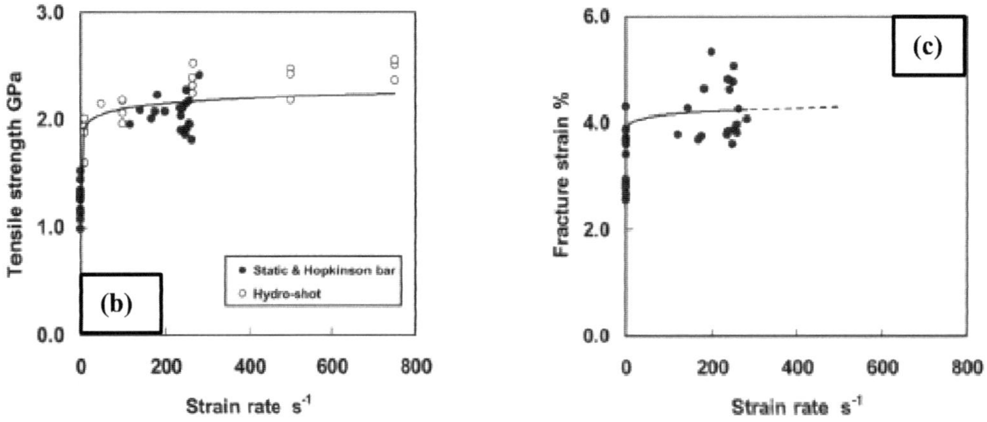

Figure 4. Dependence of strain rate on (a) Young's modulus, (b) tensile strength and (c) fracture strain in E-glass

One of the factors that tensile strength is improved by the strain rate is the slow crack-growth behavior which is the characteristic phenomenon of glass fracture. In general, the strength of the glass, σ_f is represented by the following formula in linear fracture mechanics.

$$\sigma_f = \frac{K_{IC}}{Y(a)\cdot\sqrt{\pi a(t)}} \tag{3.1}$$

where K_{IC} denotes the fracture toughness of glass, $a(t)$, the length of crack; and $Y(a)$, the shape factor.

Suratwala et al. and Freiman et al. reported that there is the slow crack-growth behavior, where the crack has grown slowly at a rate that depends on the stress intensity factor, before reaching the fracture toughness [7-8]. In equation (3.1), as the crack length is a function of time, it is possible that the crack length before reaching the fracture toughness is different under quasi-static loading and high strain rate.

For this reason, the tensile strength of the glass might be improved under high strain rates. The relationships between the slow crack-growth behavior and the strain rate dependency of E-glass on the tensile strength.

3.2. *Effect of fiber diameter on the strain rate dependency*

The tensile strength of glass fiber is generally dominated by micro-cracks present in the surface of the fiber. When varying fiber diameter of glass fiber, the surface properties of the fibers might change, and it is possible that the tensile strength of glass fiber significantly changes. It is also possible to change the strain rate dependency as well. Therefore, E-glass fiber bundles having five types different fiber diameter were prepared and evaluated the strain rate dependency of the tensile strength by Hopkinson bar method. Table 1 shows the specification of E-glass used in this study. The fiber diameter was varied by adjusting the winding speed of the glass fibers after the melting step.

Table 1. Specification of E-glass fiber

Specimen	EG17	EG13.8	EG9.2	EG7.5	EG5.3	EG3.2
Average diameter μm	17.0	13.8	9.2	7.5	5.3	3.2
Number of fiber	1000	906	200	200	200	1600

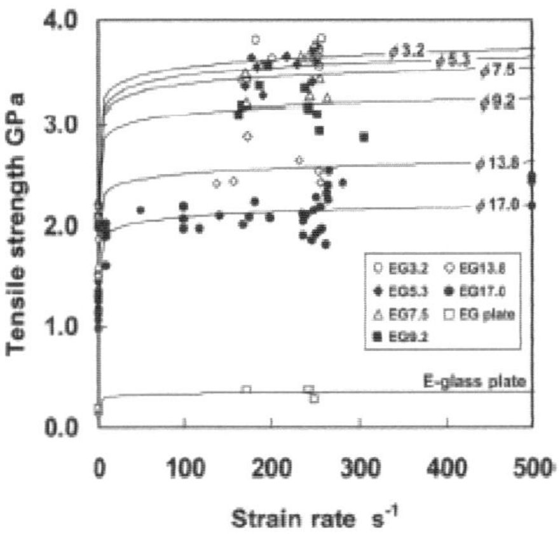

Figure 5. Relationships between tensile strength and strain rate obtained for each specimen

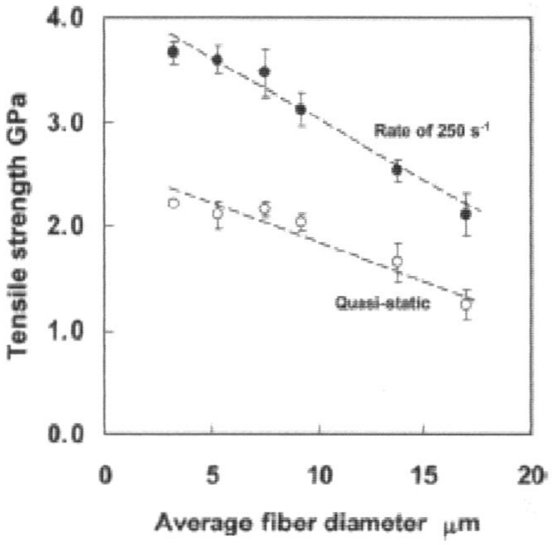

Figure 6. Relationships between tensile stress and the fiber diameter under quasi-static loading condition and strain rate of $250s^{-1}$

Figure 5 shows the relationships between tensile strength and strain rate obtained for each specimen. In all specimens, the tensile strength increased with increasing strain rate. Further, it was confirmed that E-glass fiber with smaller diameter tends to exhibit higher strength. Therefore, E-glass with smaller diameter should be less affected by micro-cracks present in the surface of the fiber. In addition, Figure 5 also shows the results of impact tensile test of the flat glass plate (the gage length, the width and the thickness of the specimen is 20mm, 5.0mm and 0.5mm. the aluminium tabs were bonded at both ends of the specimen) having same component with E-glass fiber. Although strain rate dependency of tensile strength can be confirmed in the flat glass plate specimen, the strength is very low value compared to the glass fiber specimens.

Figure 6 shows the relationships between tensile stress and the fiber diameter under quasi-static loading condition and strain rate of $250s^{-1}$. It was confirmed that E-glass fiber with smaller diameter exhibits higher strength, and E-glass fiber under strain rate of 250s-1 exhibits higher strength than under quasi-static loading condition. As a result, it is concluded that the size of the fiber diameter is a factor which affect the strain rate dependency of the tensile strength on E-glass. In addition, the surface properties of the fibers might change when varying diameter of glass fiber. So, the surface properties might have some influence on the strain rate dependency of the tensile strength.

4. Prediction of glass fiber strength

The tensile strength of glass fiber was estimated by employing the numerical simulation based on the slow crack-growth model (SCG) and using material values obtained in the double cleavage drilled compression (DCDC) test. In prediction of glass fiber strength, following assumptions were applied.

(1) The material constants of the glass fibers are not changed by the test environment.
(2) The glass fiber is treated as an elastic body.
(3) The glass fiber is cylindrical and has a penny shaped crack.
(4) The strength of the glass fiber dominated by the crack length.

First, using assumptions (1) and (2), the applied stress $\sigma_{app}(t)$ is calculated by multiplying the Young's modulus E (72.6GPa (Silica glass, E-glass), 84.5GPa (T-glass)) in the strain ε (t).

$$\sigma_{app}(t) = E\varepsilon(t) \tag{4.1}$$

Then, the stress intensity factor K_I (t) at any given time t is calculated by the following equation,

$$K_I(t) = Y(a(t))\sigma_{app}(t)\sqrt{\pi a(t)} \tag{4.2}$$

where, Y denotes the shape correction factor, $a(t)$, the length of crack. The shape correction factor Y is calculated by the following equation using assumption (3) [9].

$$Y(a(t)) = 0.661 - 0.011\left[\frac{a(t)}{r}\right] + 0.415\left[\frac{a(t)}{r}\right]^2 \tag{4.3}$$

When the stress intensity factor K_I is greater than the lower limit of the stress intensity factor K_S and crack growth rate V is calculated from the following empirical formula.

$$V(t) = \frac{\mathrm{d}a(t)}{\mathrm{d}t} = A[K_I(t)]^n \tag{4.4}$$

The crack length a is obtained by time integrating the crack growth rate V.

$$a(t) = \int_0^t V(\tau)\mathrm{d}\tau \tag{4.5}$$

Further, the apparent fiber strength $\sigma_{as}(t)$ for load time can be obtained by using the fracture toughness K_{IC}.

$$\sigma_{as}(t) = \frac{K_{IC}}{Y(a(t))\sqrt{\pi a(t)}} \tag{4.6}$$

The fiber is assumed to break when the applied stress σ_{as} reaches the apparent fiber strength σ_{app}, and then, the applied stress σ_{as} denotes the fiber strength σ_b.

Table 2 shows the calculated fiber strength σ_b. It is confirmed that the strength is T-glass > E-glass > Silica glass in the same environment. In addition, there is a tendency that the glass fiber in lower humidity environment has higher strength. However, the tensile strength of each glass fiber is at least 5GPa in silica glass (fiber diameter 125μm), 3.43GPa in E-glass (13 μm), 4.65GPa in T-glass (13 μm), so it is significantly different from the result of strength prediction. These are the strength of the glass with small defects immediately after manufacture. Therefore, it is necessary to consider the initial defect a_0. Figure 7 shows the relationship between calculated strength and initial crack length using the results of DCDC tests (Relative humidity (RH) = 2%). From Figure 7, in order to obtain the fiber strength immediately after manufacture, it is necessary that the initial defects in Silica glass, E-glass and T-glass is less than 6nm, 14nm and 9nm, respectively. Although these values may be as initial defect, its validity must be investigated further. Thus, the single fiber tensile test of E-glass was performed at normal temperature and humidity, and the validity of the strength prediction was investigated. As the result, the breaking stress σ_0 of 522MPa was obtained. In addition, observing the fracture surface (Figure 8) by the optical microscope, the initial defect size a_0 was measured to be 500nm. Strength prediction was performed using the data in DCDC test of E-glass (RH = 65%) and the initial defect size a_0. The calculated strength σ_b was 595MPa, and almost coincide with the value of the breaking stress σ_0 of 522MPa. Therefore, the validity of this strength prediction model was confirmed.

Comparing the result of the strength prediction with V-K_I diagram, to indicate low crack growth rate and high stress intensity factor is a key of high strength grass fiber. In other words, it is concluded that the slow crack-growth behaviour in the glass fiber was an essential for the increase in the strength of this material.

Table 2. Calculated fiber strength σ_b

Glass	S	E	T	S	E	T	S	E	T
σ_b GPa	1.51	1.74	1.98	1.62	1.82	2.04	1.82	1.80	2.04
RH %	65±5			25±5			2±2		
Environment	In air						In Ar(g)		

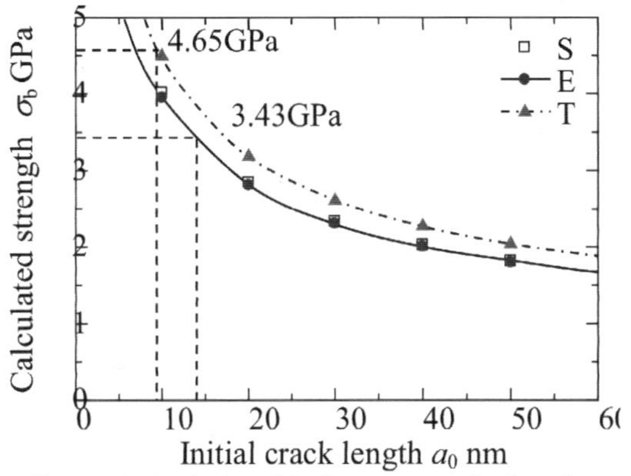

Figure 7. Relationship between calculated strength and initial crack length

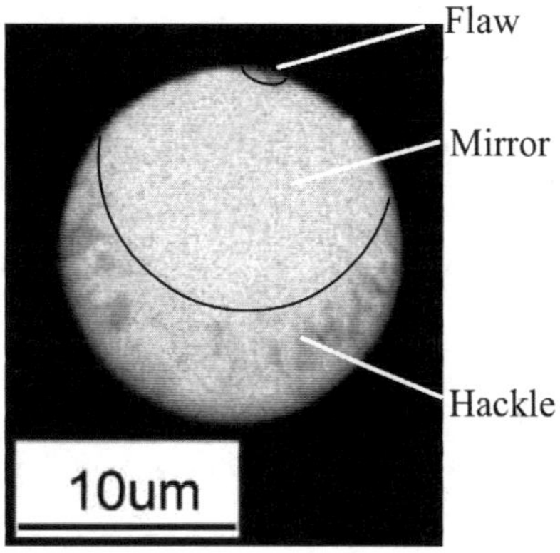

Figure 8. Fracture surface of E-glass fiber

References

[1] Harding J and Welsh L M 1983 *J. Mater. Sci.* **18** 1810-1826
[2] Harding J 1993 *Composites* **24** 4 323-332
[3] Shokrieh M M and Omid M J 2009 *Compos. Struct.* **88** 595-601
[4] Taniguchi N, Nishiwaki T, Hirayama N, Nishida H and Kawada H 2009 *Transactions of the Japan Society of Mechanical Engineers Series A* **75** 758 1284-1289
[5] Xia Y, Yuan J and Yang B 1994 *Compos. Sci. Technol.* **52** 499-504
[6] Wang Z and Xia Y 1998 *Compos. Sci. Technol.* **57** 1599-1607
[7] Suratwala T I and Steele R A 2003 *J. Non-Crystalline Solids* **316** 174-182
[8] Freiman S W, Wiederhorn S M and Mecholsky J J, Jr. 2009 *J. Am. Ceram. Soc.* **92** 7 1371-1382
[9] Okubo K and Hujii T 2000 *Journal of the Society of Materials Science* **49** 7 819-824

Shear fracture of jointed steel plates of bolted joints under impact load

M Daimaruya[1], H Fujiki[1], H Ambarita[2], H Kobayashi[3] and H-S Shin[4]

[1] Graduate School of Engineering, Muroran Institute of Technology, Mizumoto, Muroran 050-8585, Japan
[2] Department of Mechanical Engineering, University of Sumatera Utara, Jl. Almamater Kampus USU, Medan 20155, Indonesia
[3] Department of Mech. Sci. and Bioengineering, Graduate School of Eng. Sci., Osaka University, Toyonaka, Osaka 560-8531, Japan
[4] Department of Mechanical Design Engineering, Andong National University, Andong, Kyungbuk 760-749, Korea

E-mail: masashi@mmm.muroran-it.ac.jp

Abstract. The present study is concerned with the development of a fracture criterion for the impact fracture of jointed steel plates of bolted joints used in a car body, which contributes to crash simulations by CAE. We focus our attention on the shear fracture of the jointed steel plates of lap-bolted joints in the suspension of a car under impact load. Members of lap-bolted joints are modelled as a pair of steel plates connected by a bolt. One of the plates is a specimen subjected to plastic deformation and fracture and the other is a jig subjected to elastic deformation only. Three kinds of steel plate specimens are examined, i.e., a common steel plate with a tensile strength of 270 MPa and high tensile strength steel plates of 440 and 590 MPa used for cars. The impact shear test was performed using the split Hopkinson bar technique for tension impact, together with the static test using a universal testing machine INSTRON 5586. The behaviour of the shear stress and deformation up to rupture taking place in the joint was discussed. The obtained results suggest that a stress-based fracture criterion may be developed for the impact fracture of jointed steel plates of a lap-bolted joint.

1. Introduction

In the vehicle industries collision safety and reliability of cars under impact loading become more and more important in recent years, together with weight saving. The present study is concerned with the development of a fracture criterion for the impact fracture of jointed steel plates of bolted joints used in a car body, which contributes to crash simulations by computer-added engineering (CAE). For the accurate prediction of crash characteristics of car bodies by CAE, it is also necessary to examine the strength and failure of bolted joints subjected to impact loads, together with those of welded joints such as spot welds and tailored blank welds. The present study concentrates on the impact behaviour and fracture of connected steel plates of lap-bolted joints used in cars. There has been much work on bolted joints used in machine elements, in which loosening processes and tightening torque of bolted joints are investigated in detail [1][2]. However there remain many uncertainties in the impact behaviour and fracture of bolted joint parts. In particular, we can hardly find investigations regarding the fracture of jointed steel plates of lap-bolted joints under impact loading.

In this study the impact behaviour and fracture of jointed steel plates of lap-bolted joints used in cars are examined by experimental and numerical simulation. The object is the parts of lap-bolted joints connecting the suspension frame and the bracket of the suspension frame rear mounting in a car. They are modelled as a pair of two steel plates connected by a bolt. One of the plates is a specimen subjected to plastic deformation and fracture and the other is a jig subjected to elastic deformation only. Attention is given to the impact shear deformation and fracture of the specimen plate, even though the actual behaviour of the bolted joint is complicated. Three kinds of steel plate specimens are examined, i.e., a common steel plate with a tensile strength of 270 MPa rank and high tensile strength steel plates of 440 and 590 MPa rank used for cars. The jig plate is made of high tensile strength steel of 980 MPa rank in order to make sure that the shear fracture occurs only in the specimen plates. The impact shear test of the specimens was performed using the Split Hopkinson bar technique for tension impact, together with the static test using a universal testing machine INSTRON 5586. The variation of stresses with time and the fracture characteristics in the specimen plates jointed by a bolt were measured and examined. This paper reports only about the bolted joints in hand-tightened, even though the experiments under other fastening conditions of bolted joint, such as torque and axial force, were carried out [3], [4]. The obtained results suggest that a stress-based fracture criterion may be developed for the impact fracture of jointed steel plates of a lap-bolted joint.

2. Lap-bolted joints

Although the actual fracture of lap-bolted joints in cars on collision is complicated, it might be classify into shear fracture and extraction of a bolt from the members [3][4]. The shear deformation and fracture of the specimen steel plate jointed by a bolt under impact loadings are focused on in this study. First, the static tests are conducted to obtain the static characteristics of the specimen of lap-bolted joints. Figure 1 shows a lap-bolted joint specimen with a M6 bolt used in the tests. Figure 2 shows the configuration of the specimen made of steel sheet with a tensile strength of 270 MPa (CR270) and high tensile strengths of 440 MPa (CR440) and 590 MPa (CR590) used for a car body. The jig plate is made of a high tensile strength steel of 980 MPa (CR980). The thickness of CR270 plate is 2.0 mm and the thickness of CR440, CR590 and CR980 plates is 1.2 mm. In both static and impact tests the specimen plates are plastically deformed and finally ruptured under loadings, while the jig plate remains within elastic. In Figure 2 the letter "*e*" denotes the distance between centreline of a joint and the end of a jointed steel plate. Three kinds of *e* (6.0 mm, 7.5 mm and 9 mm) were used in the tests. The static test was done using a universal testing machine INSTRON 5586 with a cross-head speed 0.01 mm/s.

Figure 3 shows the static stress-strain diagrams of the four kinds of steel plates used in the present experiments.

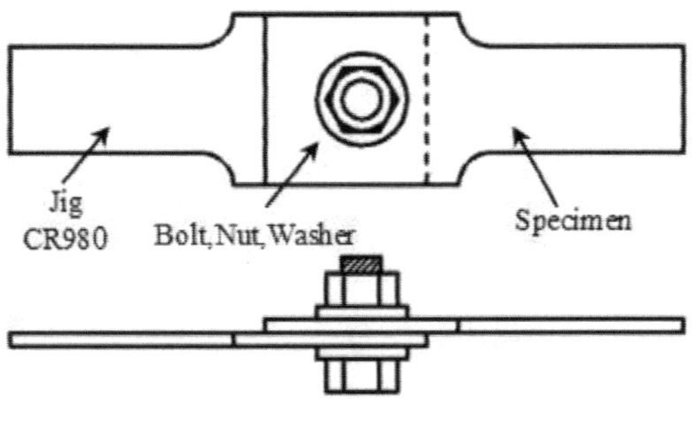

Figure 1. Lap-bolted joint specimen.

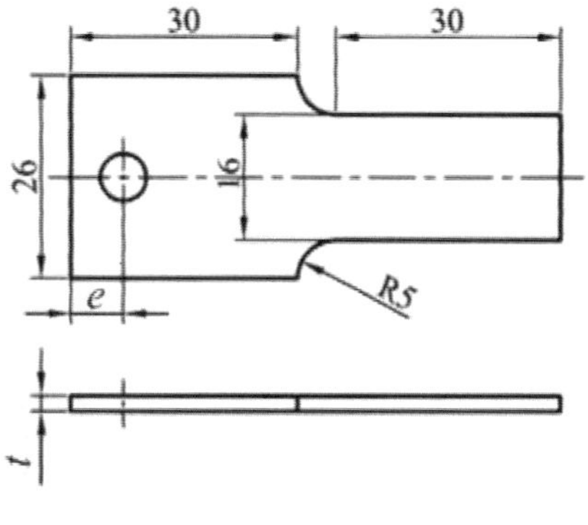

Figure 2. Dimension of steel plate specimen.

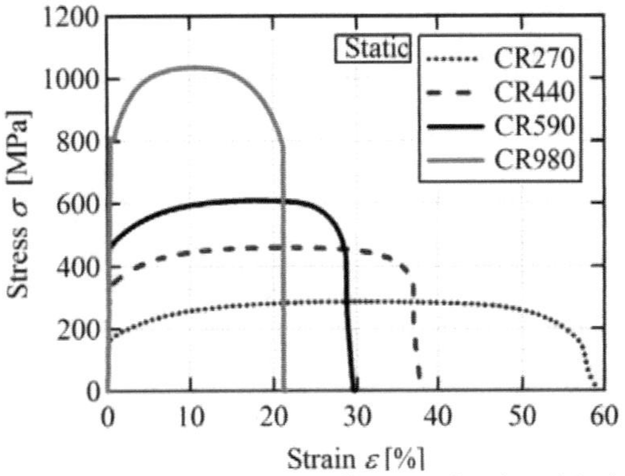

Figure 3. Stress-strain diagram for four kinds of steels in static tensile tests.

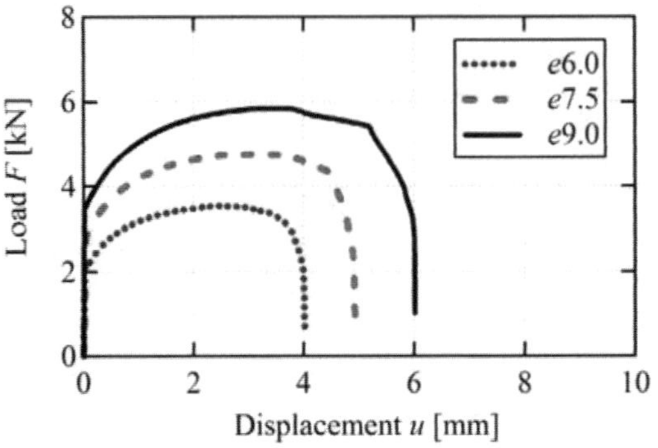

Figure 4. Load (F)-displacement (u) diagram of CR440 for each e in static tests.

Figure 4 shows the curves of load (F) versus displacement (u) of CR440 steel plates obtained by static shearing tests. It may be seen that the shearing loads increase with increasing e. The similar diagrams are also obtained regarding the plates of CR270 and CR590.

3. Impact experiments

3.1 Apparatus of impact experiments and procedure

The measurement for impact tests was performed using the vertical type of modified split Hopkinson tension bar [3]. The schematic diagram of the apparatus is shown in Figure 5.

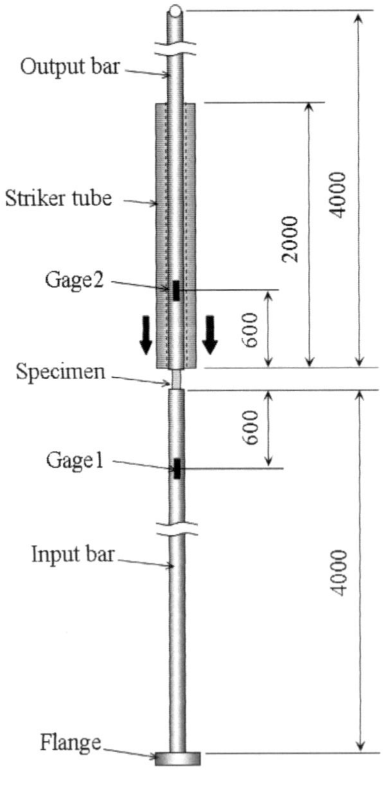

Figure 5. Schematic diagram of split Hopkinson bar.

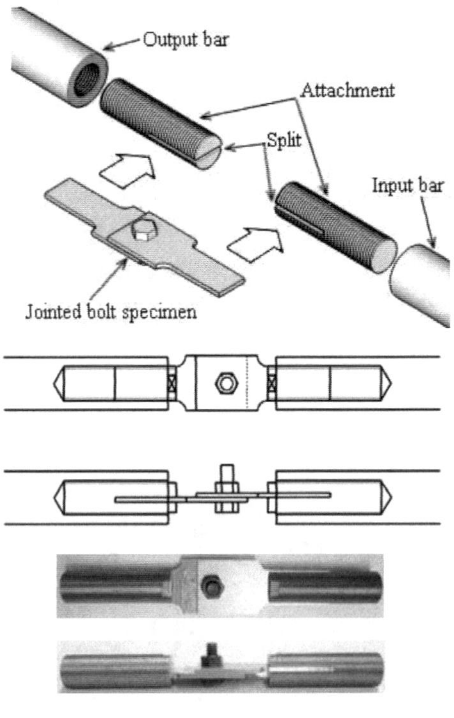

Figure 6. Bolted joint specimen with threaded attachments screwed into Hopkinson bars.

Both input and output bars are made of SUS304 stainless steel and have the same diameter of 25.0 mm and length of 4000 mm. The Young's modulus of the bars is 194 GPa, the mass density is 7.76x10^3 kg/m^3 and the velocity of elastic wave is 5.02x10^3 m/s. The steel striker tube, 42.7 mm in outside diameter, 35.5 mm in inside diameter and 2000 mm in length, was dropped from 4.0 m in height (impact velocity is about V=8.85 m/s). Two semiconductor strain gages (KYOWA, KSP-2-120-E4) were glued at three locations axial-symmetrically to capture the stress waves. The signals from the strain gauges were passing through the bridge boxes and amplifiers (KYOWA CDV-700A), and recorded by a digital oscilloscope (Nicolet INTEGRA10) with sampling time of 1 μs. The fastening method of bolted joint specimens to the Hopkinson bar is illustrated in Figure 6.

The displacement u, the rate of displacement $\dot{u}$ and the load F were calculated by Equation (1), following the Hopkinson bar theory [3].

$$u = \frac{1}{\rho c}\int(\sigma_i - \sigma_r - \sigma_t)dt \qquad \dot{u} = \frac{1}{\rho c}(\sigma_i - \sigma_r - \sigma_t) \qquad F = \frac{A}{2}(\sigma_i + \sigma_r + \sigma_t) \tag{1}$$

Here σ_i is the incident stress wave propagated to the input bar, σ_r is the reflected stress wave to the input bar and σ_t is the transmitted stress wave propagated to the output bar. And ρ is the mass density, c the elastic wave velocity, A the cross-sectional area of input and output bars.

Figure 7(a) and (b) show the curves of impact load (F)–displacement (u) for CR270 and CR440 obtained by the impact tests, which were calculated from Equation (1). The vibration on the load waves seems to be caused by the propagation of complicated stress wave in the specimen of an lap-bolted joint that assembled four kinds of elements, namely, a testing steel plate, a jig plate, a jointed bolt and washers. The impact loads increase with increasing of the value of e, same as in static tests. The fractured steel plate specimens of CR270 after the impact test are shown in Figure 8.

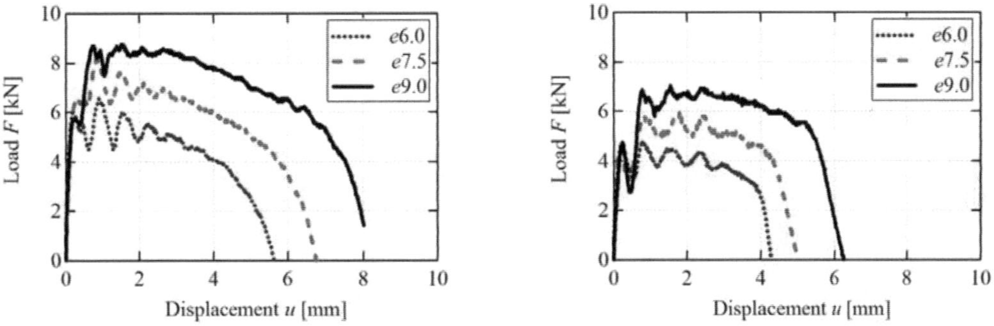

(a) CR270 (t=2.0) (b) CR440 (t=1.2)

Figure 7. Load-displacement diagrams for each e in impact tests.

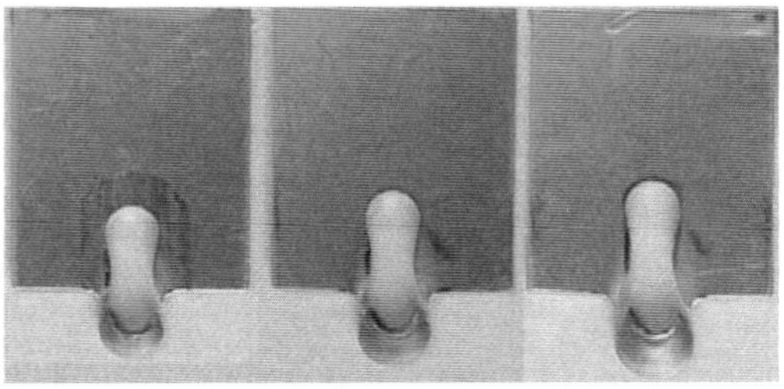

(a) e6.0 (b) e7.5 (c) e9.0

Figure 8. Fractured specimens of CR270.

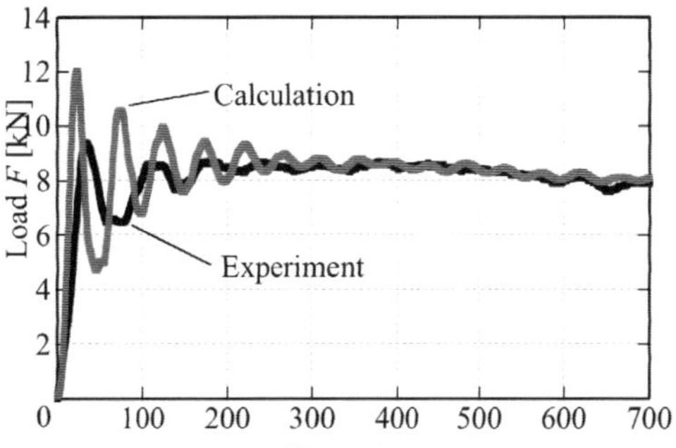

Figure 9. Variation of impact loads with time in measured and numerical results.

3.2 Transformation of impact loads into stresses

Numerical simulations were carried out for CR270 using a FEM code LS-DYNA in order to understand the fracture mechanism of jointed steel plates of lap-bolted joints under impact loadings in the previous paper [3], [4]. Figure 9 shows the variation of impact loads with time that was measured at the gauge 2 on the output bar, and a corresponding numerical simulation. The simulation may approximate the vibration on load waves observed in impact tests. Figure 10 shows the simulated, sequence distributions of impact shear stress τ_{yx}. The deformation process presented in Figure 10 is compatible with the experimental results shown in Figure 8. From consideration of the numerical simulation and the configuration of fractured specimens after impact tests, the shearing fracture process seems to be dominant, and then the area of shearing cross section, A, illustrated in Figure 11 is assumed to normalize the impact loads into shearing stresses [3], [4], [6].

Figure 12 (a) and (b) show the relations of (F/A) and (u/e) of CR270 and CR440 for each e, which correspond to Figure 7(a) and (b), respectively. For the sake of convenience, we should refer to the shearing stress and the moving ratio as (F/A) and (u/e), respectively. It may be seen that the three curves of the shearing stress and the moving ratio become similar to each other in spite of the difference of e values. The same relationship is also obtained in the static results.

Figure 13 shows the relation of the shearing fracture loads and the distance e obtained in the present static and impact tests. The outline marks denote the static fracture loads, while the filling marks the impact fracture loads. The fracture loads increase with increasing of e and the impact results increase

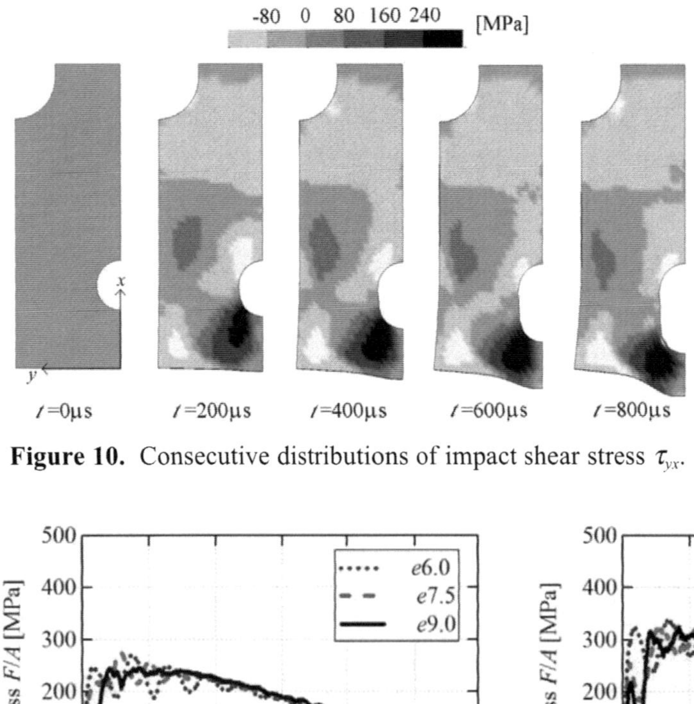

Figure 10. Consecutive distributions of impact shear stress τ_{yx}.

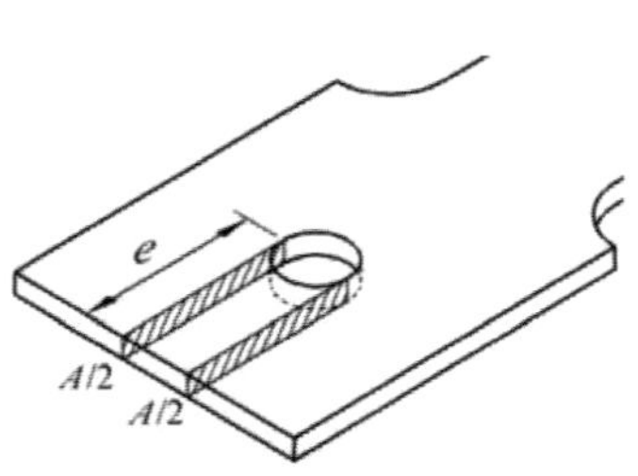

Figure 11. Shearing cross -sectional areas.

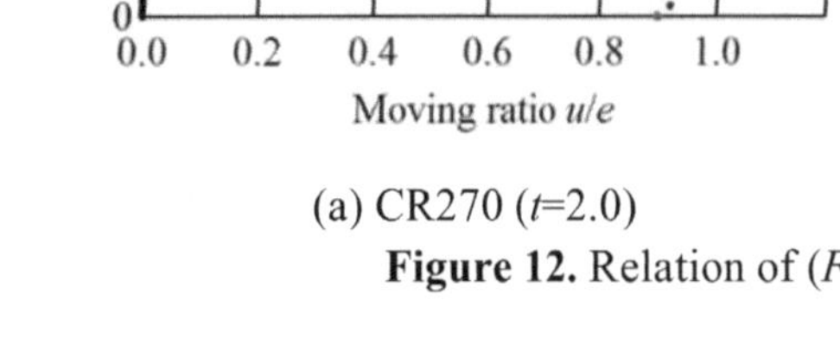

(a) CR270 (t=2.0)

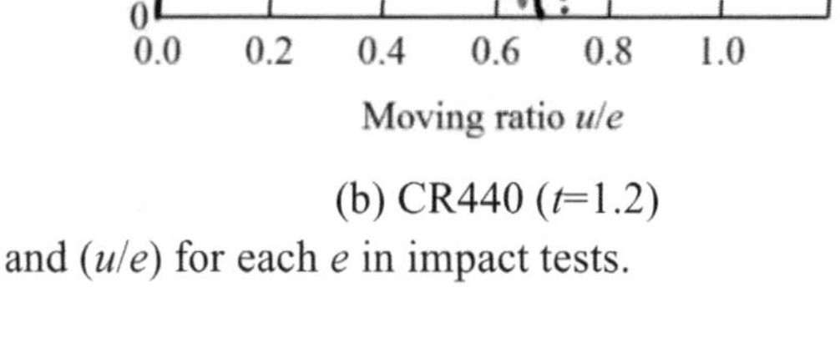

(b) CR440 (t=1.2)

Figure 12. Relation of (F/A) and (u/e) for each e in impact tests.

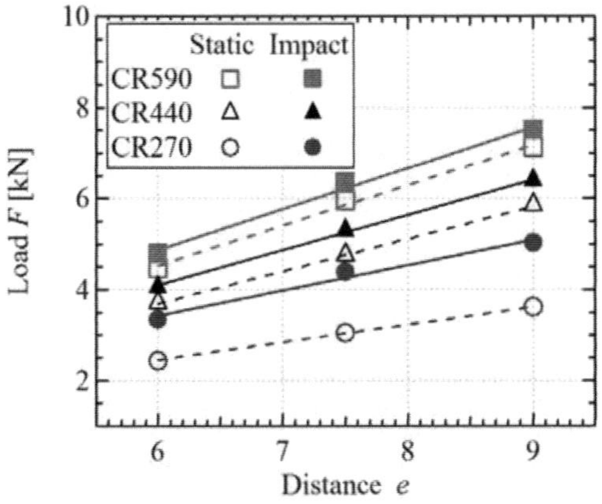

Figure 13. Relation of fracture load and distance of *e* for each steel plate.

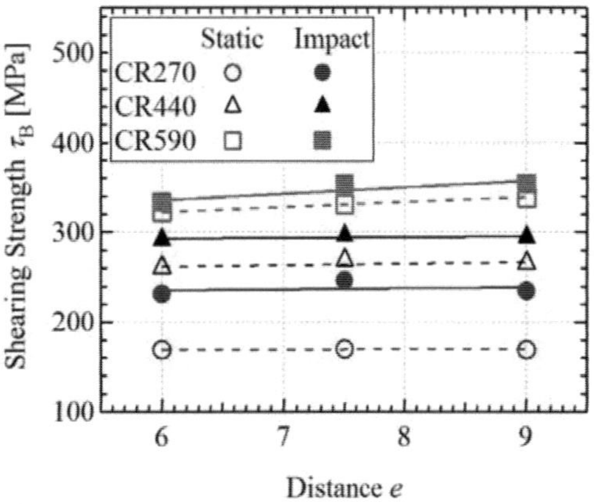

Figure 14. Relation of shearing strength and distance of *e* for each steel plate.

by comparison with the static results due to the effect of strain rate.

Figure 14 shows the relation of the average shearing strength and the distance of *e* for each steel plate. It can be seen that the shearing strength becomes much the same for each steel plate in spite of the difference of *e*. This fact suggests that a stress-based fracture criterion may be developed for the impact shearing fracture of jointed steel plates of a lap-bolted joint.

4. Concluding remarks

The behaviour and fracture of jointed steel plates of lap-bolted joints used in the suspension parts of cars are examined in static and impact experiments. In this study the shear deformation and shear fracture of the testing specimen plates are considered, even though the actual behaviour of the bolted joint parts is complicated. The specimens are made of three kinds of steel plates used for a car body. The impact shear test of the specimens was performed using the Split Hopkinson bar technique for tension impact, while the static test was done using a universal testing machine INSTRON 5586.

The shearing fracture loads increased with increasing of the distance *e* between centerline of a joint and the end of a jointed steel plate in both of static and impact tests. And the impact fracture loads increased by comparison with the static ones due to the effect of strain rate. However, the shearing strength, defined in this study, became much the same for each steel plate regardless of the difference of *e*. This fact suggests that a stress-based fracture criterion may be developed for the impact shearing fracture of jointed steel plates of lap-bolted joint used in a car body.

Acknowledgments
This work was supported by JSPS KAKENHI Grant Number 23560077. I would like to express my gratitude to Japan Society for the Promotion of Science.

References

[1] Yamamoto A 1995 *Design and Principle of Bolted Joints* (Tokyo:Yokendo)
[2] Izumi S, Yokoyama T, Iwasaki A and Sakai S 2005 *Trans. Japan Society. Mech. Eng.* **71** 204
[3] Daimaruya M, Fujiki H and Uemura U 2009 *Proc. DYMAT2009 (Brussels)* 867
[4] Uemura U, et al. 2011 *Society Mate. Science Japan* **60** 662
[5] Daimaruya M, Fujiki H and Himsar A 2012 *Proc. DYMAT2012 (Freiburg)* 01013-p.1
[6] Architectural Inst. of Japan 2006 *Recommendation for Design of Connections in Steel Structures* p46

Evaluation of dynamic collapse in thin-walled composite members

N Nishimura[1], K Murase[1], T Onda[2] and T Watanabe[3]

[1] Department of Vehicle and Mechanical Engineering, Meijo University, 1-501 Shiogamaguchi, Tempaku-ku, Nagoya 468-8502, JAPAN
[2] Iida Industry Co .Ltd., 1-5 Numa Kitaohmi, Inazawa, Aichi 492-8547, JAPAN
[3] Graduate Student, Department of Transportation Engineering, Meijo University, 1-501 Shiogamaguchi, Tempaku-ku, Nagoya 468-8502, JAPAN

E-mail: nisimura@meijo-u.ac.jp

Abstract. Many of steel thin wall closed section members are used by automobile industry. The local buckling of these elements that occur under compressive stress remains the main technological problem. To improve the performance of the members, the filling the inside of closed section with the low-density foaming material are usually used. In this study, we present the usefulness of axial compression and 3-point bending tests for the composite members filled with two kinds of epoxy resin. The difference between composite member and thin-walled member was investigated concerning their elastic and plastic collapse characteristics while tested with various deformation velocities. The adherent strength between thin-walled member and foaming material, and filling effect of the foaming material in the energy absorption characteristics was clarified.

1. Introduction

Steel thin wall closed section member is mainly used for traffic applications (e.g., automobile). The main problem here is a local buckling under compressive stress that was already thoroughly investigated [1-4]. To resolve this problem, the elements filled with the low-density foaming material are used nowadays [5-8]. In this report, we call such a member the "thin-walled composite member" or "composite member" in short. Recently, it was found that the filling effect is useful for cost and performance improvement [9,10]. The advantages of thin-walled composite member include: increase of yield load, elastic rigidity and loading-plateau in plastic region. Composite member exhibit high strength, improved rigidity and enhanced energy absorption in the plastic region. However, homogeneity and stability of the foaming in the closed section appears here the main obstacle due to employment of the polymer material [6]. Therefore, mechanical properties of the foaming material are seem very unstable, and the adherent strength, synthesis characteristics and deformation velocity dependence, remain here unsolved issue. Consequently, it is necessity of independent examination of these construction elements [11].

The aim of the present paper is to reveal the difference of thin-walled member and composite member under the dynamic compression loading. As a test piece, we used two kinds of composite member filled with the foaming material of the different density for the hat type thin-walled cross section member. Axial compression and 3-point bending tests were carried out on the composite member, the hat type thin-walled member and the epoxy resin foaming material. The difference in adherent strength and the energy absorption of composite member and thin-walled member was carefully investigated from the point of view of elasticity and plastic collapse at various deformation velocities.

2. Experimental

2.1. Test pieces

The thin-walled member contains the hat type made of cold rolled steel sheet, SPCC, as shown in Figure 1(a). The foaming material - thermosetting epoxy resin (figure1(b)) was prepared by multiple foaming procedure: 2 times foaming material, specific gravity 0.77g/cm^3 and 3 times foaming material, specific gravity 0.55g/cm^3). The composite member filled the hat type thin-walled member with 2 and 3 times foamed material (figure 1(c)). The foam – steel interface was adhered by filling. Constitution of the foaming material, specification of the test pieces and test conditions are given in tables 1, 2 and 3, respectively.

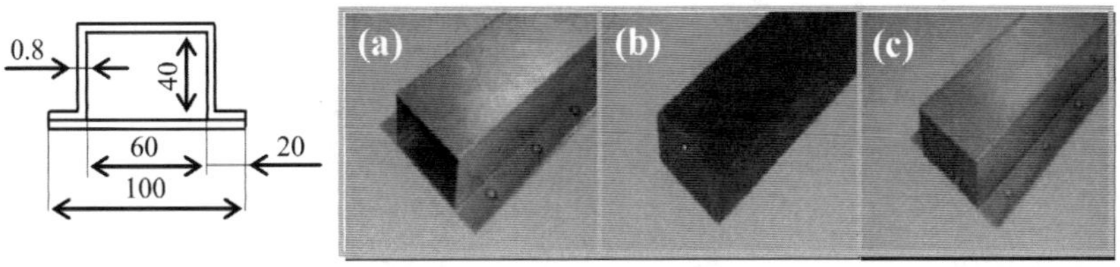

(a)Hat type (b)Foam (c)Composite member

Figure 1. Test pieces for axial compression and 3-point bending tests.

Table 1. Constitution of foaming material.

Material	Weight ratio (%)	Material	Weight ratio (%)
Epoxy resin	34.5	Carbon black	1.6
Synthetic rubber	8.1	Increasing agent	45.8
Thermoplastic elastomers	5.8	Hardening agent / accelerator	2.6
Foaming agent	1.6	Total	100

Table 2. Specification of test piece.

Test piece	Foam		Thin-walled member
	Two times*	Three times*	(Hat type)
Cross-section area A	2.40×10^{-3} (m^2)		2.24×10^{-4} (m^2)
Young's modulus E	1.14(GPa)	659(MPa)	206 (GPa)
Moment of inertia of cross -section area I	3.20×10^{-7} (m^4)		6.78×10^{-8} (m^4)
Yield stress σ_Y	13.7 (MPa)	7.73 (MPa)	288 (MPa)
Ultimate tensile stress σ_U	14.9 (MPa)	8.54 (MPa)	394 (MPa)

*The result as length of member is 100mm.

Table 3. Types of test piece and test.

Test piece	Foaming material	Thin-walled member	Composite member
Length of member	25, 50, 100, 200, 400mm, 500*mm (*only bending test)	400mm, 500*mm (*only bending test)	
Type of test	Axial compression, Three-point bending (span 400mm)		
Deformation velocity	0.05mm/s, 5mm/s, 400mm/s		
Foaming rate	Two times (0.77g/cm^3), Three times (0.55g/cm^3) (Relative density of foaming material)		

2.2. Experimental procedure

Axial compression test and three-point bending test, as shown in figure 3, were carried out using the hydraulic dynamic testing machine (figure 2). The maximum displacement in the axial compression test was 150mm on the thin-walled member and composite member, and a 50% of the member length on the foaming material. The three-point bending test concerned, the maximum displacement, and the contact radius at loading and supporting point were 50mm, 5mm and 1.5mm, respectively. The employed deformation velocity equalled 0.05mm/s, 5mm/s and 400mm/s during our experiments.

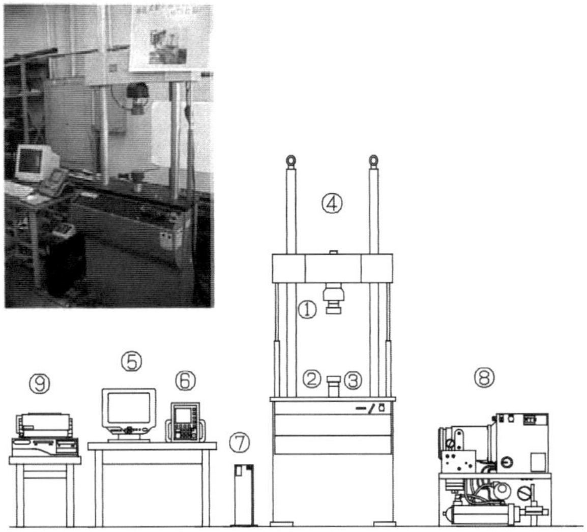

①Load Cell ②Base ③Actuator ④Cross Head
⑤Monitor ⑥Remote Station Control⑦Digital Controller (TestStar II s)
⑧Hydraulic Power Supplies⑨Printer
⑩Personal Computer

Figure 2. Hydraulic dynamic testing machine (MTS Co., Ltd).

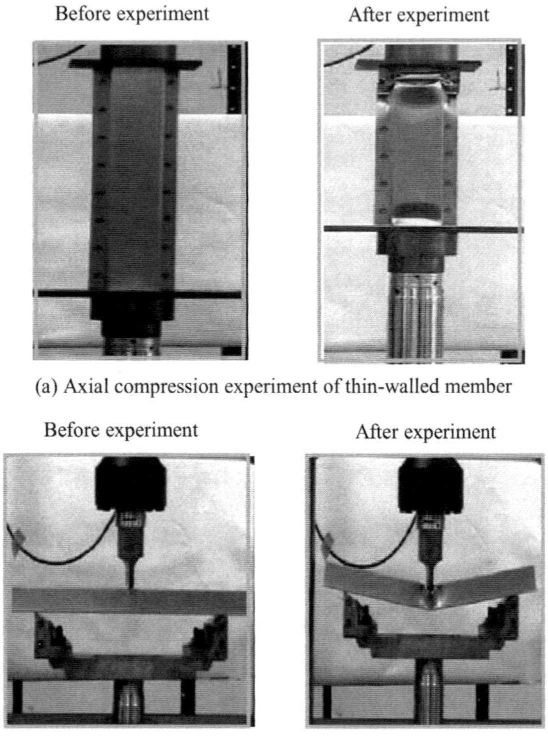

(a) Axial compression experiment of thin-walled member

(b) Three-point bending experiment of thin-walled member

Figure 3. Example of deformation of the member after each test.

2.3. Experimental result

In the axial compression test (figure 4), the maximum load in the composite member was increased due to the "filling effect". For example, maximum load of the 2 times foaming material (36kN) are 1.8 times for maximum load of the 3 times foaming material (20kN). Furthermore, the maximum load of the composite member (48kN) which filled the thin-walled member with the 3 times foaming material increased about 1.8 times for that of the thin-walled member (27kN). However, in composite member which filled the 2 times foaming material, maximum load (44kN) increased about 1.6 times. We attribute the observed effect to peeling generated at the interface of the constructional member (thin-walled and 2 times foaming) as well as to the rigidity of the composite member lowered.

The loading history of composite member, as shown in figure 4, fluctuates with displacement, after it reaches the maximum load. Plastic buckling deformation of the thin-walled member by the axial compression load is difficult to generate, because the foaming material is inside the rigid structure, something described in section 5.3. The thin-walled part of composite member is deformed outside the member by the pressure of the inside foaming material. The interface of thin-walled member and foaming material is adhered by the filling, and the rigidity of the member increases. We consider that the fluctuation behaviour becomes prominent due to these deformation constrains. Moreover, the three-point bending test (figure 5) confirmed the effect of filling on construction strength.

Maximum load P_{max} was carefully recorded for all values of strain rate in the axial compression test (figure 6). The strain rate was determined by dividing deformation velocity (0.05, 5, 400mm/s) by the length of member (foaming material:100mm, thin-walled and composite member:400mm).

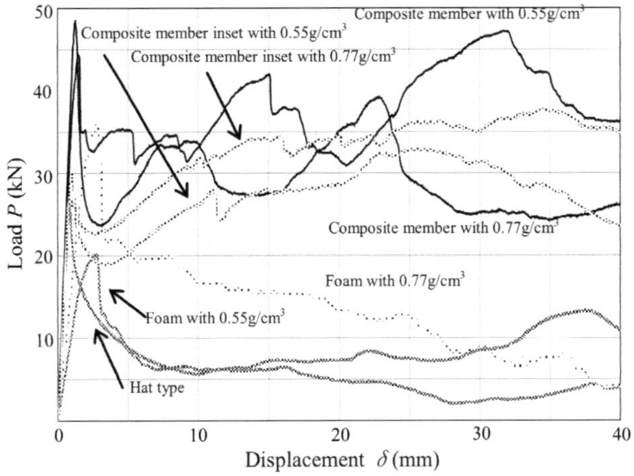

Figure 4. Load-displacement curve on axial compression test at deformation velocity 0.05mm/s.

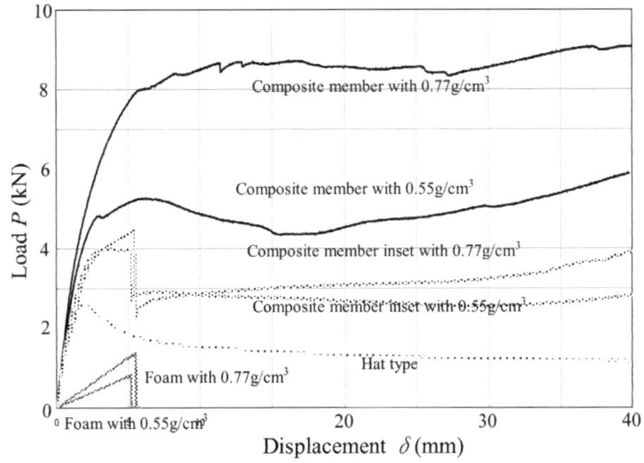

Figure 5. Load-displacement curve on three-point bending test at deformation velocity 0.05mm/s..

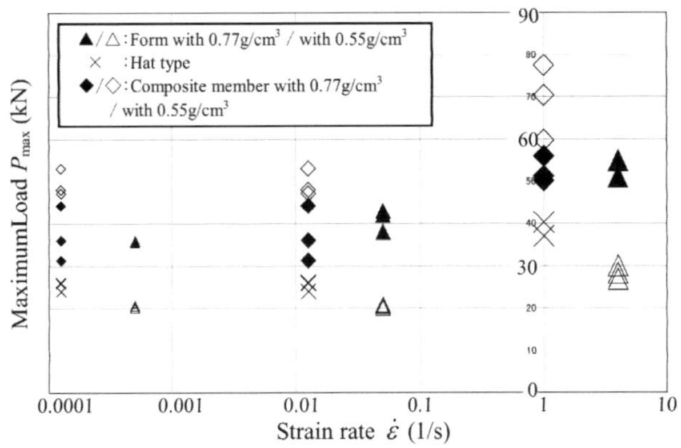

Figure 6. Maximum load achieved during axial compression test. The symbol in the figure shows 2 / 3 times foaming material (▲/△), 2 / 3 times composite member (◆/◇) and thin-walled member (×).The size of the symbol shows the deformation velocity 0.05mm/s, 5mm/s and 400mm/s in turn from the small.

3. Adherent strength

Filling effect became evident from our comparison of maximum loads. Insertion of the foaming material was made in order to examine the effect of filling-member adherence. Using insertion composite member without the adherent strength, the axial compression test and three-point bending test were carried out, and the results compared with that of the composite member (compare figures 4 and 5).

One can notice the maximum loads 0.6 times higher for foaming material, and 0.7 times in the another specimen (refer to figure 4). In the case of the thin-walled member, maximum loads were about 1.0 to 1.1 times in the 3 and 2 times foaming materials. The maximum load does not increase by the insertion of the foaming into thin-walled member, though there is the possibility of generation of the frictional force during loading. Thus, the adherence by the filling results in improved strength. In the 3-point bending test (figure 5), the foaming material in the insertion composite member was fractured during the deformation, which cased load-drop. The displacement of fracture was same as during to the test result for the foamed material (about 5.4mm at 2 times foaming material). Thus, in both of axial compression and three-point bending tests, the maximum load of composite member filled the foaming material inside is higher than that of composite member inserted the foaming material inside, which clarifies the effect of the adherent strength.

4. Characteristics of energy absorption

Energy absorption and its dependence of strain rate were examined as shown in figure 7. The relationship between energy absorption and curvature change speed of each member was examined in the three-point bending test. The curvature change speed was calculated by dividing curvature of the member which generated by displacement δ by the time. The curvature was calculated from equation $\kappa = 4\delta/l^2$ [12], where l defines span (400mm).

In the axial compression test, the energy absorption of composite member increased about 1.1 times (on both of 2 and 3 times) for that of insertion composite member. In addition, the energy absorption of insertion composite member increased about 3.8 times (2 times) and about 3.0 times (3 times) for that of thin-walled member. The energy absorption of composite member also increased about 1.5 times (2 times) and about 1.4 times (3 times) for that of insertion composite member in the three-point bending experiment. Moreover the energy absorption of insertion composite member increased about 1.5 times (2 times) and about 1.2 times (3 times) for that of thin-walled member.

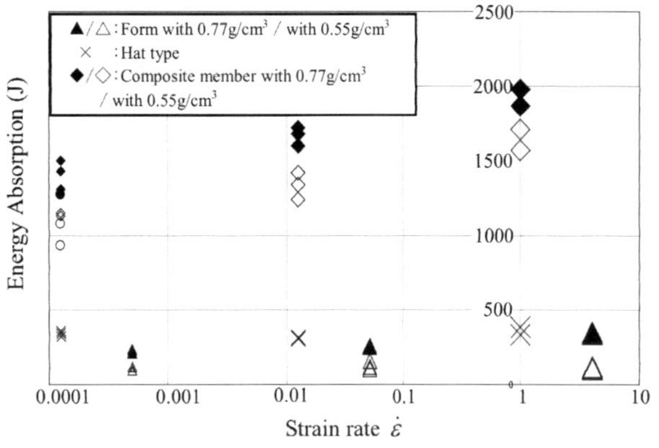

Figure 7. Characteristic of energy absorption on axial compression test. The symbol condition is the same as figure 6.

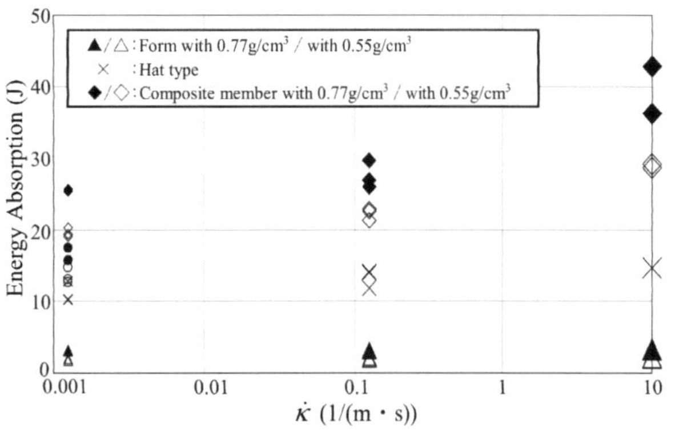

Figure 8. Characteristic of energy absorption on three-point bending test. The symbol condition is the same as figure 6.

Although our experiment explores low strain rate range, the strain rate dependence was noted (figure 6). Strain rate dependence on energy absorption is shown in figure 7, while the energy absorption increases linearly in the whole strain rate range. The energy absorption is generated by deformation resistance in the whole plastic region and it increases linearly, since viscous resistance (the plastic flow stress) in plastic region remains constant [13]. In the three-point bending test (figure 8), the energy absorption of composite member have an effect on the deformation velocity, and it increases almost linearly, in spite of that of the foaming independent of deformation velocity. Thus, the foaming does not affect strength and plastic deformation resistance in the thin-walled member.

5. Plastic collapse characteristics of each member

5.1. Plastic collapse of foaming material

In axial compression test of the foaming material, the collapse was investigated for 5 kinds of member length (see Table 3). The maximum load of the foaming material decreased with the member length (figure 9). The foaming material collapsed due to shearing stress (see figure 10②, ③). In this experiment, the axial angle with normal line concerning this shear collapse was 40 to 45 degree. The edge in surface of collapse by shearing stress contacts the surface which gives the load on the base and load cell in the equipment, when the member length is short such as 50 and 100mm. The maximum load seemed to increase, since a shear collapse is restricted by this friction. In the case of member length 100 and 200mm, the maximum load did not change. The buckling deformation was generated in the member, and it collapsed by the bending in the centre of the member, in member length 400mm. Though the 400mm length member should be considered the middle column, because the slenderness ratio is about 35, the Euler's buckling load was calculated with 13kN, when it was assumed both free end for the confirmation, and this value was closed to the experimental result (15.4kN). The bending stress takes maximum value at the central of the member length, and the crack is generated in the tension side surface, by the bending deformation supported at both end. It is considered that the crack was propagated perpendicularly to the loading direction by the bending deformation, and that the member was collapsed by it (as the example, shown in figure 11).

5.2. Plastic collapse characteristics of thin-walled member

When thin-walled member undergoes progressive plastic buckling, (1) hat type plate and thin plate are deformed between the spot welding points, as shown at arrowⓐ in figure 12B. Further, (2) side wall of member shown at arrowⓑ is deformed to the thin-walled member inside, and (3) surface part of member shown at arrowⓒ is also deformed to the member inside, similarly with the side wall. The deformation of these (1) to (3) is generated simultaneously. This deformation is terminated at figure 12A ⑤ and ⑥, and the similar deformation is generated after ⑥ continuously, then it is finished at ⑩. By repeating this deformation, the thin-walled member progressively generates plastic buckling.

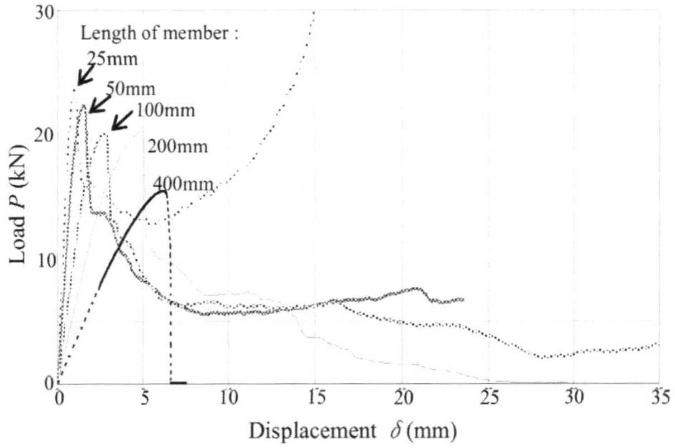

Figure 9. Load-displacement curve for epoxy resin foaming materials. Load of each 3 times foaming material on 5mm/s deformation velocity was examined on the dependence of displacement.

Figure 10. Deformation history (collapse) of foaming material. As example of the deformation process, the sequence photographs (δ=5mm (1s/1Frame)) of 2 times foaming material of the 100mm length in case of 5mm/s deformation velocity.

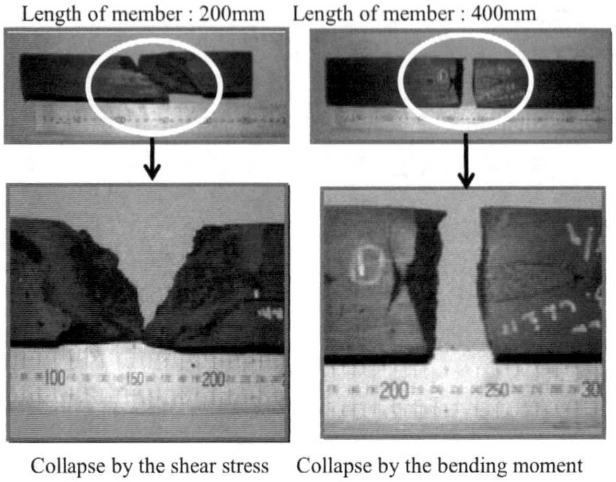

Figure 11. Collapse of foaming material after the test.

5.3. Plastic collapse characteristics of composite member

The collapse characteristics of composite member (figure 13) are dependent on the collapse mode of thin-walled member. Deformation process of the composite member is similar to the thin-walled member. However, in case of the thin-walled member, the surface part (figure 12B©) was deformed. However, the surface part of composite member could not be deformed to the member subject inside and are pushed outside, since the foaming material is filled inside. The side wall of member (figure 12Bⓑ) also could not be deformed inside by the effect of the foaming material. Moreover, by the adherence of foaming material and thin-walled member, the strength of the composite member was improved.

6. Conclusion

This study, clarifythe difference in mechanical characteristics of the thin-walled composite member and the constructional member, axial compression and 3-point bending tests were carried out on the composite member which is filled 2 kinds of epoxy resin foaming material in hat type thin-walled cross section member. The adherent strength between thin-walled member and foaming material was clarified by comparing the strength of insertion composite member and filling composite member.
The energy absorption improved by filling the member inside with the foaming material.

Acknowledgement

We would like to thank Iida Industry Co .Ltd. for invaluable comments and joint research.

References

[1] Reid S R, Reddy T Y and Gray M D 1986 *Int. J. Mech. Sci.* **28-5** 295
[2] Takatori O 1999 *Proc. Sympo. the Society of Automotive Engineers of Japan* **88-99** 5
[3] Reddy T Y and Walla R J 1988 *Int. J. Impact Eng.* **7-2** 151
[4] Mitsuishi H, Nakatani T and Matsukawa F 1999 *Proc. Sympo .the Society of Automotive Engineers of Japan* **88-99** 13

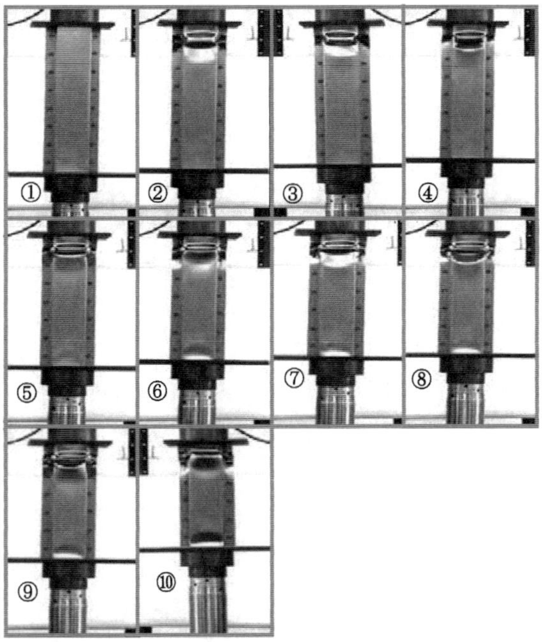

A : Deformation history (collapse) of thin-walled member.

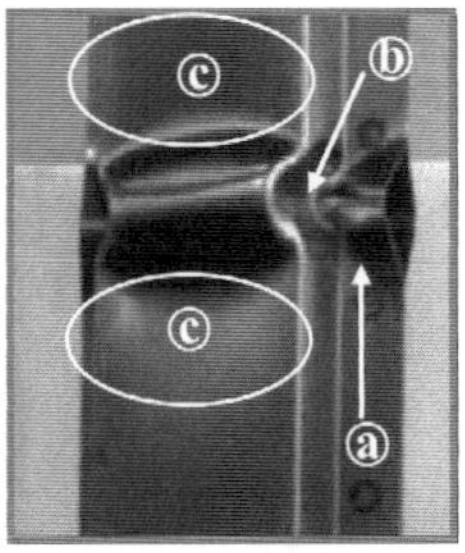

B : Enlarged photograph of A ②.

Figure 12. Collapse of thin-walled member. These figures shows the sequence photographs (δ=4mm (0.8s/1Frame)) of deformation of the thin-walled member in case of 5mm/s deformation velocity.

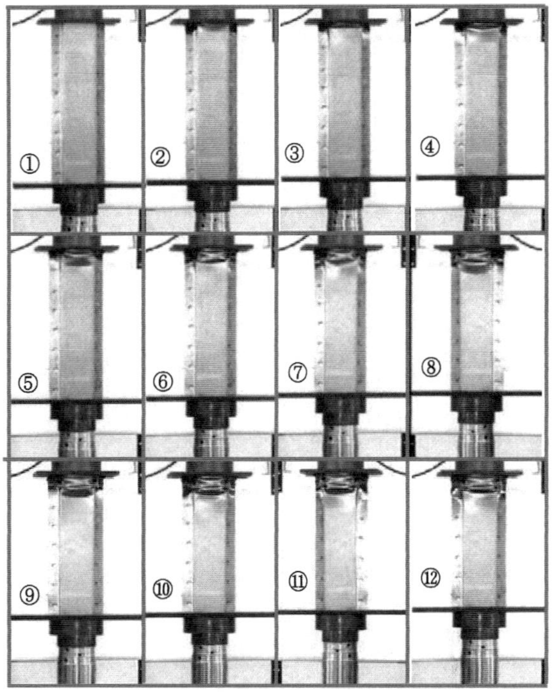

A : Deformation history (collapse) of composite member.

B : Deformation of composite member after test.

Figure 13. Collapse of composite member. These figures shows the sequence photographs (δ=5mm (1s/1Frame)) of deformation of the 3 times composite member in case of 5mm/s deformation velocity.

[5] Onda T, Tomada A, Watanabe A, Wada H and Murase K 2001 *Proc. Sympo. the Japan Society of Mech.Eng.Tokai branch* **50** 151
[6] Nakazato K, Fukudome H and Iwao A 1995 *J. Soc. Automotive Eng. Japan* **49-6** 39
[7] Reyes A, Hopperstad O S and Langseth M 2003 *Proc. 8th Int. Sympo. IMPLAST 2003* 395
[8] Reys A, Hopperstad O S, Hanssen A G and Langseth M 2004 Int. J. Impact Eng. **30** 805
[9] Lilley K and Mani A 1996 *SAE Trans. J. Passenger Cars* **6-105-960435** 374
[10] Ishida K, Fukahori M, Hanakawa K, Tanaka H, Matsuda K and Sunakawa T 2001 *J. Soc. Automotive Eng. Japan* **55-4** 70
[11] Tanimura S 1997 *Trans. Japan Soc. Mech. Eng.* **A-63-616** 2466
[12] Murase K, Katori H and Nishimura T 1983 *Bulletin of Japan Soc. Mech. Eng.* **26-220** 1696
[13] Kojima N, Mimura K, Umeda T and Nakazawa Y 2010 *J. Soc. Mat. Sci. Japan* **59-7** 505

Quasi-static and dynamic deformation behaviour of Zr-based bulk metallic glass

V Nekouie[1], U Kühn[2], A Roy[1] and V Silberschmidt[1]

[1]Wolfson School of Mechanical and Manufacturing Engineering, Loughborough University, Leicestershire, UK
[2]IFW Dresden, Institute for Complex Materials, Helmholtzstraße 20, D-01069, Dresden, Germany

E-mail: v.nekouie2@lboro.ac.uk

Abstract. Nano- and micro-indentation studies were carried out to characterise a plasticity mechanism through the evolution of localised shear bands that drive material's deformation at sub-micron length scale. Initial deformation of Zr-based bulk metallic glass (BMG) was investigated with nanoindentation tests using a spherical indenter. The indentation cycle reflects an elastic deformation with the yielding load of approx. 3 mN. For designed cycling indentation, hardening and softening phenomena were observed in nano- and micro-indentations, respectively. High-precision dynamic mechanical relaxation measurements were performed using a Dynamic Mechanical Analyzer (DMA), on decreasing frequency from 160 Hz to 0.1 Hz. A mechanical response of the BMG surface to a concentrated impact load was also studied. The obtained results indicated that the studied Zr-based BMG behaved as an elastic-perfectly plastic material at macroscale with discrete plasticity events at smaller length scales.

1. Introduction

Metallic glasses are a relatively new class of materials used in many functional and structural applications ranging from golf club to nano devices [1]. They were initially manufactured with rapid quenching techniques by exposing the molten alloys to very high cooling rates, ~ 10^6 K/s [1]. In the late 1980s, Inoue and co-workers [2-4] found a large number of new bulk metallic glasses in a variety of multi-component alloy systems comprising the rare-earth-based systems that have cooling rates less than 100 K/s and thicknesses reaching several centimeters. The first commercial BMG, $Zr_{41.2}Cu_{12.5}Ni_{10}Ti_{13.8}Be_{22.5}$ alloy, was fabricated by Johnson and Peker in 1992 with a critical cooling rate of 1 K/s. Metallic glasses are well-known for their distinctive mechanical properties, which can be exploited for a wide range of technological applications [2]. Unlike crystalline materials, amorphous metals lack an orientation long-range order. However, there is evidence of a crystal-like short-range order, which leads to unusual structural properties and non-conventional deformation mechanisms [6]. As a result, metallic glasses show excellent engineering properties such as high values of the Young`s modulus and elasticity limit, higher fracture toughness when compared to inorganic glasses and ceramics, high yield stress and high specific strength. Typically, inorganic glasses are brittle at room temperature, exhibiting a smooth fracture surface as a results of mode-I brittle fracture. BMGs, to the contrary, deform by localized shear deformation at room temperature, and are ductile in bending and rolling deformation. Under uniaxial tension or compression, BMG specimens undergo shear fracture immediately after yielding without appreciable macroscopic plastic strain.

For over a century, indentation has been conducted to probe the mechanical behaviour of materials for a wide range of engineering applications. The main reason for its ubiquitous use is its intrinsic experimental simplicity since indentation needs minimal specimens preparation and can be performed several times on single specimens, and can probe various volumes of materials via appropriate choice of load and tip geometry [7]. Instrumented nano- and micro-indentation on flat plates of BMG is ideal for the study of plastic deformation in amorphous materials because of limited global ductility in uniaxial testing conditions [8, 9]. In current study, purely elastic deformation of BMGs is analysed by using nano-indentation with a spherical indenter. Spherical indenters are finding popularity, as this type of indenter provides a smooth transition from elastic to plastic contacts. Instrumented indentation experiments are capable of showing some mechanical properties directly reflecting the inhomogeneous BMGs without the adverse effect of macroscopic casting defects. The frequency-dependent modulus is measured using Dynamic Mechanical Analyzer (DMA) to evaluate storage and loss moduli (E_1) and (E_2) and loss tangent$\left(\tan\delta = \frac{E_1}{E_2}\right)$. Additionally, the dynamic response of materials under concentrated impact loading at elevated rates is also investigated.

2. Experimental procedure

In this study, rectangular beam-shaped samples with a rectangular cross-section of a Zr-based metallic glass with nominal composition of $Zr_{48}Cu_{36}Al_8Ag_8$ (at %) with length of 40 mm, width of 10 mm and thickness of 2 mm were prepared at IFW Dresden, Institute for Complex Materials, Germany. The indentation tests were conducted with a Nano Test (Micro Materials Ltd.) system equipped with a spherical indenter with a radius of 5 μm and 50 μm. The samples with length of 10 mm, width of 10 mm and 2 mm thickness were adapted for these measurements. The samples were polished to mirror-like appearance with average roughness of around 5 nm. For the polishing process, the samples were initially mounted in epoxy resin (Epofix, Struers Ltd.) and then polished using a semi-automatic polishing machine (TegraPol-25, Struers Ltd.) with a rotating sample holder (TegraForce-5, Struers Ltd.) and an automatic feeding system for the polishing media (TegraDoser-5, Struers Ltd.). As a typical polishing cycle, mounted samples were initially ground using a polishing disc with grit size of 400 (35 μm), followed by 600 (25 μm), 1200 (15 μm), and finally 5000 (3 μm). Subsequently, polishing was continued by using a diamond paste consisting of particles of 1 μm in size. After finishing each polishing step and before starting the next one, the mounted samples were removed from the specimen holder, cleaned with deionised water and dried with a heat gun. When the final stage was completed, in order to get a clean surface without any contamination all the samples were subjected to ultrasound in deionised water for two minutes, sprayed with ethanol and dried using a heat gun. All the tests were conducted in a displacement control mode, and single loading as well as multiple unloading-reloading experiments were performed. The loading and unloading rates were 0.1 mN/s and 2 mN/s for single loading and the incremental loading-unloading experiments respectively, followed by the holding time of 1 s for all experiments. The maximum applied depth ranged between 5000 nm and 20000 nm for micro-indentation, and 200 nm and 2000 nm for nano-indentation. Three-point bending (3PB) was also used to determine the Young`s Modulus $(E)$ and Poisson`s ratio (ν) which are 86 GPa and 0.35, respectively. The load was applied by the mechanical test system operating with the displacement rate of 2 mm/min. Additionally, high-precision dynamical mechanical relaxation measurements were performed with DMA system using SD TA861.The experimental frequency range was from 0.2 Hz and 160 Hz at constant temperature. The elastic modulus can be derived from the measurements. A specimen in the shape of rectangular beam with length of 40 mm, width of 10 mm and thickness of 2 mm was tested in a 3PB configuration. The accuracy of frequency-dependent modulus measuring for DMA was better than 2%. For impact measurements, a pendulum-based impact was performed using the Nano Test system with a diamond spherical probe of radius $R = 5$ μm. The operating principles of the experimental set are outlined in [10].

3. Result and discussion

3.1. Elastic deformation in indentation
To investigate the initial elastic deformation, a maximum load of 6 mN was applied to the samples in the nano-indentation tests in this study, which is around the estimated yielding load of approx. 3 mN for the typical Zr-based BMGs corresponding to 5 μm spherical tip based on the work of Packard *et al.* [11]. Figure 1(a) and (b) show the response to nano-indentation loading, holding and unloading cycle

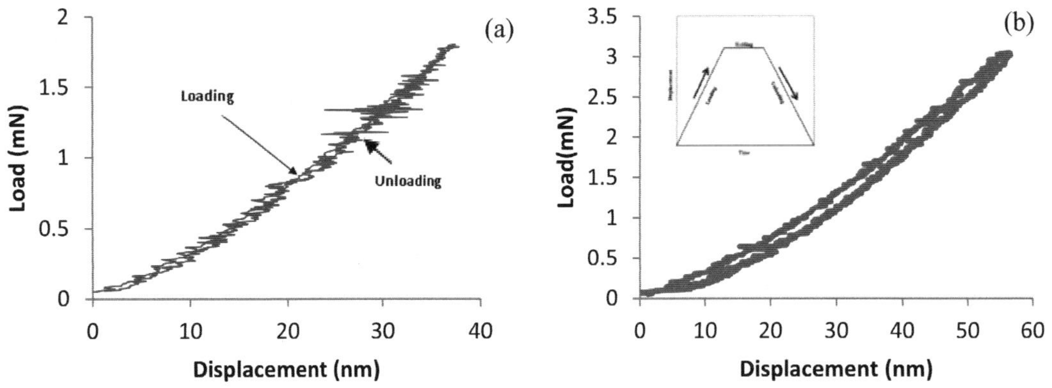

Figure 1. Typical load-displacement plot for Zr-based BMGs at loading rate of 0.1 mN/s: (a) purely elastic deformation; (b) initial plastic deformation

with maximum loading of 2 mN and 3.5 mN, respectively. It can be seen from Figure 1(a) that the loading and unloading responses are similar, with no obvious disparity between the two. This indicated an overall elastic response of the material under nano-indentation.

Next, we investigate the variation of hardness and elastic modulus with depth in the sample. The 'load/partial-unload' technique allows for hardness and modulus measurements at different indentation depths in the sample at the same location. A large plastic zone was formed under the indenter tip during indentation, commensurate with prior experiments [12-13]. The plastic zone contained a high density of shear bands, which is ideal for the investigation of deformation-induced hardening and softening effects [13]. The results show the dependence of hardness on the penetration depth. As shown in Figure 2(a) and (b), particular hardening and softening behaviors were observed when the multiple unloading-reloading tests were conducted during the indentation experiment with the same loading rate. In Figure 2(b), the onset of yielding upon each reloading is noted to take place always at a higher load than the load immediately before the previous unloading, apparently suggesting hardening effect. In addition, the hardness increased from 3.7 GPa to 5.7 GPa with a sharp enhancement in first 10 cycles and a slight rise in the following 10 cycles.

This also indicates that there is a work hardening behavior in the metallic glass at nanoscale due to increasing the hardness. However, the micro-indentation results (Figure 2(a)) indicate that there is large-scale softening associated with each load-unloading step, and hardness decreased from approximately 4.6 GPa to some 3 GPa. Such strain softening occurs in the development of shear transformation zones, which bring about an increase of free volume. At room temperature, the deformation-induced creation of free volume cannot be compensated by thermally-assisted structural relaxation and its accumulation results in formation of shear bands. Hence, this phenomena proves that plastic deformation at microscale in metallic glass is accompanied by dilatation, i.e. creation of free volume [13].

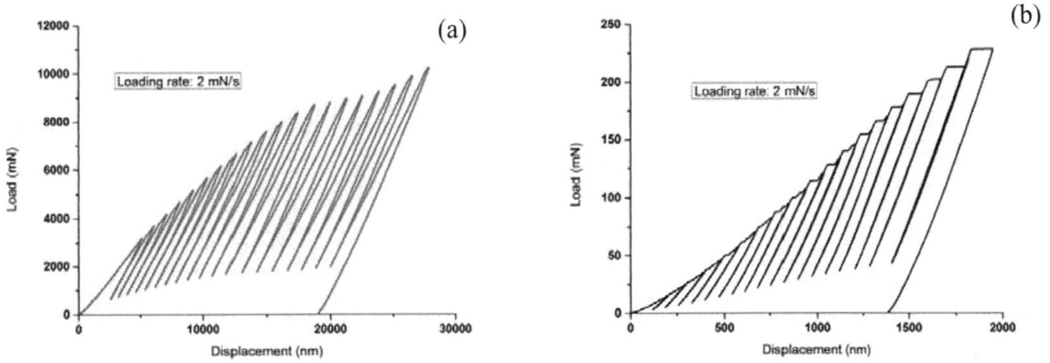

Figure 2. Indentation load-displacement curves of Zr-based BMG for incremental loading-unloading at loading rate of 2 mN/s (a) micro-indentation; (b) nano-indentation

3.2. Dynamic Mechanical Analysis (DMA)
Dynamic Mechanical Analysis (DMA) is one of the useful techniques to measure mechanical and viscoelastic properties as a functional of temperature, frequency and time when they are subjected to periodic stress cycles. The types of materials that can be analysed with this technique include thermoplastic, thermoset, metals, composites and ceramics. DMA provides quantitative and qualitative information such as the Young`s and shear moduli, damping characteristics, polymer structure and morphology, flow and relaxation behaviour. Figure 3 shows the frequency dependence of storage modulus (E_1), loss modulus (E_2) and damping ($tan\delta$) of the studied Zr-based metallic glass determined with DMA for a frequency range from 0.2 Hz to 160 Hz. It can be seen that the dynamic characteristics of the metallic glass are effectively independent of the frequency. The dynamic properties of the metallic glass are directly related to the microstructure and it can be concluded that the BMG behaves like a purely elastic material. The maximum value of storage modulus, loss modulus and the loss factor are 129 GPa, 3.57 GPa and 0.025, respectively.

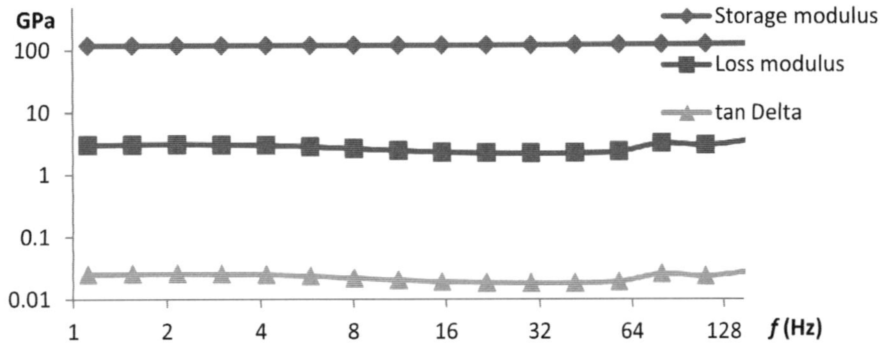

Figure 3. Frequency-dependence of storage and loss moduli of metallic glass measured with DMA at room temperature

3.3. Probe impact mode for low cycle fatigue
The probe impact mode can be used for investigating low cycling fatigue, work hardening and dynamic hardness. In this experiment, the pendulum is moved away from the specimen by a known distance and then released to produce a single impact. Successive impacts can be made at a single

point until failure occurs. In most cases, an initial period of fatigue damage generation occurs, in which cracks develop and expand, but in which no appreciable increase in penetration depth in observed. Figure 2(a) and (b) illustrate that there is an evidence of strain hardening (as shown in nano-indentation results, Figure 2(b)) as there is a significant increase in displacement (penetration depth) in first 30 seconds, where the cracks coalesce, and a slight enhancement afterwards. Additionally, these results indicate that the Zr-based BMG was an elastic-perfectly plastic material since the material is linear-elastic and as the load increases to the yield point its stress remained constant and strain increased.

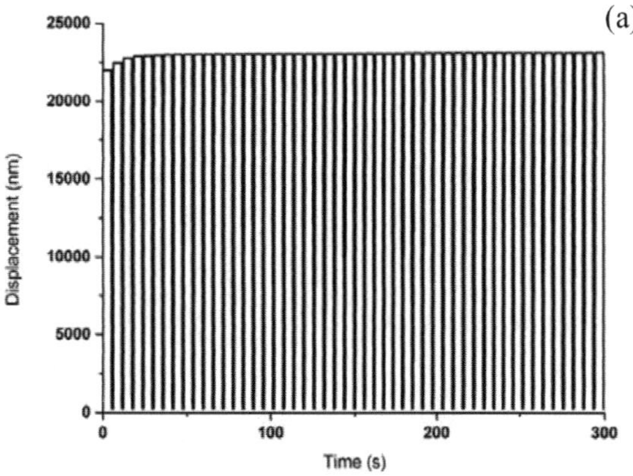

(a) (b)

Figure 4. Typical displacement-time plot of impact results for Zr-based BMGs (a) for 300 s; (b) for 60 s. One impact every five second and failure of the metallic glass occurred after 30 s.

4. Conclusion

The initial stage of elastic deformation of BMG was investigated via nano-indentation test with a spherical indenter. It was found that the atomic structure has an effect on increasing initial elastic deformation. Strain-hardening and -softening were also observed for the Zr-based BMG during plastic deformation using controlled instrumented nano- and micro-indentation techniques. The dynamic elastic response of the BMG samples was studied at various frequencies at constant temperature showing that the dynamic characteristics of the metallic glass are effectively independent of the frequency. The indentation impact characterisation findings indicated that the BMG shows pure elastic behaviour.

References

[1] Klement W, Willers R H and Duwez P 1960 Non-crystalline structure in solidified gold-silicon alloys *Nature* **187** 869-870
[2] Yavari A R 2006 Metallic glasses: The changing faces of disorder *Nature* **439** 405-06
[3] Inoue A, Ohtera K, Kita K and Masumoto T 1988 New amorphous Mg-Ce-Ni alloys with high strength and good ductility *JJAP* **27** 2248-51
[4] Inoue A, Zhang T and Masumoto T 1989 Al-La-Ni Amorphous Alloys with a Wide Supercooled Liquid Region *Mater. Trans., JIM* **30** 378-81
[5] Inoue A 2000 Stabilization of metallic supercooled liquid and bulk amorphous alloys *Acta Matter.* **48** 279-306
[6] Vincent S, Basu J, Mutry B S and Bhatt J 2012 Micro indetation study on Cu60Zr20Ti20 metallic glass *Mater. Sci. Eng., A* **550** 160-66
[7] Gouldstone A, Challacoop N, Dao M, Li J, Minor A M and Shen Y L 2007 Indentation across size scales and disciplines:Recent developments in experimentation and modelling *Acta*

Mater. **55** 4015-4039

[8] Zhao L, Ma C L, Fu M W and Zeng X R 2012 Inversitagation on the inhomogeneous structure of metallic glassesbased on the initial elastic deformation in nanoindetation *Intermetallica* **30** 65-71

[9] Wang K, Chen M W, Pan D, Fujita T, Zhang W and Wang X M 2008 Plastic deformation energy of bulk metallioc glasses *Mater. Sci. Eng., B* **148** 101-4

[10] Constantinides G, Tweedie C A, Holbrook D M, Barragan P, Smith JF and Van Vliet KJ 2008 Quantifying deformation and energy dissipation of polymeric surfaces under localized impact *Mater. Sci. Eng., A* **489** 403-12

[11] Packard C E and Schuh C A 2007 Initiation of shear bands near a stress concentration in metallic glass *Acta Mater.* **55** 5348-58

[12] Yang B, Riester L and Nieh T G 2006 Strain hardening and recovery in a bulk metallic glass under nanoindenttaion *Scr. Mater.* **54** 1277-80

[13] Schuh C A and Nieh T G 2003 A survey of instrumented indentation studies on metallic glasses *Acta Mater.* **51** 46-56

Mechanical and thermal characterisation of poly (l-lactide) composites reinforced with hemp fibres

A Shakoor[1, 3], R Muhammad[2], N L Thomas[1] and V V Silberschmidt[2]

[1]Department of [1]Materials, Loughborough University, Leicestershire, LE11 3TU
[2]Wolfson School of Mechanical and Manufacturing Engineering, Loughborough University, Leicestershire, LE11 3TU

Email: a.shakoor@lboro.ac.uk

Abstract. Polylactic acid (PLA) is the most promising in the bio-derived polymer's family. But its use can be constrained by its poor mechanical properties, poor thermal stability and processing difficulties. The objective of this research is to investigate and improve mechanical and dynamic thermal properties of PLA by developing PLA composites reinforced with natural fibres (hemp). Composites were prepared by melt blending of PLA with hemp fibres. Their properties were investigated using mechanical and dynamic thermal analysis. The elastic modulus increased significantly - from 4.1 ± 0.74 to 9.32 ± 0.86 (GPA) - when the weight fraction of hemp increased from 0 to 30(wt %). The storage modulus obtained by dynamic mechanical analysis increased from 2.20 to 4.58 (GPA) for the same change in the volume fraction of hemp. FE simulation of tensile testing and DMA were carried out to investigate the effect of strain rate and temperature on the observed properties respectively. The model was developed in the commercially available code MSC Marc mentate. The model validated all experimental results.

1. Introduction

The growing usage of synthetic polymers during the last decade has led to apprehension about the environmental impact caused by plastic waste, which has been a widespread concern. The ecological aspect of both production and disposal of standard oil-based plastics is presently of concern worldwide. This has driven the search for alternatives that are bio-derived and eco-friendly. In order to be a competitive alternative, bio-plastics must have the same desirable properties as obtained in conventional plastics.

Poly (lactic acid) (PLA) is chemically synthesized from lactic acid derived from corn starch [1], the most commonly used aliphatic polyesters exhibiting comparable properties to conventional plastics derived from corn starch [2, 3].

PLA has a poor mechanical and thermal properties which restricting its widespread applications. Shortcomings in mechanical properties of PLA may be overcome and sustained by several means like reinforcement and blending. For example, PLA has been blended with poly(caprolactone) (PCL) to increase its flexibility [4, 5] and with poly (hydroxybutyrate) (PHB) to improve tensile properties and

[3] To whom any correspondence should be addressed.

biodegradability [6]. There is much interest in PLA nanocomposites because of the potential of montmorillonite nanofillers to improve the mechanical, barrier and biodegradation properties of PLA [7-10].Most recently PLA – mineral filler composites are in focus to increase the heat distortion temperature (HDT) [11].

Natural fibres reinforced polymer composites have been remained a centre of attention for academics and industrial research for the last decade or so. Automotive, construction and decking industry have shown their interests in natural fibre polymer composites due their competitive advantages, such as low cost, high strength –to- weight ratio, recyclability and low density when compared to synthetic fibres reinforced composites. [12]

This research investigates the development of PLA composites reinforced with natural fibres (hemp). Composites have been processed and the effects of hemp content on the mechanical and thermo-mechanical properties of PLA-hemp composites have been studied [13].

2. Experimental

2.1. Materials
General purpose Polylactic acid (PLA), a trade name is HM_{1011}, was supplied by Hycail Company, Netherlands. HM_{1011} is completely amorphous. The polymer was supplied in the form of granules with specific gravity of 1.24 g/cm^3 and melting range 150 – 175 °C. The glass transition (T_g) range is 55 – 65 °C supplied by the manufacturer. Vegetable based dry retted hemp of variable dimensions was supplied by Hempcore Technology Ltd UK in the chopped form.

2.2. Composite preparation
Dried PLA is melt blended with chopped hemp fibres in different ratios of 90/10, 80/20, 70/30 (weight %). Mixing of PLA with hemp was carried out in Haake OS-Polylab rheomix at 170 ^{0}C for 10 minutes with constant rotor speed of 60 rpm. For each run, total mass (PLA + Hemp) was about 58 (g) with 70 (%) filling capacity of the Haake Polylab rheometer. Samples for tensile and dynamic mechanical analysis of the Composites were compression moulded at 180°C for three minutes at 10~12 Tons pressure followed by quenching for three minutes at the same pressure level. This was carried out in 20 Ton laboratory hot press.

2.3. Tensile testing
The tensile testing of the composites was performed using Tinius Olsen H50 KS with a clip-on extensometer, used to measure the modulus more precisely. The tensile machine was equipped with a load cell of 5kN. Compression moulded dumbbell – shape tensile specimens (width ~10mm, thickness ~ 2mm, length ~30 mm) were extended at a crosshead speed of 10 mm / min.

2.4. Dynamic Mechanical Thermal Analysis (DMTA)
The viscoelastic properties of the PLA/Hemp composites were investigated using dynamic mechanical analysis (DMA). DMA Q800 apparatus (TA Instrument Inc, USA) was used to measure the storage modulus, loss modulus and Tan Delta (Tan δ) of each composite sample as a function of temperature. All samples were tested in the flexure (dual-cantilever bending) mode. The rectangular specimens (width 12mm, thickness 3mm, length 64mm) were heated at a constant rate of 3/min from room temperature to 140 and tested at a frequency of 1Hz.

3. FEA Simulation

Finite element simulations of tensile testing and DMTA of PLA/HEMP composites were carried out using commercially available code MSC Marc/Mentat. A schematic of the specimens used in the FE simulation of PLA/Hemp composite material for both tensile testing and DMA is shown in Figure 1. Eight nodes 3D element (element type 7) were used to model specimens. The tensile specimen was discretised to 5600 elements with number of element of 70, 10 and 8 in X-axis, Y-axis and Z-axis, respectively, whereas in DMA simulation, the specimen was discretised to 4608 element with 64, 12 and 8 number of elements in X-axis, Y-axis and Z-axis, respectively. In tensile testing simulations, the left hand side of the specimen was kept fixed, restricting the movement of nodes in all three directions, whereas, the right hand side of the specimen was allowed to move in the x-direction. The overall length of the specimen was 90 mm, width of 20 mm and thickness of 2 mm. The material data used in the simulations were achieved from a tensile testing of PLA/Hemp composites. The ambient temperature is selected as 20°C for the specimen. The number of simulation steps was kept at 500 with a total time of 40 Sec. For material modelling of tensile specimen, the response of specimen measured in tensile testing was incorporated to the model. The data obtained from the machine was in the form of Force (N) Vs. Elongation (mm), which was converted into the Stress (MPa) Vs Strain (%) at the end to perform the simulation.

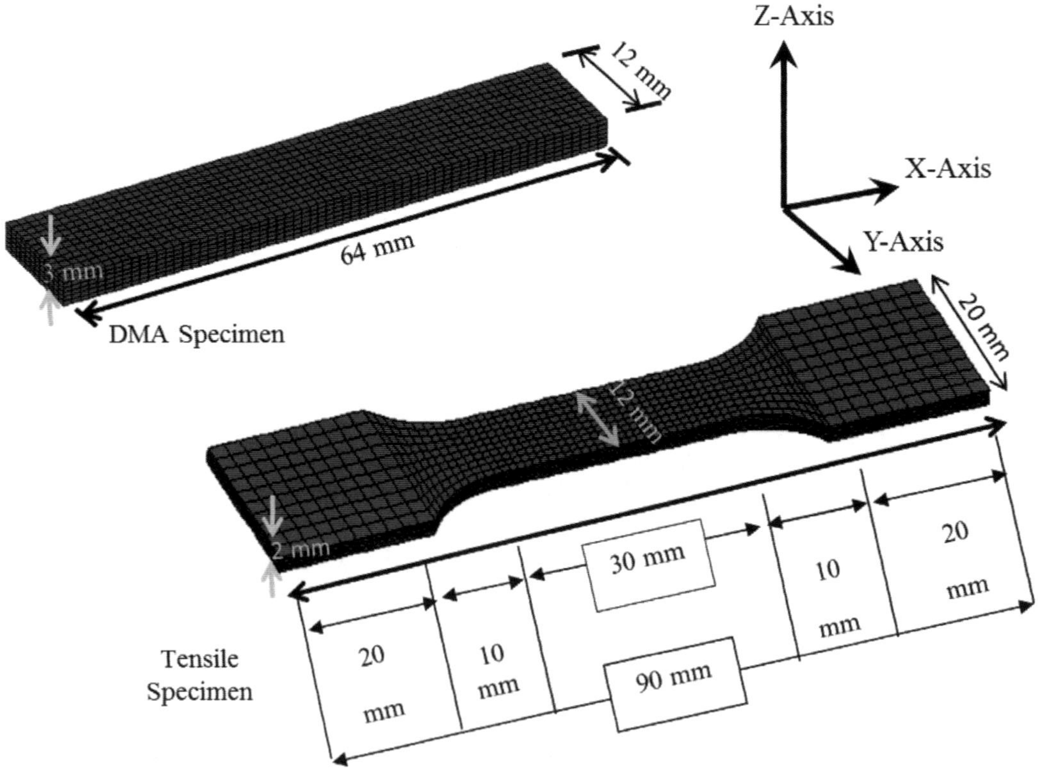

Figure 1. Schematic of the specimens used in the FE simulations

In DMA simulation, the two ends of the specimen were fixed and a fluctuating displacement was applied to the through the centre region of the specimen. The new deflection of the specimen was resembled to the double cantilever beam. The fluctuating load applied to the two rigid surfaces glued to the specimen at the centre point has amplitude of 50 μm and frequency of 1 Hz. The load was applied in the vertical direction (Z-axis). A ramp temperature of 3°C /min was applied to the material to study the dynamic response of material and various temperatures.

4. Results and discussions

4.1. Tensile testing

Table 1 presents summary of the effect of different contents of reinforcement on tensile properties of PLA. It is clearly evident that incorporation of the hemp fibre brought a stiffening effect into the PLA-hemp composites and causing a significant improvement in the elastic modulus of the PLA. The elastic modulus increased from 4.1 ± 0.3 to 9.3 ± 0.62 (GPA) when the contents of hemp fibre increased from 0 to 30 (wt %). It was also observed that tensile strength is gradually increased with increasing contents of hemp fibre, but overall the tensile strength of the composites decreased as compared to pure PLA. The same kind of behaviour for PLA- hemp composites also reported by Nina Graupner [14]. He observed that the incorporation of hemp fibre increases the young's modulus and reported the increase in tensile strength in machine direction while decrease in cross direction. Robert Masirek et al. [13] have also reported a decrease in tensile strength of the amorphous PLA-hemp composites and improvement in tensile modulus is observed. From the table it is also clear that elongation to break is also affected with hemp fibres. Sawpan et al. [14] have also studied the properties of PLA- hemp composites and reported a good adhesion between the fibres and matrix. They also reported that the as PLA and hemp fibres both are brittle, so that the higher amount of fibre loading in composites can cause the reduction in strain failure as shown in table 1.

Table 1. Tensile properties of PLA- Hemp composites

PLA / Hemp	Tensile Strength	Tensile Modulus	Elongation @ break
(wt %)	**(MPa)**	**(GPa)**	**(%)**
100/0	43 ± 1.3	4.1 ± 0.28	4.1 ± 0.6
90/10	24 ± 1.5	4.8 ± 0.34	2.0 ± 0.7
80/20	30 ± 0.8	6.9 ± 0.43	2.0 ± 0.5
70/30	38 ± 1.7	9.3 ± 0.62	1.0 ± 0.5

4.2. Morphological analysis

Morphology of hemp fibre is investigated using a field emission gun (FEG) scanning electron microscope (SEM), LEO 1530 VP prior to incorporate in the PLA. Figure 2 (a) shows the morphology of the pure chopped hemp fibres. It is clear from the figure that the hemp fibre has damage marks on the surface. The plants have natural waterways for food and water transportation to different parts of the plants, therefore the internal structure of the hemp fibre indicates hollow structure. Figure 2 (b) & (c) shows the SEM analysis of the fracture surface of the tensile samples for PLA- chopped hemp composites. It is observed from fractured surface morphology that hemp fibre is not properly dispersed in the matrix and some fibre lumps are observed. Some of the fibre pull out is also observed shows the weak mechanical interface. It is also observed that the structure of the fibre within the composites remained hollow and it seems that PLA did not able to penetrate effectively into the fibre structure,

Journal of Physics: Conference Series **451** (2013) 012010 doi:10.1088/1742-6596/451/1/012010

thus resulting a hollow structure in the composites. The creation of the hollow structure in the composites could be one the reason for reduce tensile strength of the composites with incorporation of the chopped fibres into the matrix.

Figure 2. SEM of fractured surfaces PLA - Hemp composites: (a) pure hemp fibre; (b) 90/20; (c) 70/30

4.3. Dynamic mechanical thermal analysis

In DMA analysis a sinusoidal stress is applied to the materials and the viscoelastic properties of the materials is measured under periodic deformation. The sinusoidal wave is applied in a periodic motion and the amplitude of the stress and resulting strain is used to measure the energy absorbed by the elastic portion (storage modulus, E'), energy dissipated by the viscous portion (loss modulus, E'') and the loss factor (tan δ) as a function of temperature. Figure 3(a) shows the dynamic mechanical spectra for storage modulus as a function of fibre loadings and indicates that the storage modulus increases relative to the amount of fibre loading. Table 2 summarizes the effect of hemp fibre contents on storage modulus at various temperatures. It is observed that the storage modulus of the composite increased with the increasing contents of the hemp. Storage modulus increased from 2.28 to 4.6 (GPa), when the contents of hemp increased from 0 to 30 (wt %) at 30 °C. Improvement in the storage modulus relative to fibre loadings is a reconfirming the tensile test results. Figure 3 also shows that the storage moduli of the PLA and PLA/hemp composites remain almost constant at temperatures below the glass transition temperature (up to about 60 °C) and then drop at the Tg, but seen increasing again around 90 °C due to stabilizing capability of the hemp fibres which implies that the reinforcement can be useful to increase heat distortion temperature (HDT).

The Figure 4 shows the damping factors (tan δ) for PLA and PLA-hemp filler composites as a function of temperature. The oscillation frequency sweep was set at 1Hz with a temperature ramp of 3 °C/min. Damping (tan δ) is highly sensitive to the structural transformation of the polymers. It was observed that incorporation of hemp fibre caused reduction in the area under the *tan-delta* curve thus reducing its damping capability of the composites. Also the peak of the tan delta curve lies in the same temperature range, suggesting that the glass transition temperature of the composites remains unchanged.

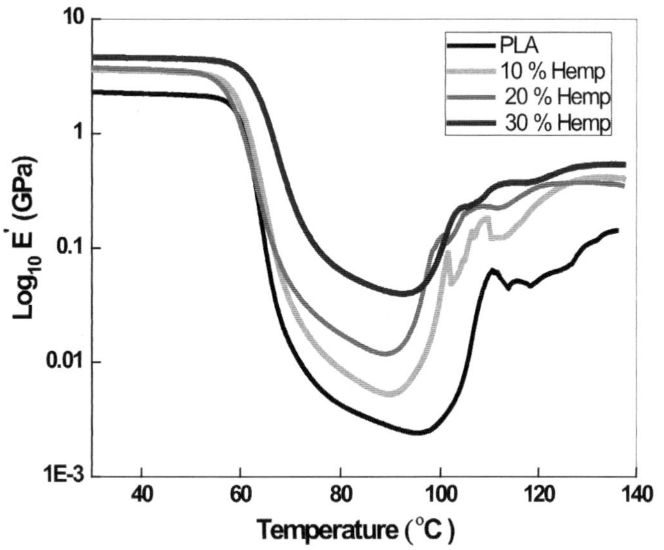

Figure 3 . Storage modulus of PLA and PLA – chopped Hemp composites as a function of temperature

Table 2. Glass tranition temperature (T_g) and storage modulus of PLA and composites

PLA / Hemp	Storage Modulus (30°C)	Storage Modulus (40°C)
100 / 0	2.28	2.18
90 / 10	3.72	3.63
80 / 20	3.60	3.52
70 / 30	4.60	4.55

4.4. Simulation analysis

Figure 5 shows the stress-strain curves of PLA/Hemp composites. The stress and strain levels were observed at the centre point of the specimen. A significant increase in stiffness of a hemp fibre composite was observed in the analysis. A close agreement between experimental and numerical results was achieved in the analysis. Incorporation of content of hemp fibres reduced the tensile strength of PLA/Hemp composites. However, an increase in tensile strength was observed with increasing level of hemp content in PLA composites. The model was simulated for different content of hemp in PLA composites and the deformation process in tensile tests was observed (see Figure 6 & 7). An inhomogeneous deformation in the specimen behaviour was observed at the start tensile testing in the FE simulations. The simulation was carried out for lower strain rate level and no significant changes in the deformation behaviour of the specimens were observed. The model was used to verify the experimental results and good agreement between experimental and numerical results were observed.

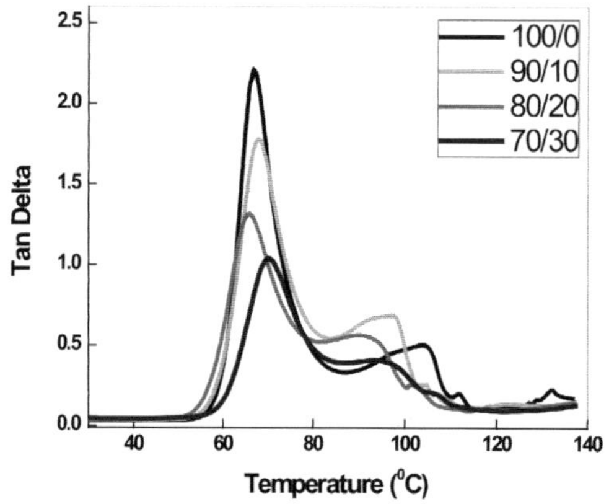

Figure 4. Damping factors (tan δ) of PLA and PLA / hemp composites as a function of temperature.

A DMA simulation of the developed PLA composites was carried out and response of the studied materials was observed at various temperatures as shown in Figure 8. A nearly linear increased in plastic deformation was observed in PLA/Hemp composites. The data were used to calculate the Storage modulus of the composites at various temperatures. The Modulus is the ratio of periodic stress to periodic strain i.e. ($E^{|*}$ = stress /strain) and the measure of the energy absorbed by the elastic contents in the sample i.e. storage modulus ($E' = E^{*}Cos\ \delta$) and the measured of the energy dissipated i.e. loss modulus can be calculated using ($E'' = E^{*}Sin\ \delta$) where the tan delta is $tan\ \delta = E'' / E'$, where δ is the phase difference between stress - strain and equal to $0^{o} < \delta < 90^{o}$.

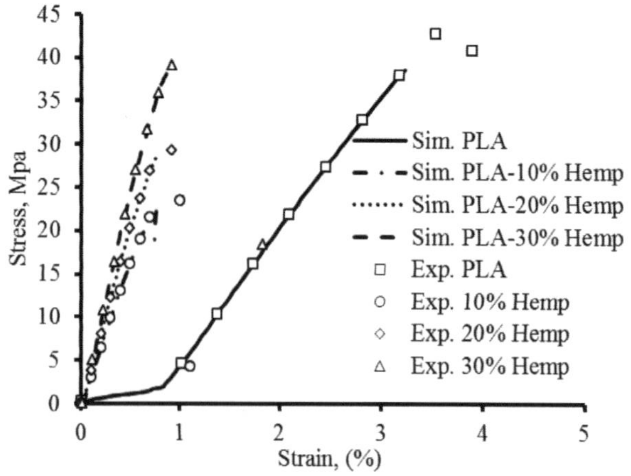

Figure 5. Stress strain curves of pure PLA and its composites obtained from experiments and simulation tests

Journal of Physics: Conference Series **451** (2013) 012010 doi:10.1088/1742-6596/451/1/012010

It is observed that experimental results have good agreement with numerical results obtained from the simulation for the dynamic mechanical spectra as shown in Figure 8. The differences in calculated magnitudes for storage moduli between experimental and FE simulation, can be subjected to assumption incorporated in the FE simulations and the expressed chemical and physical differences among the fibres type, that are, shape and hardness of the fibres, surface area of the fibres and orientation of the fibres.

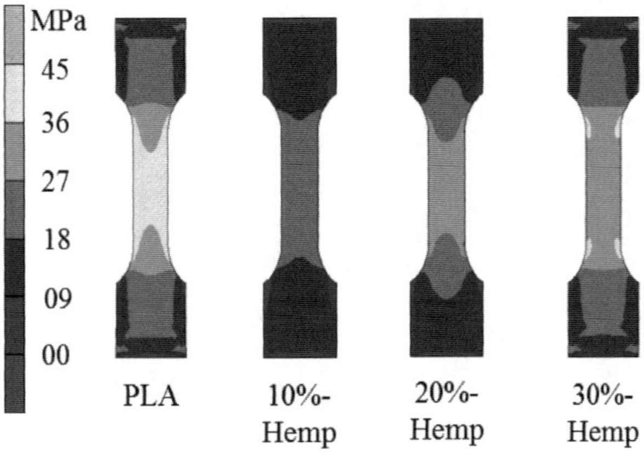

Figure 6. Stress distribution in the specimens

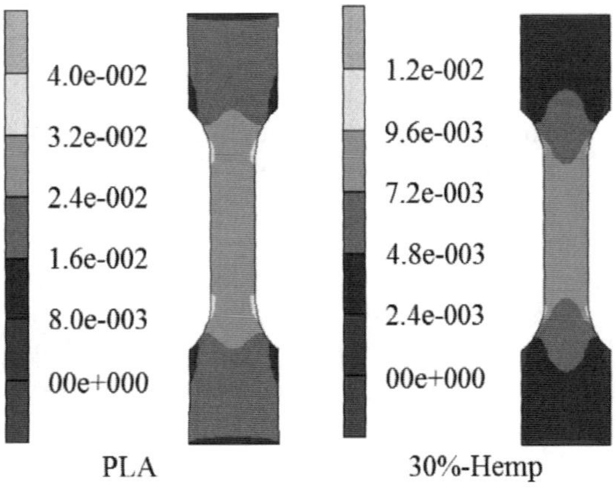

Figure 7. Calculated plastic strain level in tensile testing

5. Conclusion

The PLA – hemp composites were developed and analysed using standard tensile and thermo mechanical techniques to investigate their mechanical and visoelastic properties. The reinforcement fibres increased the mechanical stiffness and caused to reduce the tensile strength. The storage modulus is evidently increased with fibres loading in the composites reinforcing the tensile properties. In order to compare the experimental results, the FE simulation models were developed for tensile and DMA analysis. A close agreement was observed between the experimental and simulation results.

Further studies are required to understand the structural properties of biocomposites.

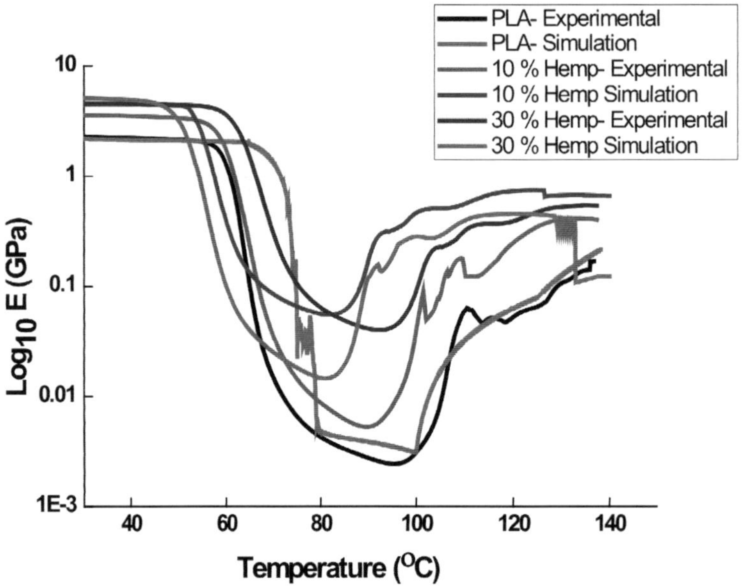

Figure 8. Storage moduli obtained from FE simulation of PLA and PLA- hemp composites.

References

[1] Auras R, Harte B and Selke S 2005 Polylactides: a new era of biodegradable polymers for packaging alication *SPE ANTEC Proc. Int. Conf. (USA)* 3240

[2] Oksman K, Skrifvars M and Selin JF 2003 Natural fibres as reinforcement in polylactic acid (PLA) composites *Comp. Sci and Tech.* **63 (9)** 1317

[3] Mohanty A K, Mishra M and Hinrichsen G 2000 Biofibres- biodegradable polymers and biocomposites- An Overview *Macro-mol. Mater. and Eng.* **276 / 277** 1-24

[4] Simoes C L, Viana J C and Cunha A M 2009 Mechanical properties of poly(ε-caprolactone) and poly(lactic acid) blends *J. Al. Poly. Sci.* **112(1)** 345

[5] Jain S, Reddy M, Mohanty A, Misra M and Ghosh A K. 2010 A new biodegradable flexible composit sheet from poly(lactic acid)/poly(ε-caprolactone) blends and micro-talc. *Macro-mol. Mater. and Eng.* **295(8)** 750

[6] Zhang M. and Thomas N L 2011 Blending polylactic acid with polyhydroxybutyrate: The effect on thermal, mechanical, and biodegradation properties *Adv. in Poly. Tech.* **30(2)** 67

[7] Ray S S, Yamada K., Ogami A, Okamoto M and Ueda K. 2002 Polylactide-Layered Silicate Nanocomposite: A Novel Biodegradable Material *Nano Lett.* **2**(10) 1093

[8] McLauchlin A R and Thomas N L 2009 Preparation and thermal characterisation of poly(lactic acid) nanocomposites prepared from organoclays based on an amphoteric surfactant *Poly. Deg. and Stab* **94(5)** 868

[9] Ray S S and Ueda K 2002 New Polylactide/Layered Silicate Nanocomposite: Nanoscale Control Over Multiple Properties *Macro-mol. Rapid Comm.* **23(16)** 943

[10] Sinha R S and Okamoto M 2003 New Polylactide/Layered Silicate Nanocomposites, *Macro-mol. Mater. and Eng.* **288(12)** 936

[11] Shakoor A and Thomas N L 2013 Talc as a nucleating agent and reinforcing filler in poly(lactic acid) composites *Poly. Eng. & Sci.* Doi 10.1002/pen.23543

[12] Garlotta D 2001 A literature review of Poly(Lactic Acid). *J. of Poly. and Envirn.* **9(2)** 63
[13] Auras R, Lim L K, Selke S and Tsuji H 2010 *Poly(lactic acid) - Synthesis, Structures, Properties, Processing, and Alications* (New Jersey John Wiley & Sons) 522
[14] Graupner N 2009 Improvement of the mechanical properties of biodegradable hemp fiber reinforced poly(lactic acid) (PLA) composites by the admixture of man-made cellulose fibers *J. Comp. Mater.* **43(6)** 689

Analysis of interfacial crack propagation under asymmetric loading in anisotropic materials

L Pryce[1,3], L Morini[2] and G Mishuris[1,3]

[1] Institute of Mathematics and Physics, Aberystwyth University, Aberystwyth, Ceredigion. SY23 3BZ

[2] Department of Civil, Environmental and Mechanical Engineering, University of Trento, Via Mesiano 77, 38123 Trento, Italy

[3] Enginsoft Trento, Via della Stazione 27 - fraz. Mattarello, 38123, Trento, Italy

E-mail: `lep8@aber.ac.uk`, `Lorenzo.Morini@unitn.it`, `ggm@aber.ac.uk`

Abstract. This paper considers a steady-state crack propagating along an interface between dissimilar orthotropic materials under an asymmetric load. Although most of the known results so far deal with symmetric loading, it has been shown recently that a significant asymmetry in the applied loading may lead to a pronounced effect in terms of the values of the SIFs. The aim of the paper is to extend these results from the static case to a moving crack. In particular, we show the significance of the asymmetry of the loading for computing the energy release rate.

1. Problem formulation and preliminary results

As a model, a semi-infinite crack propagating at a constant, subsonic speed, v, along a perfect interface between two semi-infinite anisotropic materials is considered. In our model the crack is said to occupy left hand portion of the interface whereas the materials are bonded along the remaining region of the interface. The separation point (crack tip) is moving from left to right. The material occupying the region above the interface will be referred to as material I and the material occupying the region below the interface as material II. When applying the theory to orthotropic materials the axes of orthotropy are assumed to be oriented in such a way that one axis is parallel and another perpendicular to the interface between the two materials. The results shown in this paper also hold for monoclinic materials if the plane of symmetry is the plane spanned by the x_1 and x_2 coordinates. In the absence of body forces, the applied tractions on the crack faces are self-balanced but distributed arbitrarily (continuously or as point forces) and move with the same speed, v, in the direction of the crack propagation. An example of this system for a given loading is shown in Fig. 1. For the purpose of this paper it is not necessary for $\sigma^{\pm}$ to vanish at infinity but we do have the condition that $\sigma^{\pm}$ has to vanish in a region of the crack tip.

To begin with we consider the static case, which has the governing equation defined by Hooke's law

$$\sigma_{ij} = C_{ijkl}\epsilon_{kl} = C_{ijkl}\frac{\partial u_k}{\partial x_l}, \qquad \text{for } i, j, k, l = 1, 2, \tag{1}$$

where the tensors of stress, σ, and strain, ϵ, are related via the stiffness tensor, C, for each

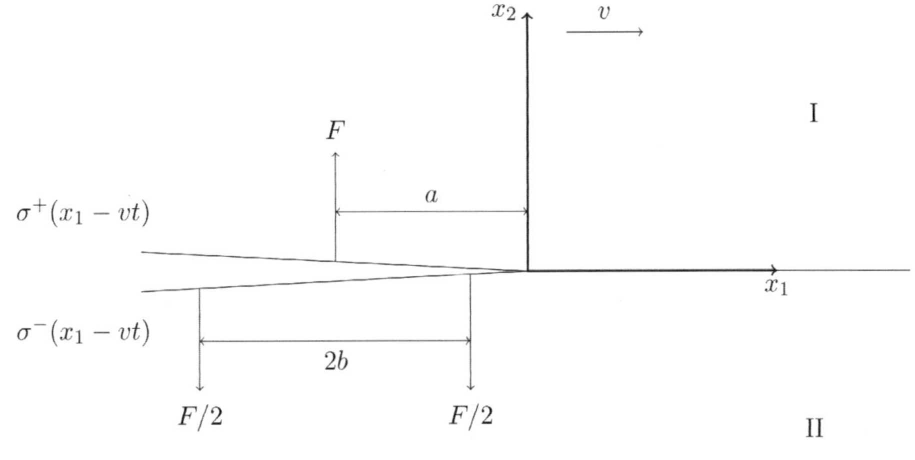

Figure 1. Geometry

material. In the absence of body forces, the balance equation reads

$$\sum_{j,k,l=1}^{2} \frac{\partial \sigma_{ij}}{\partial x_j} = 0. \tag{2}$$

Combining (1) and (2) gives the Navier-Stokes equation for an arbitrary anisotropic material

$$C_{ijkl} \frac{\partial^2 u_k}{\partial x_j \partial x_l} = 0. \tag{3}$$

Following [3], a solution for the displacement field is sought in the form $u_i = A_i f(x_1 + px_2)$ and this, in conjunction with (3), gives the following eigenvalue problem to find the respective eigenvalues p:

$$(\mathbf{Q} + p(\mathbf{R} + \mathbf{R}^T) + p^2\mathbf{T})\mathbf{A} = 0, \tag{4}$$

where $\mathbf{Q}$, $\mathbf{R}$ and $\mathbf{T}$ all depend on elastic constants for the given material. The matrix $\mathbf{A}$ consists of the two eigenvectors A_i. Knowing this result, Suo in [11] constructed the solution for an interface crack problem and found expressions for the traction along the interface ahead of the crack tip and the jump in displacement over the crack. These results were found under the assumption that the load applied on the crack faces was symmetric.

A similar method has been used to study the steady-state problem. Now the governing equation also contains an inertial term:

$$C_{ijkl} \frac{\partial^2 u_k}{\partial x_j \partial x_l} = \rho \frac{\partial^2 u_i}{\partial t^2}, \tag{5}$$

where ρ is the material density. Introducing the respective moving coordinate system: $(\tilde{x}_1 = x_1 - vt, \tilde{x}_2 = x_2)$, this can be rewritten as

$$\tilde{C}_{ijkl} \frac{\partial^2 u_k}{\partial \tilde{x}_j \partial \tilde{x}_l} = 0, \tag{6}$$

in the new frame of reference, where $\tilde{C}_{ijkl} = C_{ijkl} - \rho v^2 \delta_{ik}\delta_{1j}\delta_{1l}$. This allowed Yang [14] to solve the interface crack problem in an anisotropic bimaterial under symmetrical load prescribed along the crack surfaces.

From this stage on, for convenience, the moving coordinate system will be written as $\tilde{x}_1 = x$ and $\tilde{x}_2 = y$. Searching for a solution in the form $u_i = A_i f(x + py)$, Yang obtain the eigenvalue problem (4), where the matrix $\mathbf{Q}$ now relies on both the material constants and the crack velocity, while matrices $\mathbf{R}$ and $\mathbf{T}$ are identical to those in (4). Using the same method as seen in [11], it was possible to find the physical solution of the problem under the assumption that the loading on the crack faces was symmetric.

The theory has recently been developed further in [4] where the results were obtained for asymmetric loading on the crack faces, for a static crack. The method used in this analysis originates from the approach used in [12] where special weight functions were incorporated into the fundamental Betti formula. These results have been then extended to a moving crack under asymmetric loading in [9] and it is the application of these results that we will discuss in the remainder of this paper.

In order to perform further work, certain information regarding the asymptotic expansions of the traction, $\mathbf{t}(x)$, and displacement jump, $\mathbf{d}(x)$, at the crack tip is required. The expansions, as $x \to 0$, are given by

$$\mathbf{t}(x) = \frac{x^{-\frac{1}{2}}}{2\sqrt{2\pi}}\mathbf{T}(x)\mathbf{K} + \frac{x^{\frac{1}{2}}}{2\sqrt{2\pi}}\mathbf{T}(x)\mathbf{Y} + \mathcal{O}((x)^{\frac{3}{2}}), \tag{7}$$

$$\mathbf{d}(x) = \frac{(-x)^{\frac{1}{2}}}{\sqrt{2\pi}}\mathcal{U}(x)\mathbf{K} + \frac{(-x)^{\frac{3}{2}}}{\sqrt{2\pi}}\mathcal{U}(x)\mathbf{Y} + \mathcal{O}((-x)^{\frac{5}{2}}), \tag{8}$$

where $\mathbf{K} = [K, \bar{K}]$ and $\mathbf{Y} = [Y, \bar{Y}]$. Here, $K = K_1 + iK_2$ is the complex stress intensity factor of the system [1, 13]. The matrices $\mathcal{U}(x)$ and $T(x)$ are given by the following equations

$$\mathcal{U}(x) = \frac{2(H + \bar{H})}{\cosh \pi\epsilon}\left[\frac{\mathbf{w}(-x)^{i\epsilon}}{1 + 2i\epsilon}, \frac{\bar{\mathbf{w}}(-x)^{-i\epsilon}}{1 - 2i\epsilon}\right], \qquad \mathbf{T}(x) = 2\left[\mathbf{w}x^{i\epsilon}, \bar{\mathbf{w}}x^{-i\epsilon}\right].$$

The matrix $\mathbf{H}$ is a bimaterial matrix defined as $\mathbf{B}_I + \bar{\mathbf{B}}_{II}$, where $\mathbf{B}$ is the surface admittance tensor, $\mathbf{B} = i\mathbf{A}\mathbf{L}^{-1}$, for each material. The vector $\mathbf{w}$ and Dundurs paramater ϵ are found from the eigenvalue problem [11]

$$\bar{\mathbf{H}}\mathbf{w} = e^{2\pi\epsilon}\mathbf{H}\mathbf{w}. \tag{9}$$

Follwing the work of [2] an expression was also found for the energy release rate at the crack tip, G, for the static case in [11] in the form

$$G = \frac{\bar{\mathbf{w}}^T(\mathbf{H} + \bar{\mathbf{H}})\mathbf{w}|K|^2}{4\cosh^2(\pi\epsilon)}. \tag{10}$$

It was noted in [15] that the same method can be used to find the energy release rate for a moving crack as long as it moves at subsonic speeds. As this paper only considers cracks with sub-Rayleigh velocities this expression can also be used in our analysis. For orthotropic materials equation (10) can be simplified further

$$G = \frac{H_{11}(1 - \beta^2)|K|^2}{4}. \tag{11}$$

The Dundurs parameter, β, and H_{11} are both obtained from the bimaterial matrix $\mathbf{H}$ (see [9]).

The following section of the paper considers a specific example of loading and materials to calculate energy release rates using the methods developed in [9]. We concentrate on the consequence of the asymmetry in the applied load for the major fracture mechanics paramaters, that is the stress intensity factors and energy release rate, highlighting some effects which cannot be observable in a dissimilar structure when only symmetric loading is applied.

2. Specific example

The loading on the crack faces is given by a point force of magnitude F acting on the upper crack face a distance a behind the crack tip and two point forces, both of magnitude $F/2$, acting on the lower crack face a distance b away from the point force acting upon the upper crack face. All forces move at the same speed and in the same direction as the crack propagation. This system is shown in Fig. 1.

In order to compare the contribution of the different parts of the loading to the overall energy release rate, we decompose the stress intensity factor, K, into its symmetric and asymmetric parts, denoted K^S and K^A respectively. Both components are complex numbers.

We now compute the energy release rate for two given materials. The piezoceramic Barium Titanate is set as material I, with material properties $C_{11} = 120.3$GPa, $C_{22} = 120.3$GPa, $C_{12} = 75.2$GPa, $C_{66} = 21$GPa and $\rho = 6,020$kgm^{-3}. Material II is set as the Aluminium which has material properties $C_{11} = 107.3$GPa, $C_{22} = 107.3$GPa, $C_{12} = 60.9$GPa, $C_{66} = 28.3$GPa and $\rho = 2,700$kgm^{-3}. For the purpose of the calculations in this paper we set $a = 1$. The velocity has been normalised by dividing by c_R, the lowest Rayleigh wave speed for either material, which in this case is the Rayleigh wave speed of Barium Titanate. We present a dimensionless form of the ERR in our results, where G is normalised in the following manner: $GC_{66}^{(1)}/F^2$. For this normalisation we take our value of $C_{66}^{(1)}$ as that of the material above the crack.

Fig. 2 shows the normalised ERR and Fig. 3 indicates the normalised energy release rates, G^S and G^A, corresponding to K^S and K^A, respectively. Both components are normalised by the total energy release rate corresponding to $K = K^S + K^A$, given by G.

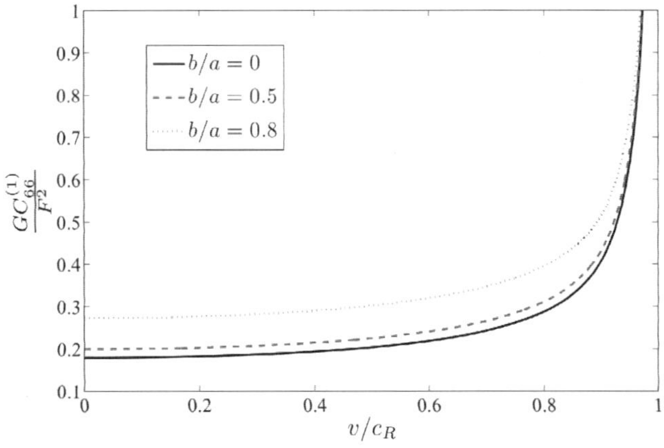

Figure 2. The normalised ERR, as a function of the velocity, for different positions of the self-balanced point forces applied to the crack surfaces, described by the ratio b/a in Fig. 1.

The graph in Fig. 2 clearly shows that ERR increases as the velocity increases for all values of b/a and approaches infinity as the velocity approaches the Rayleigh wave speed , as expected. Moreover, the energy release rate is higher for larger values of b/a, that is, as the asymmetry of the loading increases. Therefore, symmetric loading is energetically more beneficial than any choice of asymmetric one.

Fig. 3 show that the contribution of the asymmetric part of the loading to the overall energy release rate is small compared to the contribution of the symmetric part for the lower crack velocities. However, when the crack is moving at velocities close to the Rayleigh wave speed the asymmetric nature of the loading begins to play a more crucial role in the calculations. This fact is also highlighted in the left hand graph of Fig. 4. Interestingly, the energy release rate corresponding to the symmetric part of the loading is higher than the overall ERR for the physical loading, apart from when the crack velocity is very close to the Rayleigh wave speed. The behaviour of the symmetric part of G near the Rayleigh wave speed is particularly noteworthy as it can be seen that for asymmetric loading (when $b/a > 0$) the ERR stops increasing and starts decreasing. Note that, for the symmetrical load ($b/a = 0$), the energy release does not depend on the crack speed. This was previously observed for isotropic and anisotropic bimaterials in [13, 14].

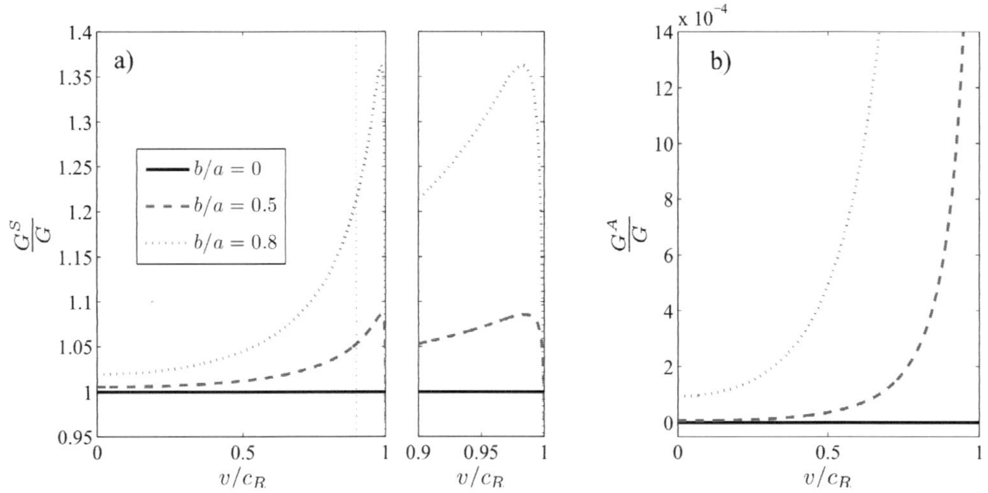

Figure 3. a) The ratio G^S/G, as a function of velocity, for different values of b/a, with particular attention being given to the behaviour near c_R. b) The ratio G^A/G with a different scale on the vertical axis.

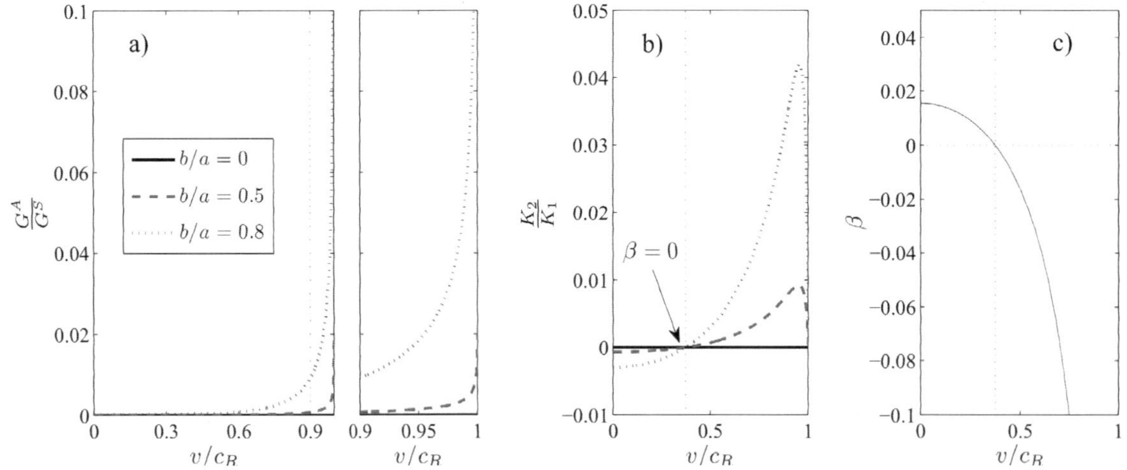

Figure 4. a) The ratio G^A/G^S, as a function of the velocity, for different values of b/a with particular attention being given to the behaviour near c_R. b) The ratio K_2/K_1, as a function of thevelocity, for different values of b/a. c) The value of β as a function of the velocity.

Fig. 4a shows the ratio of the mode 2 contribution of the stress intensity factor, K, to the mode 1 component. For the symmetric case the mode 2 component is 0 for all values of the velocity, as one would expect, whereas for asymmetric symmetry the ratio is initially negative and then at a certain velocity the sign of the ratio changes. This is connected to the fact that for a determinate value of the propagation speed there is a change in the sign of the Dundurs parameter, β (which only depends on the elastic constants of the materials). It is therefore possible to obtain a characteristic value for v, depending only on the material properties. This may lead to a possible change in the fracture mechanism or indicate a possible redirection of the interface crack propagation, for example kinking or branching. Note that there are other investigations, both theoretical and experimental, demonstrating an existence of a specific sub-Rayleigh velocity which is related to the stability of the crack propagation [5, 6].

3. Conclusions

In this paper we have analysed some results concerning the ERR and SIF for a crack propagating along the interface between anisotropic materials with a constant speed, under asymmetric, self-balanced loading acting upon the crack faces. For small velocities the results agree with those obtained for the stress intensity factors for the static crack in [4]. In particular, the load asymmetry influences the critical loading magnitude. Moreover, it has been shown that the asymmetry may play an important role for a moving crack, where the SIF changes its sign at a certain velocity depending on the material parameters but independent of the asymmetry of the load. However,this effect could only be observered in the presence of the asymmetry and may be important when analysing possible crack kinking or branching.

Acknowledgments

LP, and GM acknowledge support from the FP7 IAPP project 'INTERCER2', project reference PIAP-GA-2011-286110-INTERCER2. LM gratefully thanks financial support from the Italian Ministry of Education, University and Research in the framework of the FIRB project 2010 "Structural mechanics models for renewable energy applications.

References

[1] Hwu C 1993 Explicit solutions for collinear interface crack problems *Int. J. Solids Struct.* **30** 301-312
[2] Irwin G 1957 Analysis of stresses and strains near the end of a crack traversing a plate *J. Appl. Mech.* **24** 361-364
[3] Lekhnitskii S G 1963 *Theory of elasticity of an anisotropic body* (San Francisco: Holden-Day)
[4] Morini L, Radi E, Movchan A B and Movchan N V 2012 Stroh formalism in analysis of skew-symmetric and symmetric weight functions for interfacial cracks *Math. Mech. Solids.* 1-18
[5] Obrezanova O, Willis J R and Movchan A B 2002 Stability of an advancing crack to small perturbation of its path *J. Mech. Phys. Solids* **50** 57-80
[6] Obrezanova O, Willis J R and Movchan A B 2002 Dynamic stability of a propagating crack *J. Mech. Phys. Solids* **50** 2637-2668
[7] Piccolroaz A, Mishuris G and Movchan A B 2007 Evaluation of the Lazarus-Leblond constants in the asymptotic model for the interfacial wavy crack *J. Mech. Phys. Solids.* **55** 1575-1600
[8] Piccolroaz A, Mishuris G and Movchan A B 2009 Symmetric and skew-symmetric weight functions in 2D perturbation models for semi-infinite interfacial cracks *J. Mech. Phys. Solids.* **57** 1657-1682
[9] Pryce L, Morini L and Mishuris G 2013 Weight function approach to study a crack propagating along a bimaterial interface under asymmetric loading in orthotropic solids *To be submitted for publication. arXiv No.1305.0486*
[10] Stroh A 1957 Steady state problems in anisotropic elasticity *Math. Phys.* **41** 77-103
[11] Suo Z 1990 Singularities, interfaces and cracks in dissimilar anisotropic media *Proc. R. Soc. Lond* **427** 331-358
[12] Willis J R and Movchan A B 1995 Dynamic weight function for a moving crack I Mode I Loading *J. Mech. Phys. Solids* **43** 319-341
[13] Wu K C 1990 Stress intensity factors and energy release rate for interfacial cracks between dissimilar anisotropic materials *J. Appl. Mech.* **57** 882-886
[14] Yang W, Suo Z and Shih C F 1991 Mechanics of dynamic debonding *Proc. R. Soc. Lond* **433** 679-697
[15] Yu H H and Suo Z 2000 Intersonic crack growth on an interface *Proc. R. Soc. Lond* **456** 223-246

Kinetics of interfacial crack bridged zone degradation

M Perelmuter
Institute for Problems in Mechanics of RAS
prospect Vernadskogo 101-1, Moscow, 119526 Russia

E-mail: perelm@ipmnet.ru

Abstract. A kinetic model of bonds destruction, interfacial adhesion strength and durability of interface junction is developed. It is assumed that a zone of weakened bonds exists at materials interface at an initial time instant. This weakened zone is considered as a crack-like defect filled with distributed spring-like bonds. Bonds mechanical properties are time-dependent under external loads. Bonds properties kinetics analysis is performed on the basis of Zhurkov's thermofluctuational approach of bonds destruction. The computational results those obtained on the basis of the developed numerical algorithm are presented.

1. Introduction

The model of a crack with bonds in the bridged zone allows one to determine the crack resistance and the adhesive strength of joints between different materials on the basis of micromechanical properties of the bonds [1–3]. The development of this model with consideration of time-variation in the physical and mechanical properties in the crack bridged zone allows one to estimate the long-term strength and time characteristics of material crack resistance.

In the present paper, Zhurkov's kinetic model of thermal fluctuation fracture [4,5] together with the crack bridged zone model, is used for bonds destruction and interfacial adhesion strength modelling. It is assumed that at least one of the materials is a polymer and the crack part occupied by the bridges (the bridged zone) is not small compared with the crack length. The estimate of the bonds destruction in the crack bridged zone and kinetic characteristics of the interfacial junction are based on the following assumptions:

- at the initial time, on the interface there is a region of weakened bonds between materials (this may be a technological defect or a weakened region caused, for example, by the diffusion activity of the medium);
- the bond density in that region varies in time according to the thermal fluctuation mechanism;
- the bond rigidity is proportional to their density at each point of the crack bridged zone;
- the defect nucleation occurs near the center of the weakened bond region;
- the condition for the crack-defect formation is the decreasing in the average bond density to the critical value on the corresponding part of the weakened bond region.

2. Kinetics of bonds in the crack bridged zone

The durability of materials under the action of the tensile stress σ satisfies the following experimentally established formula [4,5]

$$\tau = \mu\tau_0 e^{\frac{U(\sigma)}{kT}} \qquad (1)$$

where τ is the specimen fracture time, k is the Boltzmann constant, T is the absolute temperature, $\tau_0 \approx h/(kT)$ is a constant of the order of the atomic thermal vibration period ($10^{-13} - 10^{-12}$ s), h is the Planck constant, μ is a dimensionless coefficient depending on the type of the material (polymer, metal, or ceramics), and $U(\sigma)$ is the bond destruction energy (fracture activation energy).

For a sufficiently wide interval of external loads and temperature, the function $U(\sigma)$ is linear,

$$U(\sigma) = U_0 - qT - A(\sigma), \quad A(\sigma) = \gamma\sigma \tag{2}$$

where U_0 is the interatomic bond breakage activation energy and q and γ are structure-sensitive parameters characterizing the material thermal and strength properties.

It follows from expression in Eq. (2) that $A(\sigma)$ decreases the bond breakage energy barrier and can be treated as the work done by external stresses in the body fracture.

A formula similar to that in Eq. (1) also holds for the average life time (durability) τ_b of a loaded interatomic bond [4,5]. The work of external stresses in this case depends on the stress value in the bond. The inhomogeneity of the stress distribution in the bonds in an actual body results in local damage accumulation and formation of defects. The most intensive damage accumulation occurs in the weakened interatomic (intermolecular) bond regions in the material. The weakened bond region in the material (or on the material interface) will be treated as a crack filled with bonds (bridges) whose properties vary in time according to the thermal fluctuation mechanism.

It is assumed that the expression in Eq. (1) also holds for bonds in the crack bridged zone but the work $A(\sigma)$ contained in Eq. (2) is the bond deformation work, which is determined with the inhomogeneous stresses distribution over the bonds in the bridged zone taken into account.

The bond deformation work A and the bond durability τ_b in the bridged zone of a rectilinear crack on the material interface (Fig. 1) depend on the tension of bond in the bridged zone σ_b (for details, see Eqs. (12)-(13) below) and the coordinate x determining the bond location along this zone, $A = A(\sigma_b, x)$ and $\tau_b = \tau_b(\sigma_b, x)$.

Assume (following [6]) that the time variation in the bond density $n(x, t)$ in the crack bridged zone is described by the chemical destruction equation

$$\frac{dn(x,t)}{dt} = -\frac{n(x,t)}{\tau_b(\sigma_b, x)} \tag{3}$$

where $\tau_b = \tau_b(\sigma_b, x)$ is the time till the breakage of the molecular bond located in the crack bridged zone at the distance $\ell - x$ from its tip (the bond durability).

It follows from Eq. (3) that

$$n(x,t) = n_0 \exp\left(-\frac{t}{\tau_b(\sigma_b, x)}\right) \tag{4}$$

where $n(x, 0) = n_0$ is the initial bond density.

The time variation in the bond density leads to a variation in the bond compliance in the crack bridged zone. Let $k_s(x)$ denote the rigidity of a single molecular bond. Then the effective rigidity of bonds per unit area of the crack bridged zone is

$$k(x,t,\sigma_b) = k_s(x)n(x,t) = k_b(x)\exp\left(-\frac{t}{\tau_b(\sigma_b,x)}\right) \tag{5}$$

where $k_b(x) = k_s(x)n_0$ is the initial bond rigidity per unit area of the crack bridged zone.
It follows from Eq. (5) that the bond compliance in the crack bridged zone can be represented as

$$c(x,t,\sigma_b) = c_b(x)\exp\left(\frac{t}{\tau_b(\sigma_b,x)}\right) \tag{6}$$

where $c_b(\cdot) = 1/k_b(\cdot)$ is a function determining the initial bond compliance in the crack bridged zone.

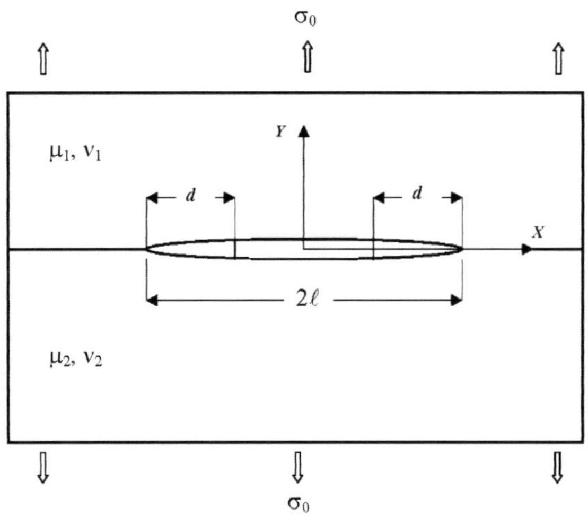

Figure 1. Bridged crack at the interface of different material

3. Bridged interface crack model

We use proposed in the previous papers [1-3] the bridged interface crack model with the assumption that the crack surfaces interact in some zones starting from the crack tips. Let us consider the planar elasticity problem on a crack ($|x| \le \ell$) at the interface $(y=0)$ of two dissimilar joint half-planes, see Fig.1. Assume that the uniform load σ_0 normal to the interface is acted at infinity. The crack surface interaction is supposed to be existing in the bridged zones, $\ell - d \le |x| \le \ell$. The size of the interaction zone depends on time $d = d(t)$ due to the possibility the time changing of the bond properties. As a simple mathematical model of the crack surfaces interaction we will assume that the linearly elastic bonds act through the crack bridged zones.

Denote by $Q(x,t)$ the bond tractions occurring under the action of the external loads such that

$$Q(x,t) = q_y(x,t) - iq_x(x,t)\,, \quad i^2 = -1 \tag{7}$$

where $q_y(x,t)$ and $q_x(x,t)$ are the normal and shear components of the traction, respectively.

By incorporating the superposition principle we'll consider the 2D-elasticity interface crack problem with the following boundary conditions at the crack surfaces $(y=0)$

$$\sigma_{yy}(x,t) - i\sigma_{xy}\ (x,t) = -\sigma_o, \quad |x| \le \ell - d(t) \tag{8}$$

$$\sigma_{yy}(x,t) - i\sigma_{xy}(x,t) = (-\sigma_o + q_y(x,t)) - iq_x(x,t), \quad \ell - d(t) \le |x| \le \ell \tag{9}$$

The opening of the interface crack, $u(x,t)$, can be written as follows

$$u(x,t) = u_y(x,t) - iu_x(x,t), \tag{10}$$

where $u_{x,y}$ are the projections of the crack opening on the coordinate axes x and y respectively.
The function $u(x,t)$ can be represented in the following form

$$u(x,t) = u_\infty(x) + u_Q(x,t), \quad u_\infty(x) = u_{\infty y}(x) - iu_{\infty x}(x), \quad u_Q(x,t) = u_{Qy}(x,t) - iu_{Qx}(x,t) \tag{11}$$

Where $u_\infty(x)$ is the crack opening caused by the loads, $-\sigma_0$, acting at the crack surfaces and $u_Q(x,t)$ is the crack opening caused by the bond induced tractions under the action of the external loads.
The relation between the crack opening and bond tractions (the bond deformation law) will be written as follows

$$u(x,t) = c(x,t,\sigma_b)(q_y(x,t) - iq_x(x,t)), \tag{12}$$

where the function $c(x,t,\sigma_b)$ is determined by formula (6) and $\sigma_b = \sqrt{q_x^2 + q_y^2}$.
For dimensionality reasons one can represent function $c(x,t,\sigma_b)$ in the formula (6) in the following form [1]

$$c_b(x) = \frac{H}{E_B}\varphi(x,t,\sigma_b) = c_0\frac{\ell}{E_B}\varphi(x,t,\sigma_b) \tag{13}$$

where H is a linear scale proportional to the bonding zone thickness, E_B is the effective Young modulus of the bonds and $\varphi(x,t,\sigma_b)$ is dimensionless function and

$$c_0 = \frac{H}{\ell}$$

is the relative bond compliance.
Let us write a system of the integral-differential equations for computation of the bond tractions $Q(x,t)$ in the crack bridged zone. From the first relation in expressions (11) we can obtain

$$\frac{\partial u(x,t)}{\partial x} = \frac{\partial u_\infty(x)}{\partial x} + \frac{\partial u_Q(x,t)}{\partial x} \tag{14}$$

The interface crack opening components caused by the external normal stresses, σ_o (or the stresses - σ_o at the crack surfaces), are equal to [7]

$$u_\infty(x) = \frac{\sigma_o\sqrt{\alpha}}{2(1+\alpha)}\left(\frac{k_1+1}{\mu_1}+\frac{k_2+1}{\mu_2}\right)\sqrt{\ell^2-x^2}\left(\frac{\ell-x}{\ell+x}\right)^{-i\beta}, \quad \alpha = \frac{\mu_2 k_1+\mu_1}{\mu_1 k_2+\mu_2}, \quad \beta = \frac{\ell n\,\alpha}{2\pi} \tag{15}$$

where $k_{1,2} = 3-4\nu_{1,2}$ or $k_{1,2} = (3-\nu_{1,2})/(1+\nu_{1,2})$ for the plane strain and plane stress, respectively; $\nu_{1,2}$ and $\mu_{1,2}$ are the Poisson ratios and shear moduli of the joint materials 1 (Y>0) and 2 (Y<0).

To obtain the expression of the derivative of the crack opening caused by the bond action ($\partial u_Q(x,t)/\partial x$) we had used the representations for the derivative of the displacements of the interface crack surfaces under the action of arbitrary loads given in [7], see details of transformations in [1].

Finally, by substituting the expressions for the derivatives (those obtained on the basis of Eqs. (12),(13) and (15)) into Eq. (14) we arrive to system of the integral-differential equations for unknown functions $p_{x,y}(x,t)$ (see details in [8])

$$\frac{\partial}{\partial s}\left[\varphi(s,t,\sigma_b)(q_y(s,t)-iq_x(s,t))\right] - \frac{\pi\varepsilon(1-\alpha)}{2(1+\alpha)}(q_x(s,t)+iq_y(s,t)) +$$
$$\frac{\varepsilon}{\sqrt{1-s^2}}\left(\frac{1-s}{1+s}\right)^{-i\beta}\int_{1-d/\ell}^{1} R(s,\eta)\left[sp_y(s,t)-i\eta p_x(s,t)\right]d\eta = \frac{\pi\varepsilon\sqrt{\alpha}\,(2\beta i - s)}{(1+\alpha)\sqrt{1-s^2}}\left(\frac{1-s}{1+s}\right)^{-i\beta} \tag{16}$$

where

$$q_y(s,t)-iq_x(s,t) = \left(p_y(s,t)-ip_x(s,t)\right)\left(\frac{1-s}{1+s}\right)^{-i\beta}$$

$$s=\frac{x}{\ell},\ t=\frac{\xi}{\ell},\ q_{y,x}(s,t)=\frac{q_{y,x}(x,t)}{\sigma_o},\ p_{y,x}(s,t)=\frac{p_{y,x}(x,t)}{\sigma_o}$$

The singular kernel $R(s,\eta)$ and the parameter ε in Eqs. (16) are defined as

$$R(s,\eta)=\frac{\sqrt{1-\eta^2}}{\eta^2-s^2}, \quad \varepsilon = \frac{E_B\ell}{2\pi H}\left(\frac{k_1+1}{\mu_1}+\frac{k_2+1}{\mu_2}\right)$$

Parameter ε characterizes the ratio of the compliance of the joint part containing a crack without bonding to the bond compliance. Equations (16) represent the system of two singular integral-differential equations with the Catchy type kernels. This system can be solved numerically.

4. Bonds deformation work and stress intensity factors

The bond deformation work (per unit width of the body) on the crack bridged zone of size dx is given by the expression [1, 8]

$$dU(x) = \left[\int_0^{u_x(x)} q_x(u_x)du_x + \int_0^{u_y(x)} q_y(u_y)du_y\right]dx \tag{17}$$

where $u_{x,y}(x)$ are the opening components in the crack bridged zone at the point with coordinate x and $q_{x,y}(x)$ are the bonds tractions. The number of molecular bonds on the interval dx is

$$dn = n(x,t)dx$$

The work per one intermolecular bond is estimated under the assumption that the adhesion junction is formed by chains of polymer molecules of size of the monomer link a_m. It is assumed that the bond elongation (of the chain molecule) in deformation within the limits of linear elasticity is much less than its initial length and the crack opening in the bridged zone at the distance $\ell - x$ from the crack tip is approximately equal to the length of the loaded segment of the chain molecule bridging the crack surfaces. In this case, the number N_m of monomer links between the crack surfaces on an interval of length dx is

$$N_n = \omega dn, \quad \omega = \sqrt{u_x^2(x,t) + u_y^2(x,t)} \Big/ a_m \tag{18}$$

where ω is the number of monomer links of the polymer chain making a bond between the crack lips. The work per one intermolecular bond is equal to

$$A(\sigma_b, x) = \frac{dU(x)}{\omega dn} \tag{19}$$

Finally, expression (1) can be rewritten in the form convenient for computations,

$$\tau_b(\sigma_b, x) = \tau_0 \exp\left(\frac{U_M - A_M(\sigma_b, x)}{RT}\right) \tag{20}$$

where U_M is the energy of activation of one mole of intermolecular bonds, $U_M = U_0 N_A$, $A_M(\sigma_b,\, x) = A(\sigma_b,\, x) N_A$, $R = kN_A$ is the gas constant, and N_A is the Avogadro number.

Having the solution of system (16) one can calculate the stress intensity factors (SIF) K_I and K_{II} following to [9]

$$K_{\mathrm{I}} + iK_{\mathrm{II}} = \lim_{\delta \to 0} \sqrt{2\pi\delta}\left(\sigma_{yy}(\delta,t) + i\sigma_{xy}(\delta,t)\right)\delta^{-i\beta}, \tag{21}$$

where $\sigma_{yy}(\delta,t)$ and $\sigma_{xy}(\delta,t)$ are the stresses ahead the crack tip caused by the external loads and bond stresses, δ is the small distance to the crack tip.
On the other hand, the SIF can be written as follows

$$K_{\mathrm{I}} + iK_{\mathrm{II}} = (K_{\mathrm{I}}^{ext} + K_{\mathrm{I}}^{\mathrm{int}}) + i(K_{\mathrm{II}}^{ext} + K_{\mathrm{II}}^{\mathrm{int}}), \tag{22}$$

where $\mathrm{K}_{\mathrm{I,II}}^{ext}$ and $K_{\mathrm{I,II}}^{\mathrm{int}}$ are the SIF caused by the external loads and bond stresses.
Taking into account relations (21) and (22) we can obtain (see details in [1])

$$K_{\mathrm{I}} + iK_{\mathrm{II}} == \frac{\sqrt{\pi\ell}}{(2\ell)^{i\beta}}\left[\sigma_o(1+2i\beta) - \frac{2\cosh(\pi\beta)}{\pi}\int_{1-d/\ell}^{1}\frac{q_y(s,t) + itq_x(s,t)}{\sqrt{1-s^2}}ds\right], \quad K = \sqrt{K_{\mathrm{I}}^2 + K_{\mathrm{II}}^2} \tag{23}$$

where K is modulus of the SIFs.

Journal of Physics: Conference Series **451** (2013) 012012 doi:10.1088/1742-6596/451/1/012012

5. Numerical algorithm

The system of the integral-differential equations (16) is solved by a collocation scheme for each time-step. The crack bridged zone, $d(t)$, is divided into M quadratic finite elements. Unknown functions are represented by parametric quadratic polynomials on each element. This procedure enables us to obtain expressions for the unknown functions and their derivatives through the nodal values of these functions. Then the system of the integral-differential equations is reduced to a system of linear algebraic equations relative to the nodal values of the unknown functions [1].

For each time step bond compliances are assumed to be proportional to the density of unbroken bonds. The following condition is used as the criterion of full bonds destruction on a part of bridged zone of size $\Delta\ell$ at its trailing edge

$$\bar{N}(t_k) \leq N_{cr} \tag{24}$$

where $\bar{N}\ (t_k)$ is the average density of bonds at the time step t_k and N_{cr} is the limit value of the bond density.

The modeling further growth of the initial nucleated defect is performed by decreasing the bridged zone size according to condition (24). The time-step scheme of numerical solving the integral-differential equation for a crack-like region with the bridged zones is used at each step of the defect increasing [8].

The computations were carried out for a combination of materials typical of microelectronic devices [10]: one of the materials is metal (copper-based alloy) with elasticity modulus $E_1 = 130\,GPa$, and the other is an (epoxy-based) polymer with elasticity modulus $E_2 = 2\,GPa$, the Poisson ratios of the materials are $\nu_1 = 0.3$ and $\nu_2 = 0.35$, respectively, and the elasticity modulus of the bonds is $E_b = E_2$. The size of the weakened bond region located along the material interface is set to be equal to $2\ell = 10^{-5}m$. It is assumed that the bonds are formed by chains of polymer molecules where the size of the molecule monomer chain is equal to $a_m = 10^{-9}m$. The bond deformation law in the computation process was assumed to be linearly elastic with compliance constant along the weakened bond region at the initial time. (Estimates of mechanical parameters of bonds in the crack bridged region can be found in [1].) The kinetic dependences were calculated for the following parameter values [4,5]: $q \cdot N_A = 8.3\ J\,/\,mole{\cdot}K$, the energy of activation of the molecular bond breakage was $U_M = 150\ kJ\,/\,mole$, the initial bond density was $n_0 = 10^{18}m^{-2}$, $\tau_0 = 10^{-12}s$, and $\mu = 1$. The limit value of the bond density in the initial defect formation region was chosen as $N_{cr} = 0.1n_0$. The external load was assumed to be constant in time.

As an example of proposed model application in Fig. 2 is shown the time variation of the stress intensity factor module in the process of growth of the bond-free crack part for different values of temperatures. In the bond breakage process, their strengthening effect decreases, and the stress intensity factor increases. The largest increase in the stress intensity factor occurs at the last stages of the defect growth when the stress intensity factor tends to the corresponding value for the bond-free crack. Assuming (for the conservative estimate) that the ratios of the time intervals in which the stress intensity factor attains the largest value are proportional to the durability of the joint, one can see that the increase in the temperature (in the range under study) by $\Delta T = 25K$ decreases the joint durability by more than one order of magnitude. The values of the stress intensity factor modulus are normalized by the value K_0, the stress intensity factor modulus due to the action of the external load for cracks without bonds.

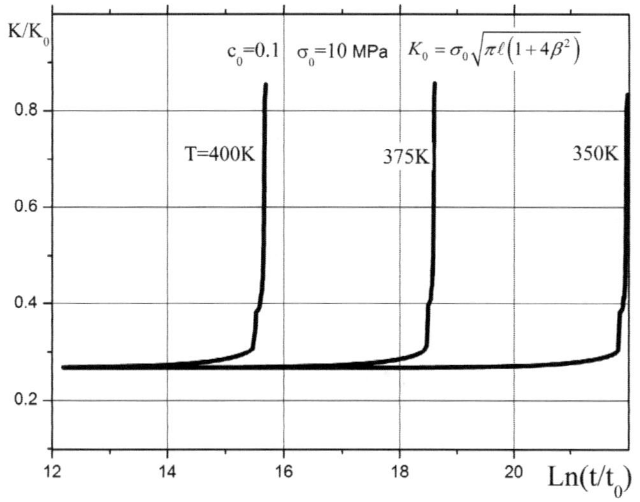

Figure 2. Variation of SIF module due to bonds destruction

6. Summary

The method of estimation of bonds degradation in the weakened region on the interface between materials is proposed. The theory of thermal fluctuation fracture and the interfacial crack bridged zone model are combined and numerical algorithm based on singular integral-differential equations is proposed. The results of computations allow to estimate the increasing the SIF module due to bonds degradation over time. These results might be helpful for durability analysis of adhesion joints. Since the computational parameters strongly depend on the initial data (which is caused by the exponential dependence of the durability in formula (1)), the comparative analysis of joints of different materials under appropriate loading conditions is of greatest practical interest in this method.

Acknowledgements. Financial support for this research was partly provided by RFBR (Project No. 11-08-01243).

References

[1] Goldstein R V and Perelmuter M N 1999 Modeling of bonding at the interface crack, *Int. J. Fracture* **99** 53-79

[2] Perelmuter M N 2007 A criterion for the growth of cracks with bonds in the end zone *J. Appl. Math. Mech.* **71** (1) 137-153

[3] Goldstein R V and Perelmuter M N 2009 Modeling of crack resistance of composite materials *Computational continuum mechanics* **2** (2) 22-39, in Russian

[4] Zhurkov S N 1965 Kinetic concept of the strength of solids *Int. J. Fracture Mech.* **1** 311-323

[5] Regel V G, Slutsker A I and Tomashevskii E E 1974 *Kinetic Nature of Strength of Solids* (Moscow: Nauka), in Russian

[6] Barenblatt G I, Entov V M and Salganik R L 1966 On the kinetics of crack propagation. General concepts. Cracks close to equilibrium *Mech. Solids* **1**(5) 53–59

[7] Slepijn L I 1981 *Mechanics of Cracks* (Leningrad: Sydostroenie), in Russian

[8] Goldstein R V and Perelmuter M N 2012 Kinetics of crack formation and growth on the material interface, *Mech. Solids* **47**(4) 400–414

[9] Rice J R 1988 Elastic fracture mechanics concepts for interfacial cracks *Trans. ASME J. Appl. Mech.* **55** 98-103

[10] Khanna V K 2011 Adhesion–delamination phenomena at the surfaces and interfaces in microelectronics and MEMS structures and packaged devices *J. Phys.D: Appl. Phys.* **44** 1–19

Application of imperfect interface weight function techniques for modelling of glued structures containing cracks and small defects

A Vellender[1,2], G Mishuris[1,2] and A Piccolroaz[3]

[1] Institute of Mathematics and Physics, Aberystwyth University, Physical Sciences Building, Aberystwyth, Ceredigion, Wales, SY23 3BZ

[2] Enginsoft Trento, Via della Stazione 27 - fraz. Mattarello, 38123, Trento, Italy

[3] Department of Civil, Environmental and Mechanical Engineering, University of Trento, Via Mesiano 77, 38123 Trento, Italy

E-mail: asv2@aber.ac.uk, ggm@aber.ac.uk, andrea.piccolroaz@ing.unitn.it

Abstract. We analyse a problem of anti-plane shear in a bimaterial plane containing a semi-infinite crack situated on a soft imperfect interface. The plane also contains a small thin inclusion (for instance an ellipse with high eccentricity) whose influence on the propagation of the main crack we investigate. The problem can be considered as modelling bimaterial ceramics which are joined with a thin adhesive substance. An important element of our approach is the derivation of a new weight function (a special solution to a homogeneous boundary value problem) in the imperfect interface setting. The weight function is derived using Fourier transform and Wiener-Hopf techniques and allows us to obtain an expression for an important constant σ_0 (which may be used in a fracture criterion) that describes the leading order of tractions near the crack tip for the unperturbed problem. We present computations that demonstrate how σ_0 varies depending on the extent of interface imperfection and contrast in material stiffness. We then perform perturbation analysis to derive an expression for the change in the leading order of tractions near the tip of the main crack induced by the presence of the small defect, whose sign can be interpreted as the inclusion's presence having an amplifying or shielding effect on the propagation of the main crack.

1. Introduction

We consider a problem of anti-plane shear formulated in the whole plane, with different materials occupying the regions above and below the crack line. The geometry considered contains a semi-infinite crack situated along a soft *imperfect interface*; we will formulate and solve a weight function problem in such a geometry before using the newly derived weight function to describe the tractions near the crack tip. We will then conduct perturbation analysis to evaluate the effect of a small inclusion's presence, in particular whether it encourages or shields the propagation of the main crack.

Soft imperfect interfaces model a very thin layer of adhesive between two larger bodies of material. Typically such an interface is represented in the model by transmission conditions (justified for example in [1]) which impose continuity of tractions across the interface while the jump in displacement is proportional to the traction.

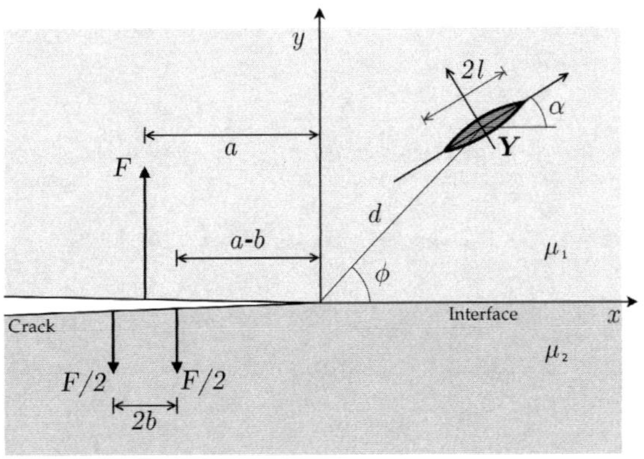

Figure 1. Geometry of the crack, imperfect interface, and small defect in a bimaterial plane.

Weight functions are special solutions to homogeneous boundary value problems that aid in the evaluation of constants describing the behaviour of physical fields near crack tips; a weight function for the perfect interface analogue of the problem considered here has been previously constructed [5] but the presence of the imperfect interface fundamentally alters many aspects of weight functions and their application [6].

Our approach uses Betti's identity to relate physical fields to the weight function; the presence of the imperfect interface introduces new challenges here. We present computations that demonstrate how the stress at the crack tip (which is finite in the imperfect interface interfacial crack setting and may be used in a fracture criterion) varies depending on the extent of interface imperfection and the contrast in stiffness between the two matierials. We also draw comparisons against the previously studied analogous perfect interface case. We then move on to consider the perturbed problem. In particular, the Betti identity again allows us to use the weight function to derive an expression for $\Delta\sigma_0$, the change in the leading order of tractions near the tip of the main crack induced by the presence of the small defect. The sign of $\Delta\sigma_0$ can be interpreted as the inclusion's presence having an amplifying or shielding effect on the propagation of the main crack.

2. Physical problem formulation

The geometry under consideration is shown in Figure 1. A crack occupies $\{(x,y) : x < 0, y = 0\}$ while an imperfect interface lies along the positive x-axis joining two materials with shear moduli μ_1 and μ_2 lying respectively above and below the crack. The methods described are applicable to arbitrary loadings on the crack faces, although we will concentrate on asymmetric self-balanced point loadings as shown in Figure 1. A small elliptic inclusion of shear modulus μ_{in} is centered at a point $\boldsymbol{Y}$, making an angle ϕ with the positive x-axis and oriented at an angle α to the horizontal.

The anti-plane shear displacement function u satisfies the Laplace equation

$$\nabla^2 u(x,y) = 0.$$

We assume continuity of tractions and impose imperfect transmission conditions ahead of the crack, that is

$$\mu_1 \frac{\partial u_1}{\partial y}\bigg|_{y=0+} = \mu_2 \frac{\partial u_2}{\partial y}\bigg|_{y=0-}, \qquad [\![u]\!](x) = \kappa \mu_1 \frac{\partial u_1}{\partial y}\bigg|_{y=0+}, \quad x > 0,$$

where κ is a parameter describing the extent of interface imperfection. On the crack faces, tractions are prescribed as follows

$$\mu_1 \frac{\partial u_1}{\partial y}\bigg|_{y=0+} = p_+(x), \qquad \mu_2 \frac{\partial u_2}{\partial y}\bigg|_{y=0-} = p_-(x), \quad x < 0,$$

where $p_\pm$ are such that the total load is self-balanced.

3. Weight function and Betti identity

The weight function problem is formulated similarly but with zero traction prescribed on the crack faces and no inclusion is present. Also, the geometry is modified; the crack occupies $x > 0$ while the imperfect interface occupies $x < 0$ (the mirror image of the physical problem setup).

The non-trivial solution to this weight function problem can be found by employing the Wiener-Hopf technique since it is a homogeneous problem of the form $Au = 0$. This process involves a suitable choice of factorisation of the Wiener-Hopf kernel that is obtained by taking Fourier transforms and applying boundary conditions. The factorisation results in a representation of this Wiener-Hopf kernel as the product of functions analytic in overlapping half-planes, which upon performing asymptotic analysis yields the imperfect interface weight function via Liouville's Theorem.

Betti's identity can then be applied in an imperfect interface setting; this gives a relationship between the hitherto unknown physical fields and the newly derived weight function. The presence of the imperfect interface introduces some changes in the application of Betti's identity; details of this derivation are given in [7]. For instance, the Fourier transform of the displacement jump across $y = 0$ is analytic in the upper complex half-plane in the perfect interface case, since the displacement jump is zero across a perfect interface. This is not true for the imperfect interface and so a new version of the Betti identity is required. The identity takes the following form, where lowercase u and σ are the physical displacement and out-of-plane stress component, while uppercase U and Σ are their weight function counterparts. Bars denote Fourier transforms (with Fourier coordinate ξ) and $^{(\pm)}$ superscripts represent the restriction of the preceding quantity to either the positive or negative semi-axis.

$$\overline{[\![U]\!]^{(+)}}(\xi)\overline{\sigma^{(+)}}(\xi) - \bar{\Sigma}(\xi)\overline{[\![u]\!]^{(-)}}(\xi) = -[\![\bar{U}]\!](\xi)\langle\bar{p}\rangle(\xi) - \langle\bar{U}\rangle(\xi)[\![\bar{p}]\!](\xi), \quad \xi \in \mathbb{R}.$$

4. Asymptotic behaviour near the crack tip

Full radial asymptotics near the tip of an interfacial crack sitting on an imperfect interface have been constructed in [3]; importantly, along the interface,

$$\sigma_{yz} \sim \sigma_0, \quad x \to 0^+, \tag{1}$$

where σ_0 is a constant we aim to find. This asymptotic behaviour is markedly different to the equivalent expression for a crack sitting on a perfect interface, where a square root singularity occurs. That is, in the perfect interface case,

$$\sigma_{yz} = \frac{K_{III}}{\sqrt{2\pi}} x^{-1/2} + O(1), \quad x \to 0,$$

where the constant K_{III} is the mode III stress intensity factor. Stress intensity factors can be used in a simple fracture criterion; if K_{III} exceeds the fracture toughness of the cracked body's material, then the crack begins to grow [2]. There is no such singularity in the imperfect case (as is readily seen in (1)), but the constant σ_0 plays an analagous role to K_{III} in the perfect interface setting.

The application of Betti's identity obtains an expression for the important constant σ_0:

$$\sigma_0 = \frac{1}{2}\sqrt{\frac{\mu_0}{\pi}} \int_{-\infty}^{\infty} \xi \left([\![\bar{U}]\!](\xi)\langle\bar{p}\rangle(\xi) + \langle\bar{U}\rangle(\xi)[\![\bar{p}]\!](\xi) \right) \mathrm{d}\xi.$$

Details of the derivation of this expression are given in [7]. We stress here that both the weight function U and the constant μ_0 depend heavily on the parameter of interface imperfection κ.

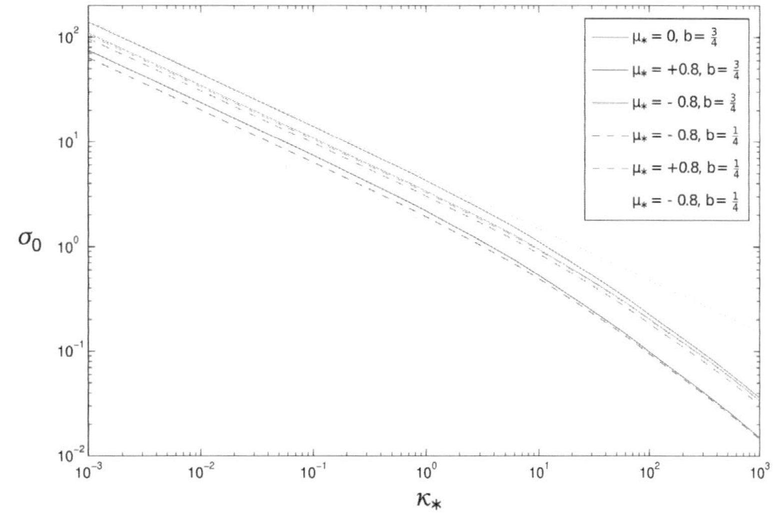

Figure 2. Log-log plot of σ_0 against κ_* for differently contrasting materials.

Figure 2 plots σ_0 as a function of a dimensionless parameter of interface imperfection κ_* which is defined as $\kappa_* = \kappa(\mu_1 + \mu_2)/a$, on a log-log plot for a range of differently contrasting materials described by the dimensionless contrast parameter $\mu_* = (\mu_1 - \mu_2)/(\mu_1 + \mu_2)$. The plot indicates that $\sigma_0 = O(\kappa_*^{-1/2})$ as $\kappa_* \to 0$ (a dotted line of slope $-\frac{1}{2}$ is included in the figure and clearly lies tangent to the curves as $\kappa_* \to 0$) and $\sigma_0 = O(\kappa_*^{-1})$ as $\kappa_* \to \infty$.

The problem considered is a singular perturbation problem; taking small values of $\kappa > 0$ gives a qualitatively different solution to the perfect case in which $\kappa = 0$ (as is seen by the different forms of crack tip asymptotics). This makes the comparison of fields near the crack tip difficult. However, given two particular pairs of materials with contrast parameters $(\mu_*)_1$ and $(\mu_*)_2$ say, we might expect the dimensionless ratio of stress intensity factors $(K_{III})_1/(K_{III})_2$ from the perfect interface case and $(\sigma_0)_1/(\sigma_0)_2$ from the imperfect interface case to be similar. The ratio

$$r(\kappa_*) = \frac{(\sigma_0)_1/(\sigma_0)_2}{(K_{III})_1/(K_{III})_2}$$

is plotted in Figure 3. From this we see that the ratio does indeed tend to 1 as $\kappa_* \to 0$, that is as the extent of interface imperfection decreases.

Figure 4 plots σ_0 as a function of the dimensionless parameter μ_* for two different configurations of point loadings. We see that the positioning of the point loadings on the lower crack face have the least effect for values of μ_* near -1; this is because the material occupying the lower half plane in such cases has a much higher shear modulus than the material occupying the upper half plane and so acts as an almost inelastic body. The exact positioning of the point loadings therefore has a smaller influence on the value of σ_0.

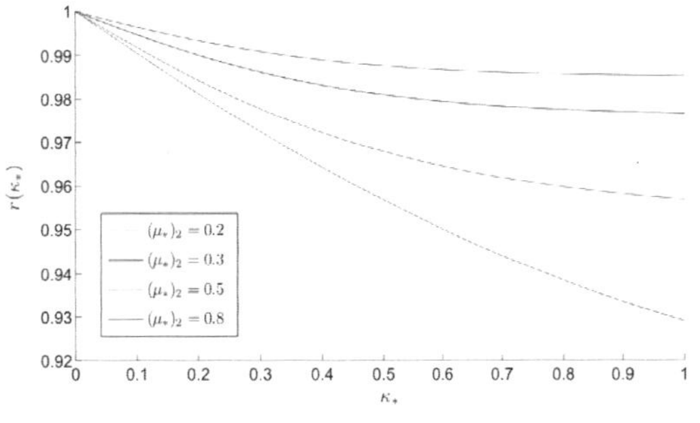

Figure 3. Plot of the ratio $r(\kappa_*)$ with $(\mu_*)_1 = 1$ for various values of $(\mu_*)_2$.

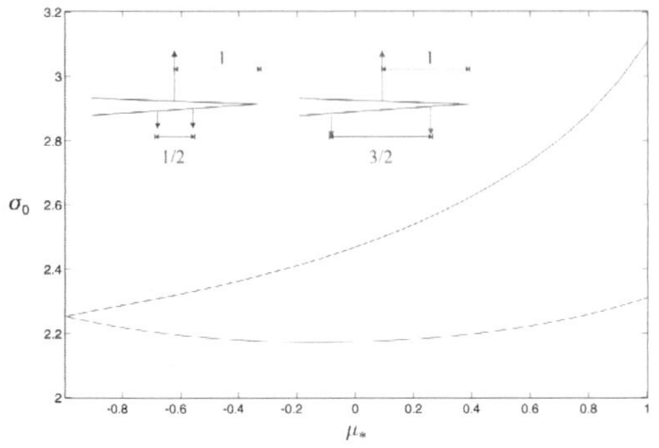

Figure 4. Plot of σ_0 as a function of μ_* for two different configurations of point loadings on the crack faces.

5. Perturbation analysis

We shall construct an asymptotic solution of the perturbed problem (with the inclusion present) using the method of Movchan and Movchan [4], that is the asymptotics of the solution will be taken in the form

$$u_{1,2}(\boldsymbol{x}, \varepsilon) = u_{1,2}^{(0)}(\boldsymbol{x}) + \varepsilon W^{(1)}(\boldsymbol{\xi}) + \varepsilon^2 u_{1,2}^{(1)}(\boldsymbol{x}) + o(\varepsilon^2), \quad \varepsilon \to 0.$$

In this expression, the leading term $u_{1,2}^{(0)}$ corresponds to the unperturbed solution while $\varepsilon W^{(1)}$ represents a boundary layer concentrated near the small defect that is needed to satisfy perfect transmission conditions on the boundary of the inclusion. The term $\varepsilon^2 u_{1,2}^{(1)}$ is introduced to fulfil the boundary conditions on the crack faces and the imperfect interface which are disturbed by the boundary layer. This term, in turn, will produce perturbations of the crack tip fields and correspondingly of the constant σ_0.

The constant σ_0 is expanded in the form

$$\sigma_0 = \sigma_0^{(0)} + \varepsilon^2 \Delta\sigma_0 + o(\varepsilon^2), \quad \varepsilon \to 0.$$

Our objective is to find the first order variation $\Delta\sigma_0$ by employing the Betti identity corresponding to the first order perturbation $u^{(1)}$. The presence of "effective" tractions induced on the crack faces by the small elastic inclusion causes additional terms to be present:

$$\begin{aligned}
&[\![\overline{U}]\!]^+(\xi)\langle\overline{\sigma}^{(1)}\rangle^+(\xi) - \langle\overline{\Sigma}\rangle(\xi)[\![\overline{u}^{(1)}]\!]^-(\xi) \\
&\quad = -[\![\overline{U}]\!](\xi)\overline{P}^-(\xi) - \langle\overline{U}\rangle(\xi)\overline{Q}^-(\xi) - \kappa\langle\overline{\Sigma}\rangle(\xi)\overline{P}^+(\xi) - \langle\overline{U}\rangle(\xi)\overline{Q}^+(\xi).
\end{aligned}$$

Here, the functions $\overline{P}^{\pm}$ and $\overline{Q}^{\pm}$ are computable functions which result from the boundary layer analysis. Similar reasoning to that employed earlier for finding the unperturbed tractions near the crack tip $\sigma_0^{(0)}$ yields an expression for $\Delta\sigma_0$ in the form

$$\begin{aligned}
\Delta\sigma_0 = -\frac{1}{2}\sqrt{\frac{\mu_0}{\pi}}\bigg\{&\int_{-\infty}^{\infty}\Big[\xi[\![\overline{U}]\!](\xi)\overline{P}^-(\xi) + \xi\langle\overline{U}\rangle(\xi)\overline{Q}^-(\xi)\Big]\,d\xi \\
&+ \int_{-\infty}^{\infty}\Big[\kappa\xi\langle\overline{\Sigma}\rangle^-(\xi)\overline{P}^+(\xi) + \xi\langle\overline{U}\rangle(\xi)\overline{Q}^+(\xi)\Big]\,d\xi\bigg\}.
\end{aligned}$$

The sign of $\Delta\sigma_0$ is interpreted as the inclusion's presence having either a shielding effect on the propagation of the main crack if $\Delta\sigma_0 < 0$ or an amplifying effect if $\Delta\sigma_0 > 0$. Figure 5 plots the sign of $\Delta\sigma_0$ for a particular configuration of materials for varying location (characterised by the angle ϕ) and orientation α of the small defect. Different analysis should be sought however when the defect lies very close to the imperfect interface (when ϕ is near 0), since we assumed that the inclusion is at a finite distance from the interface between the half-planes in the boundary layer analysis.

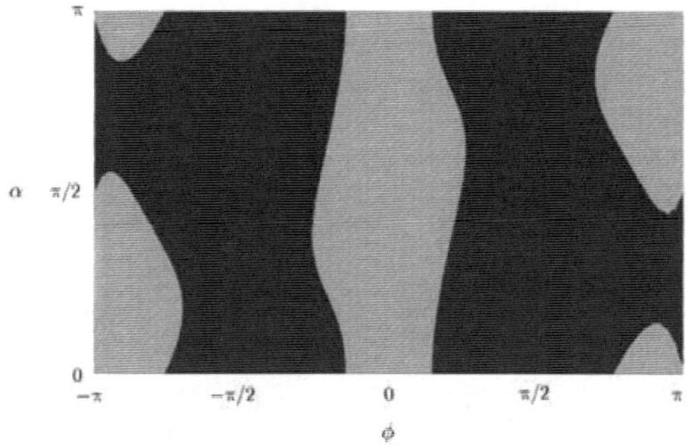

Figure 5. Plot of the sign of $\Delta\sigma_0$ for varying α and ϕ. The darker shaded areas are those (ϕ, α) for which $\Delta\sigma_0 > 0$ while paler regions have $\Delta\sigma_0 < 0$.

6. Conclusions

The method descibed uses a new imperfect interface weight function to aid in the computation of the leading order of tractions near the crack tip in an imperfect interface crack problem under arbitrary loadings on the crack faces. Further, perturbation analysis obtains the next term in the asymptotics which describes the change in the tractions near the crack tip brought about by the presenece of a small inclusion. The presence of the imperfect interface causes many significant differences in comparison to the analagous perfect interface problem; in particular the asymptotic behaviour of tractions near the crack tip is of a different form.

Acknowledgments

AV and GM acknowledge support from the FP7 IAPP project 'INTERCER2', project reference PIAP-GA-2011-286110-INTERCER2. AP would like to acknowledge the Italian Ministry of Education, University and Research (MIUR) for the grant FIRB 2010 Future in Research "Structural mechanics models for renewable energy applications" (RBFR107AKG).

References

[1] Antipov Y A, Avila-Pozos O, Kolaczkowski S T and Movchan A B 2001 Mathematical model of delamination cracks on imperfect interfaces *Int. J. Solids Structures* **38** 6665–97

[2] Irwin G R 1957 Analysis of stresses and strains near the end of a crack traversing a plate *J. Appl. Mech.*, **24**, 361–4

[3] Mishuris G S 2001 Interface crack and nonideal interface concept (Mode III) *Int. J. Frac.* **107** 279–96

[4] Movchan A B and Movchan N V 1995 Mathematical modeling of solids with nonregular boundaries CRC-Press

[5] Piccolroaz A, Mishuris G and Movchan A B 2010 Perturbation of mode III interfacial cracks *Int. J. Frac.* **166** 41–51

[6] Vellender A, Mishuris G S and Movchan A B 2011 Weight function in a bimaterial strip containing an interfacial crack and an imperfect interface. Application to Bloch–Floquet analysis in a thin inhomogeneous structure with cracks *Multiscale Model. Simul.* **9** 1327–49

[7] Vellender A, Mishuris G S and Piccolroaz A 2013 Perturbation analysis for an imperfect interface crack problem using weight function techniques *To be submitted,* `arXiv:1303.6913`

Dynamic energy release rate in couple-stress elasticity

L Morini[1], A Piccolroaz[1] and G Mishuris[2]

[1] Department of Civil, Environmental and Mechanical Engeneering, University of Trento, Via Mesiano 77, 38123 Trento, Italy
[2]Institute of Mathematical and Physical Sciences, Aberystwyth University, Ceredigion SY23 3BZ, Wales, U.K.

E-mail: `lorenzo.morini@unitn.it`

Abstract. This paper is concerned with energy release rate for dynamic steady state crack problems in elastic materials with microstructures. A Mode III semi-infinite crack subject to loading applied on the crack surfaces is considered. The micropolar behaviour of the material is described by the theory of couple-stress elasticity developed by Koiter. A general expression for the dynamic J-integral including both traslational and micro-rotational inertial contributions is derived, and the conservation of this integral on a path surrounding the crack tip is demonstrated.

1. Introduction

The explicit evaluation of the energy release rate and the analysis of its variation in function of the microstructural parameters and of the crack velocity are crucial issues for studying crack propagation stability in couple stress elastic materials [11]. For this reason, a general expression for the dynamic J-integral associated to a semi-infinite Mode III steady-state crack is obtained, and the energy release rate is derived by means of the conservation of this integral. A general procedure for the explicit evaluation of the J-integral, considering a rectangular-shaped contour surrounding the crack tip [3, 4, 5], is also illustrated.

2. Problem formulation

A Cartesian coordinate system $(0, x_1, x_2, x_3)$ centred at the crack-tip at time $t = 0$ is assumed. The micropolar behavior of the material is described by the indeterminate theory of couple stress elasticity [6]. The non-symmetric Cauchy stress tensor $\boldsymbol{t}$ can be decomposed into a symmetric part $\boldsymbol{\sigma}$ and a skew-symmetric part $\boldsymbol{\tau}$, namely $\boldsymbol{t} = \boldsymbol{\sigma} + \boldsymbol{\tau}$. The reduced tractions vector $\boldsymbol{p}$ and couple stress tractions vector $\boldsymbol{q}$ are defined as

$$\boldsymbol{p} = \boldsymbol{t}^T\boldsymbol{n} + \frac{1}{2}\nabla\mu_{nn} \times \boldsymbol{n}, \quad \boldsymbol{q} = \boldsymbol{\mu}^T\boldsymbol{n} - \mu_{nn}\boldsymbol{n}, \tag{1}$$

where $\boldsymbol{\mu}$ is the couple stress tensor, $\boldsymbol{n}$ denotes the outward unit normal and $\mu_{nn} = \boldsymbol{n} \cdot \boldsymbol{\mu}\boldsymbol{n}$. For the dynamic antiplane problem, the conditions of dynamic equilibrium of forces and moments, taking into condideration rotational inertia, and neglecting body forces and body couples, write

$$\sigma_{13,1} + \sigma_{23,2} + \tau_{13,1} + \tau_{23,2} = \rho\ddot{u}_3, \quad \mu_{11,1} + \sigma_{21,2} + 2\tau_{23} = J\ddot{\varphi}_1, \quad \mu_{12,1} + \sigma_{22,2} - 2\tau_{13} = J\ddot{\varphi}_2, \tag{2}$$

where φ_1 and φ_2 are components of the rotations vector, defined as:

$$\varphi_1 = \frac{1}{2}u_{3,2}, \quad \varphi_2 = -\frac{1}{2}u_{3,1}, \tag{3}$$

$\boldsymbol{\rho}$ is the mass density and J is the rotational inertia. The stresses and couple stresses can be expressed in terms of the out-of plane displacement u_3 [11, 12]:

$$\sigma_{13} = Gu_{3,1}, \quad \sigma_{23} = Gu_{3,2}, \tag{4}$$

$$\tau_{13} = -\frac{G\ell^2}{2}\Delta u_{3,1} + \frac{J}{4}\ddot{u}_{3,1}, \quad \tau_{23} = -\frac{G\ell^2}{2}\Delta u_{3,2} + \frac{J}{4}\ddot{u}_{3,2}, \tag{5}$$

$$\mu_{11} = -\mu_{22} = G\ell^2(1+\eta)u_{3,12}, \quad \mu_{21} = G\ell^2(u_{3,22} - \eta u_{3,11}), \quad \mu_{12} = -G\ell^2(u_{3,11} - \eta u_{3,22}). \tag{6}$$

where Δ denotes the Laplace operator, J is the rotational inertia, G is the elastic shear modulus, ℓ and η the couple stress parameters introduced by Koiter [6], with $-1 < \eta < 1$. Both material parameters ℓ and η depend on the microstructure and can be connected to the material characteristic lengths in bending and in torsion [12], namely $\ell_b = \ell/\sqrt{2}$ and $\ell_t = \ell\sqrt{1+\eta}$. Typical values of ℓ_b and ℓ_t for some classes of materials with microstructure can be found in references [7] and [8].
Substituting expressions (4), (5) and (6) in the dynamic equilibrium equation $(2)_{(1)}$, the following equation of motion is derived [11]:

$$G\Delta u_3 - \frac{G\ell^2}{2}\Delta\Delta u_3 + \frac{J}{4}\Delta\ddot{u}_3 = \rho\ddot{u}_3. \tag{7}$$

We assume that the crack propagates with a constant velocity v straight along the x_1-axis and is subjected to reduced force traction p_3 applied on the crack faces, moving with the same velocity v, whereas reduced couple traction q_1 is assumed to be zero,

$$p_3(x_1, 0^\pm, t) = \mp\tau(x_1 - vt), \quad q_1(x_1, 0^\pm, t) = 0, \quad \text{for} \quad x_1 - vt < 0. \tag{8}$$

We also assume that the function τ decays at infinity sufficiently fast and it is bounded at the crack tip. These requirements are the same requirements for tractions as in the classical theory of elasticity.
It is convenient to introduce a moving framework $x = x_1 - vt$, $y = x_2, z = x_3$. By assuming that the out of plane displacement field has the form $u_3(x_1, x_2, t) = w(x, y)$, then the equation of motion (7) writes:

$$\Delta w - \frac{\ell^2}{2}\Delta\Delta w = m^2\left(w_{,xx} - h_0^2\ell^2\Delta w_{,xx}\right) \tag{9}$$

where $m = v/c_s$ is the crack velocity normalized to the shear waves speed c_s, and $h_0 = \sqrt{J/4\rho}/\ell$ is the normalized rotational inertia defined in Mishuris et al., 2012.
According to (1), the non-vanishing components of the reduced force traction and reduced couple traction vectors along the crack line $y = 0$, where $\boldsymbol{n} = (0, \pm 1, 0)$, can be written as

$$p_3 = t_{23} + \frac{1}{2}\mu_{22,x}, \quad q_1 = \mu_{21}, \tag{10}$$

respectively. By using $(4)_2$, $(6)_{1,2}$, $(5)_2$, and (10), the loading conditions (8) on the upper crack surface require the following conditions for the function w:

$$w_{,y} - \frac{\ell^2}{2}\left[(2+\eta-2m^2h_0^2)w_{,xx} + w_{,yy}\right]_{,y} = -\frac{1}{G}\tau(x), \quad w_{,yy} - \eta w_{,xx} = 0, \quad \text{for} \quad x < 0, \quad y = 0^+. \tag{11}$$

Ahead of the crack tip, the skew-symmetry of the Mode III crack problem requires

$$w = 0, \quad w_{,yy} - \eta w_{,xx} = 0, \quad \text{for} \quad x > 0, \quad y = 0^+. \tag{12}$$

Note that the ratio η enters the boundary conditions (11)-(12), but it does not appear into the governing PDE (9).

3. Dynamic energy release rate

For couple stress elastic materials, the energy flux for a dynamic crack propagating along x_1-axis is given by:

$$F(\Gamma) = \int_\Gamma \left[(W+T)vn_1 + \mathbf{t}^T\mathbf{n} \cdot \frac{\partial \mathbf{u}}{\partial t} + \boldsymbol{\mu}^T\mathbf{n} \cdot \frac{\partial \boldsymbol{\varphi}}{\partial t} \right] ds, \tag{13}$$

where $\mathbf{n}$ is an outward unit normal on Γ. The strain-energy density W and the kinetic enrgy density T are given by

$$W = \frac{1}{2}\left(\boldsymbol{\sigma} \cdot \nabla \mathbf{u} + \boldsymbol{\mu}^T \cdot \nabla \boldsymbol{\varphi}\right), \quad T = \frac{1}{2}\left(\rho \left|\dot{\mathbf{u}}\right|^2 + J\left|\dot{\boldsymbol{\varphi}}\right|^2\right). \tag{14}$$

Considering a Mode III steady-state crack prapagating at constant velocity v along the x_1-axis, and using the expression for the the out-of-plane defined in Section 2, in the moving framework (x, y, z) the energy flux (13) assumes the special form

$$F(\Gamma) = v \int_\Gamma \left[(W+T)n_x - \mathbf{t}^T\mathbf{n} \cdot \mathbf{e}_z w_{,x} - \boldsymbol{\mu}^T\mathbf{n} \cdot \boldsymbol{\varphi}_{,x} \right] ds, \tag{15}$$

then the generalized J-integral for an antiplane dynamic steady-state crack in couple-stress elastic materials can be defined:

$$\begin{aligned} \mathcal{J} = \frac{F(\Gamma)}{v} &= \int_\Gamma \left[(W+T)n_x - \mathbf{t}^T\mathbf{n} \cdot \mathbf{e}_z w_{,x} - \boldsymbol{\mu}^T\mathbf{n} \cdot \boldsymbol{\varphi}_{,x} \right] ds = \\ &= \int_\Gamma \left[(W+T)\mathbf{n} - \mathbf{p} \cdot \mathbf{e}_z \nabla w - (\nabla \boldsymbol{\varphi})^T \mathbf{q} \right] \cdot \mathbf{e}_x ds \end{aligned} \tag{16}$$

The J-integral (16) has been proved to be path independent, the details of the demonstration are reported in the next Section, and it is the generalization of the static expressions derived by Freund and Hutchinson [2] and Lubarda and Markenscoff [9] to the antiplane dynamic steady state case. The dynamic energy release rate is then defined by the limit [1]:

$$\mathcal{G} = \lim_{\Gamma \to 0} \int_\Gamma \left[(W+T)n_x - \mathbf{t}^T\mathbf{n} \cdot \mathbf{e}_z w_{,x} - \boldsymbol{\mu}^T\mathbf{n} \cdot \boldsymbol{\varphi}_{,x} \right] ds. \tag{17}$$

4. Conservation of the J-integral

In this Section, we demonstrate that the dynamic J-integral expression (16) is path independent. Considering a closed oriented path formed by two crack tip contours Γ_1 and Γ_2 and by the segments of the crack faces of length d that connect the ends of these contours and using the notation introduced in Section 2, the energy flux integral corresponding to this entire closed path Γ_{tot} for a Mode III steady state crack in couple-stress materials is given by

$$F(\Gamma_{tot}) = F(\Gamma_2) - F(\Gamma_1) = v \oint_{\Gamma_{tot}} \left[(W+T)n_x - \mathbf{t}^T\mathbf{n} \cdot \mathbf{e}_z w_{,x} - \boldsymbol{\mu}^T\mathbf{n} \cdot \boldsymbol{\varphi}_{,x} \right] ds, \tag{18}$$

then from the definition (16) we derive

$$\mathcal{J}(\Gamma_{tot}) = \mathcal{J}(\Gamma_2) - \mathcal{J}(\Gamma_1) = \oint_{\Gamma_{tot}} \left[(W+T)n_x - \mathbf{t}^T\mathbf{n}\cdot\mathbf{e}_z w_{,x} - \boldsymbol{\mu}^T\mathbf{n}\cdot\boldsymbol{\varphi}_{,x} \right] ds, \tag{19}$$

where the notation $\mathcal{J}(\Gamma_1)$ and $\mathcal{J}(\Gamma_2)$ denotes that the dynamic J-integral (16) is evaluated respect to the crack tip contours Γ_1 and Γ_2, respectively. Applying the divergence theorem to the (19) we obtain and remembering that $n_x = \mathbf{n}\cdot\mathbf{e}_x$, we obtain

$$\mathcal{J}\Gamma_2) - \mathcal{J}(\Gamma_1) = \int_{A_{tot}} \nabla\cdot\left[(W+T)\mathbf{e}_x - \mathbf{t}^T\mathbf{e}_z w_{,x} - \boldsymbol{\mu}^T\boldsymbol{\varphi}_{,x} \right] dA, \tag{20}$$

where A_{tot} is the area within the closed area. For the antiplane steady-state problem the strain elastic energy density and the kinetic energy density are given by

$$W = \frac{1}{2}(\sigma_{13}w_{,x} + \sigma_{23}w_{,y} + \mu_{11}\varphi_{1,x} + \mu_{12}\varphi_{2,x} + \mu_{21}\varphi_{1,y} + \mu_{22}\varphi_{2,y}), \tag{21}$$

$$T = \frac{v^2}{2}(\rho w_{,x}^2 + J\varphi_{1,x}^2 + J\varphi_{2,x}^2), \tag{22}$$

the first term of the integral (20) is the given by

$$\begin{aligned}
\nabla\cdot[(W+T)\mathbf{e}_x] &= (W+T)_{,x} = v^2\left(\rho w_{,xx}w_{,x} + J\varphi_{1,xx}\varphi_{1,x} + J\varphi_{2,xx}\varphi_{2,x}\right) + \\
&+ \frac{1}{2}\left(\sigma_{13,x}w_{,x} + \sigma_{13}w_{,xx} + \sigma_{23,x}w_{,y} + \sigma_{23}w_{3,yx}\right) + \\
&+ \frac{1}{2}\left(\mu_{11,x}\varphi_{1,x} + \mu_{11}\varphi_{1,xx} + \mu_{12,x}\varphi_{2,x} + \mu_{12}\varphi_{2,xx} + \right. \\
&+ \left. \mu_{21,x}\varphi_{1,y} + \mu_{21}\varphi_{1,yx} + \mu_{22,x}\varphi_{2,y} + \mu_{22}\varphi_{2,yx}\right).
\end{aligned} \tag{23}$$

Taking into account the dynamic equilibrium conditions (2), the second term can be written as follows

$$\begin{aligned}
\nabla\cdot(\mathbf{t}^T\mathbf{e}_z w_{,x}) &= (\nabla\cdot\mathbf{t}^T)\cdot\mathbf{e}_z w_{,x} + \mathbf{t}^T\cdot\nabla w_{,x} = \\
&= \rho\ddot{u}w_{,x} + (\sigma_{13}+\tau_{13})\,w_{,xx} + (\sigma_{23}+\tau_{23})\,w_{,yx} = \\
&= \rho v^2 w_{,xx}w_{,x} + (\sigma_{13}+\tau_{13})\,w_{,xx} + (\sigma_{23}+\tau_{23})\,w_{,yx},
\end{aligned} \tag{24}$$

while the third

$$\begin{aligned}
\nabla\cdot(\boldsymbol{\mu}^T\boldsymbol{\varphi}_{,x}) &= (\nabla\cdot\boldsymbol{\mu}^T)\cdot\boldsymbol{\varphi}_{,x} + \boldsymbol{\mu}^T\cdot\nabla\boldsymbol{\varphi}_{,x} = \\
&= (J\ddot{\varphi}_1 - 2\tau_{23})\,\varphi_{1,x} + (J\ddot{\varphi}_2 + 2\tau_{13})\,\varphi_{2,x} + \mu_{11}\varphi_{1,x} + \mu_{12}\varphi_{2,x} + \mu_{21}\varphi_{1,y} + \mu_{22}\varphi_{2,y} = \\
&= \left(Jv^2\varphi_{1,xx} - 2\tau_{23}\right)\varphi_{1,x} + \left(Jv^2\varphi_{2,xx} + 2\tau_{13}\right)\varphi_{2,x} + \mu_{11}\varphi_{1,x} + \mu_{12}\varphi_{2,x} + \mu_{21}\varphi_{1,y} + \mu_{22}\varphi_{2,y}.
\end{aligned} \tag{25}$$

Substituting (23), (25) and (25) into the integral (20) and expressing φ_1 and φ_2 in function of the displacement by means of relations (3), we obtain

$$\begin{aligned}
\mathcal{J}(\Gamma_2) - \mathcal{J}(\Gamma_1) &= \int_{A_{tot}} \left[\frac{1}{2}\left(\sigma_{13,x}w_{,x} + \sigma_{23,x}w_{,y} + \frac{1}{2}\left(\mu_{11,x}w_{,yx} - \mu_{12,x}w_{,xx} + \mu_{21,x}w_{,yy} - \mu_{22,x}w_{,xy}\right)\right) - \right. \\
&- \left. \frac{1}{2}\left(\sigma_{13}w_{,xx} + \sigma_{23}w_{,yx} + \frac{1}{2}\left(\mu_{11}w_{,yxx} - \mu_{12}w_{,xxx} + \mu_{21}w_{,yyx} - \mu_{22}w_{,xyx}\right)\right)\right] dA,
\end{aligned} \tag{26}$$

finally, introducing into expressions (4) and (6), which define stress and couple-stress tensors in function of the derivatives of the displacement, we get:

$$\begin{aligned}\mathcal{J}(\Gamma_2) - \mathcal{J}(\Gamma_1) &= \frac{G\ell^2}{2}\int_{A_{tot}} [(w_{,xx}w_{,xxx} + w_{,yy}w_{,yyx} - \eta\,(w_{,yyx}w_{,xx} + w_{,xxx}w_{,yy})) - \\ &- (w_{,xx}w_{,xxx} + w_{,yy}w_{,yyx} - \eta\,(w_{,yyx}w_{,xx} + w_{,xxx}w_{,yy}))]\, dA = 0. \qquad (27)\end{aligned}$$

We have demonstrated that the J-integral (16) is path independent, as a consequence the dynamic energy release rate (17) can be evaluated by choosing an arbitrary path Γ surrounding the crack tip and making the limit $\Gamma \to 0$.

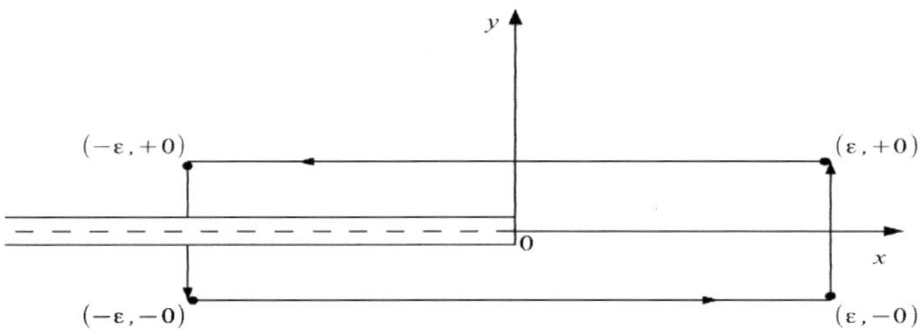

Figure 1. Rectangular-shaped contour around the crack-tip

5. Evaluation of the energy release rate

For the evaluation of the energy release rate (17), a rectangular-shaped contour contour Γ with vanishing height along the y-direction and with $\varepsilon \to +0$ reported in Fig.1. Such contour was first introduced by Freund [1] and recently used by Gourgiotis and Georgiadis [3, 4, 5] and it permits using solely the asymptotic near-tip stress and displacement fields.

Considering the moving framework in Fig.1 with the origin at the crack tip, the cartesian components of the outward unit vector normal to Γ are $\mathbf{n} = (n_x, n_y, 0)$, and the generalized J-integral (16) becomes:

$$\mathcal{J} = \int_\Gamma [(W+T)n_x - (t_{13}n_x + t_{23}n_y)w_{,x} - (\mu_{11}n_x + \mu_{21}n_y)\varphi_{1,x} - (\mu_{12}n_x + \mu_{22}n_y)\varphi_{2,x}]\, ds =$$

$$= \int_\Gamma [(W+T) - t_{13}w_{,x} - (\mu_{11}\varphi_{1,x} + \mu_{12}\varphi_{2,x})]\, dy - \int_\Gamma [t_{23}w_{,x} + (\mu_{21}\varphi_{1,x} + \mu_{22}\varphi_{2,x})]\, dx, \quad (28)$$

In order to avaluate the energy release rate, we allow the height of the rectangular path reported in Fig.1 to vanish. In this limit, the first integral of the (28) is zero. It is also important to note that anti-symmetry conditions (11) together with boundary conditions (12) provide that the reduced traction $q_1 = \mu_{21}$ is zero along the whole crack line $y = 0$, where $\mathbf{n} = (0, \pm 1, 0)$. Consequently, the energy release rate (17) becomes:

$$\begin{aligned}\mathcal{G} &= -2\lim_{\varepsilon\to+0}\left\{\int_{-\varepsilon}^{+\varepsilon} [t_{23}(x,0^+)w_{,x}(x,0^+) + \mu_{22}(x,0^+)\varphi_{2,x}(x,0^+)]\, dx\right\} \\ &= -2\lim_{\varepsilon\to+0}\left\{\int_{-\varepsilon}^{+\varepsilon} \left[t_{23}(x,0^+)w_{,x}(x,0^+) - \frac{1}{2}\mu_{22}(x,0^+)w_{,xx}(x,0^+)\right] dx\right\}. \qquad (29)\end{aligned}$$

Asymptotic expressions for the total shear stress t_{23}, the couple stress field μ_{22}, and the out-of-plane diplacement w have been derived by Radi [12] and Piccoloraz et al. [10]. These fields exhibit the following behavior near to the crack tip:

$$t_{23}(x,0^+) = Af(h_0,\eta,m)x^{-3/2}, \quad x>0, \tag{30}$$
$$\mu_{22}(x,0^+) = Ag(h_0,\eta,m)x^{-1/2}, \ x>0, \tag{31}$$
$$w(x,0^+) = A(-x)^{3/2}, \quad x<0. \tag{32}$$

where A is a constant determined by the boundary conditions (12) and depending by the characteristic of the loading applied at the faces $\tau(x)$, and $f(h_0,\eta,m)$ and $g(h_0,\eta,m)$ are factors relating the asymptotics leading term of the displacement to the leading terms of t_{23} and μ_{22}, respectively. Now, by substituting the (30) and the (32) in the general formula (29), we finally derive:

$$\mathcal{G} = -2\lim_{\varepsilon\to+0}\left\{A\left[f(h_0,\eta,m)\int_{-\varepsilon}^{+\varepsilon} x_-^{1/2}x_+^{-3/2}dx - g(h_0,\eta,m)\int_{-\varepsilon}^{+\varepsilon} x_-^{-1/2}x_+^{-1/2}.dx\right]\right\} \tag{33}$$

The two products of distributions $x_-^{1/2}, x_-^{-1/2}$ and $x_+^{-3/2}, x_+^{-1/2}$ are obtained through the use of Fisher's theorem, that's leads to the operational relation [3]:

$$(x_-)^\gamma(x_+)^{-1-\gamma} = -\frac{\pi\delta(x)}{2\sin(\pi\gamma)}, \text{ with } \gamma\neq -1,-2,-3\ldots, \tag{34}$$

where $\delta(x)$ is the Dirac delta distribution. Then, applying the relation (34) to the (33) and considering the fundamental property of the Dirac delta distribution $\int_{-\varepsilon}^{+\varepsilon}\delta(x)dx = 1$, we finally get:

$$\mathcal{G} = A\pi\left(f(h_0,\eta,m) + g(h_0,\eta,m)\right). \tag{35}$$

6. Conclusions

A general expression for the dynamic J-integral valid for antiplane crack in a couple stress elastic material has been derived. A direct procedure for the evaluation of the energy release rate, based on the assumption of a rectangular-shaped path around the crack tip and requiring only asymptotics expressions of stresses and displacement, has been illustrated. The proposed method can be used for evaluating the energy release rate for many dynamic Mode III problems in couple-stress elastic materials characterized by different loading configurations acting on the crack faces.

7. Acknowledgements

L. M. and A. P. gratefully thank financial support of the Italian Ministry of Education, University and Research in the framework of the FIRB project 2010 "Structural mechanics models for renewable energy applications", G. M. gratefully acknowledges the support from the European Union Seventh Framework Programme under contract number PIAP-GA-2011-286110-INTERCER2.

References

[1] Freund LB 1990, *Dynamic Fracture Mechanics*, (Cambridge University Press, Cambridge, UK).
[2] Freund LB and Hutchinson JW 1985, "High strain-rate crack growth in rate-dependent plastic solids", *J. Mech. Phys. Solids* **33**: 169-91.
[3] Georgiadis HG, "The mode III crack problem in microstructured solids governed by dipolar gradient elasticity: static and dynamic analysis", *J. Appl. Mech.*,**70**: 517-30.

[4] Gourgiotis PA and Georgiadis HG 2007, "Distributed dislocation approach for cracks in couple-stress elasticity: shear modes", *Int. J. Fract.***147**: 83-102.

[5] Gourgiotis PA and Georgiadis HG 2008, "An approach based on distributed dislocations and disclinations for crack problems in couple-stress elasticity", *Int. J. Solids Struct.***45**: 5521-39.

[6] Koiter WT 1964 "Couple-stresses in the theory of elasticity, I and II", *Proc. Kon. Nederl. Akad. Wetensch (B)* **67**: 17-44.

[7] Lakes RS 1986, "Experimental microelasticity of two porous solids", *Int. J. Solids Struct.*, **22**: 55-63.

[8] Lakes RS 1995, "Experimental methods for study of Cosserat elastic solids and other generalized elastic continua", in *Continuum Models for Materials with Micro-structure* , (John Wiley, New York), pp. 1-22.

[9] Lubarda VA and Markenscoff X 2000, "Conservation integrals in couple stress elasticity", *J. Mech. Phys. Solids* **48**: 553-64.

[10] Piccolroaz A, Mishuris G and Radi E 2012, "Mode III interfacial crack in the presence of couple stress elastic materials", *Eng. Fract. Mech.* **80**: 60-71.

[11] Mishuris G, Piccolroaz A and Radi E 2012, "Steady-state propagation of a Mode III crack in couple stress elastic materials" *Int. J. Eng. Sci.* **61**: 112-28.

[12] Radi E 2008, "On the effects of characteristic lengths in bending and torsion on Mode III crack in couple stress elasticity", *Int. J. Solids Struct.* **45**: 3033-58.

Effect of crack induced nonlinearity on dynamics of structures: application to structural health monitoring

D Broda[1], V Hiwarkar[2], V V Silberschmidt[3] and W J Staszewski[1]
[1]Department of Robotics and Mechatronics, AGH University of Science and Technology, Al. Mickiewicza, 30-059 Krakow, Poland
[2]Mechanical, Materials and Manufacturing Engineering, University of Nottingham, University Park, Nottingham, NG7 2RD, UK
[3]Wolfson School of Mechanical and Manufacturing Engineering, Loughborough University, Loughborough, Leicestershire LE11 3TU, UK

E-mail: dbroda@agh.edu.pl

Abstract. In recent years, there has been a growing interest in vibration-based structural health-monitoring techniques. At present, most of the vibration-based techniques consider an open crack, which is far from realistic situations. In real-life scenarios crack can open and close under the dynamic loading. This bilinear behavior of crack has a significant effect on dynamics of structures which was investigated by performing finite-element simulations of a beam with a breathing crack excited longitudinally. Its response measured at different locations with regard to the crack indicated that the crack-induced nonlinearity had a localized effect on dynamics of the beam. Measurements obtained near the crack revealed complex spectra with higher harmonics.

1. Introduction

A growing complexity of modern civil, aerospace and power-plant structures led to development of more strict safety regulations. To meet these strict standards of safety, there is a need for cost-effective and advanced reliable damage-detection techniques. At present, there are various damage-detection techniques such as visual inspection, radiographic testing, linear ultrasonic techniques etc. Most of these techniques are time-consuming and used for specific purpose. Some of the most upcoming and popular techniques are vibration-based structural health-monitoring techniques.

It was well established that damage in structures such as open/breathing cracks influence their natural characteristics such as frequency, damping and mode shapes. So far several researchers have addressed the issue of the effect of an open crack on dynamics of the structure using finite-element studies and validated the results through experimental analysis. They made an assumption of open crack to avoid the problem of nonlinearities related to its closure [1]. Some researchers have investigated the effect of the breathing crack on the dynamics of the beam. Their investigation revealed presence of higher harmonics in a frequency response [2,3].

1.1. Nonlinear vibrations in Structural Health Monitoring (SHM) applications

Physical basis of crack-induced nonlinearities has been investigated for both vibration- and ultrasonic-based techniques. In its simplest form, the breathing crack might be modelled as bi-linear spring, with different elastic moduli for 'open' and 'closed' crack. More complex models have to deal with a non-zero velocity (that leads to impacts) and roughness of crack faces during the contact. It was shown that not only contact of crack faces, but also temperature gradients near crack tips play a significant role in introducing the nonlinearity [4,5].

Numerous papers addressed the problem of vibrations of cracked structures and NDT (Non-Destructive Testing) methods based on crack–induced nonlinearity [6-9]. It is a well-known fact, that a crack affects the natural frequencies of the structure. For a breathing-crack model, a measured frequency drop was shown to be smaller than in case of the open-cracks model [9].The smaller frequency drop makes the closing crack more difficult to detect than open cracks by using the 'linear' methods. The problem of natural frequencies of a beam with the breathing crack was solved by introducing 'bilinear frequency', based on two linear configurations (with the open and closed cracks) [10]. Numerous investigations are focused on higher and sub-harmonic frequencies in the output spectra [7]. Sundermeyer and Weaver [11] presented a model, which takes into consideration appearance of difference of two driving frequencies in presence of nonlinearity. Bi-spectral analysis was proposed as efficient signal-processing method for detection of bi-linear stiffness in the structure [12]. More complex signal-processing methods are based on instantaneous frequency calculated via Hilbert Transform [13]. The methods employing modulation of an ultrasonic wave by vibrations were also suggested. In cracked specimens, the modulation might be observed in the form of sidebands around the ultrasonic wave frequency peak [14].

A 2D FE model of a cracked beam was investigated by Andreaus *et al.* [15]. The beam was excited with harmonic signal at different frequencies (multiplies of the first natural frequency) revealing significantly different spectra. The model of the system was simplified with the SDOF (Single Degree of Freedom) system. These studies were extended in [16] by taking the depth and position of the crack into consideration; a method for crack detection was proposed. In [17] a finite-element simulation was used to study the longitudinal harmonic loading to assess damage in a cracked cantilever bar. For nonlinear materials, the properties were introduced in terms of their plastic behaviour. The effect of crack size and position along the beam length was studied. It was shown that observed change in the natural frequency can be used to characterize the damage state of the component.

1.2. Earlier work

The longitudinal vibration of the bar was investigated by Hiwarkar *et al.* [18,19]. The expression for dynamic compliance (receptance) was obtained in the following form [18]:

$$L_l(l,j\omega) = \frac{l}{ES\zeta}[\tan\zeta - \frac{jX}{4\pi}\frac{\zeta + \frac{1}{2}\sin 2\zeta}{\cos^2\zeta}] \quad (1)$$

where $\zeta = \frac{\omega l}{\gamma}$, $\gamma = \left(\frac{E}{\rho}\right)^{\frac{1}{2}}$, *l*, *S* and *E* are the length, cross-sectional area and modulus of elasticity of the bar, respectively, ω is frequency of excitation and X stands for absorption coefficient. The equation was transformed using a general linear theory of integral equations for straight rods, to enable its numerical simulation in Matlab-Simulink environment [18]. The results of simulations include generation of a higher harmonics, frequency shift and their dependency on crack parameters, as well as generation of low-frequency components.

1.3 Aim and scope

In the present paper, simulations with a two-dimensional finite-element model are presented. The investigations are focused on the influence of a breathing crack on dynamics of the beam, in particular, of the impacts of crack faces. The beam was excited with a longitudinal force. Longitudinal vibrations were measured and compared at different distances from the crack.

2. Finite-element simulations

2.1. Problem formulation

The finite element model was developed using commercial FE software ABAQUS. Figure 1 shows the scheme of the beam used for simulations. The beam's dimensions are 300 mm × 25 mm × 10 mm. Simulation parameters are given in Table 1. The crack was placed in the middle of the beam, its depth was 10 mm. The beam was meshed with 2D plane-stress elements CPS4R (4-node bilinear, reduced-integration with hourglass control) and CPS3 (3-node linear). A basic element size applied was 1mm; smaller elements were used to mesh the crack's vicinity. The meshed beam is shown in Figure 2a and details of crack-tip meshing are shown in Figure 2b. One end of the beam was fixed; the other end was excited by a longitudinal sinusoidal force with frequency of the first longitudinal mode. The response was measured at eight different locations from the crack, marked with number 1 – 8, as shown in Figure1.

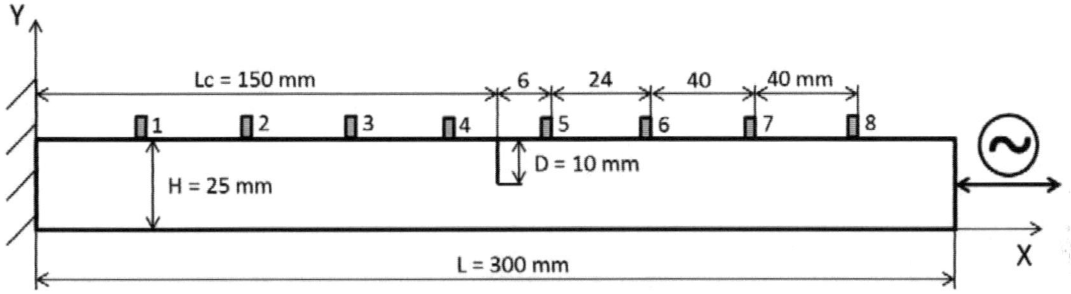

Figure 1. Scheme of cracked beam

Table 1. Simulation parameters

Material	Aluminium (Grade 6082T6)
Modulus of elasticity ($kg/mm\ s^2$)	7×10^7
Density (kg/mm^3)	2.70×10^{-6}
Poisson's ratio	0.33
Longitudinal stiffness of beam (N/mm)	5.84×10^7
Normal contact stiffness (N/mm)	7×10^9

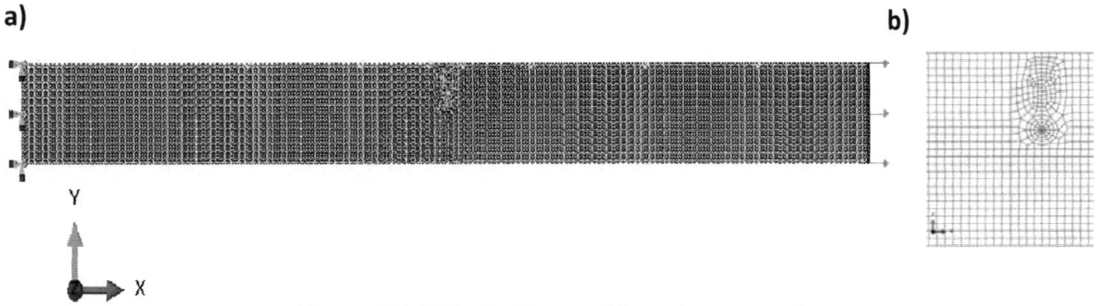

Figure 2.(a) Meshed beam, (b) mesh near crack

2.2.Results

Longitudinal natural frequencies of the un-cracked beam were extracted by modal analysis in Abaqus. It was found that the first longitudinal natural frequency of un-cracked beam was 4248 Hz (Figure 3; notation *acc* and *ampl* for axis titles in Figures 3-6 is used for *acceleration* and *amplitude*, respectively). In the figure 3, the time and frequency responses of intact beam are shown for sensors 1-4. The results obtained for sensors 5-8 are similar and thus are not shown. The cracked beam was excited at the first longitudinal natural frequency of the un-cracked beam, i.e. at 4248 Hz. The results obtained with Abaqus were post processed with Matlab software. The time and frequency responses were measured at different locations along the beam and the obtained results were compared. It was found that presence of the crack led to distortion of the time response (Figures. 4a and 5a) that is stronger near the crack. The frequency response (Figures. 4b and 5b) of the distorted time response (Figures. 4a and 5a) shows the generation of higher harmonics, which are the multiples of the frequency of excitation. Figures 6a and 6b show the magnified view of frequency response of the cracked bar at the excitation frequency measured at different location from sensors 1-8. The results demonstrate that crack-induced nonlinearity has the localised effect on dynamics of the beam as the amplitude of measured response decreases with the distance from the crack.

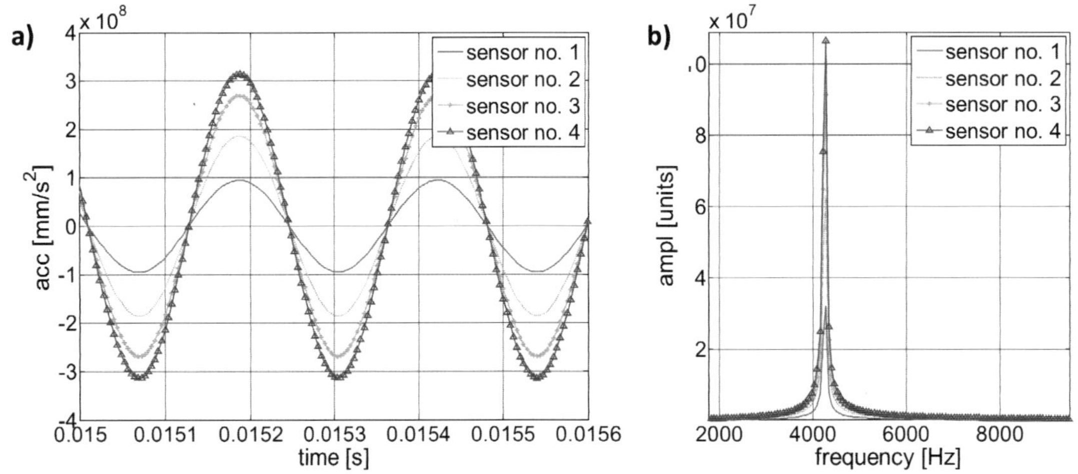

Figure 3. Measured time (a) and frequency (b) responses of un-cracked beam at sensors 1-4.

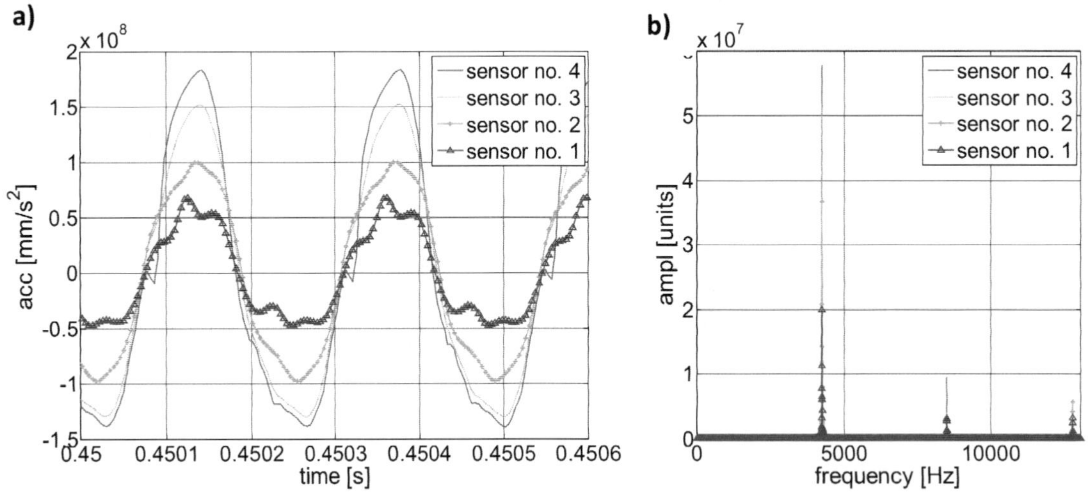

Figure 4. Measured time (a) and frequency (b) responses of cracked beam at sensors 1-4.

It was also observed that the most pronounced distortion was measured at the first harmonic. In order to compare the level of distortion (nonlinearity) of the first harmonic, the coefficient of distortion (D) in the form of

$$D = \frac{f_1}{T_1}, \tag{2}$$

wheref_1 is the first harmonic amplitude and T_1 is the first longitudinal mode excitation amplitude, was introduced. Table 2 shows the ratios D for each measurement taken by sensor at 8 different locations on the beam. The ratios measured on the sensors placed between the crack and the fixed end are higher than on the sensor between the crack and free end of the beam.

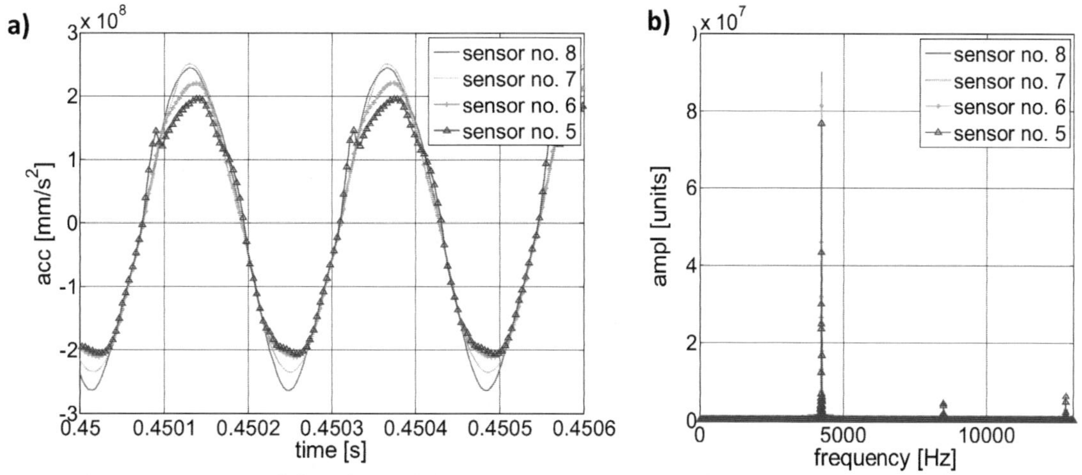

Figure 5. Measured time (a) and frequency (b) responses of cracked beam at sensors 5-8.

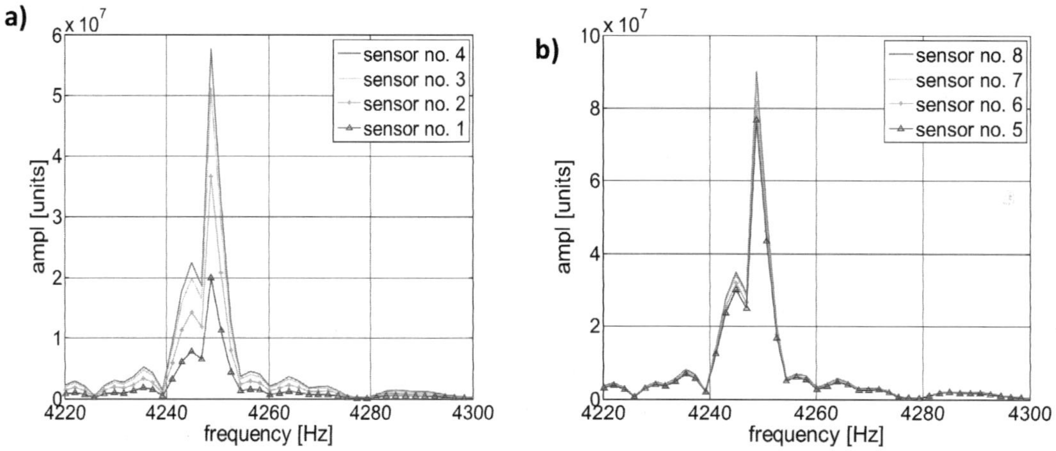

Figure 6. Magnified view of frequency response of cracked bar at excitation frequency: (a) sensors 1-4, (b) sensors 5-8.

The two highest values of distortion ratio were obtained at responses measured in the vicinity of the crack (sensors 4) and near the fixed end (sensor 1). High ratio was also measured on sensor 5, i.e. in the vicinity, but on the free end side, of the crack. From the values of distortion coefficient in Table 2 it is clear that the crack-induced nonlinearity in localized as the distortion coefficient decreases for responses measured far away from the crack with exception of the one measured at sensor 1 which needs further investigation.

3. Conclusions

From the results of two-dimensional finite-element simulations of longitudinal vibrations of the cracked beam it was established that crack-induced nonlinearity had localized effect on dynamics of the beam as the amplitude of measured response decreases when measured far away from the crack.

Further to this, it was also found that the distortion coefficient of the first harmonic decreased for the responses measured far away from the crack which can also be useful tool for monitoring the structural health.

Table 2. Coefficient of distortion D

Sensor	1st Harmonic
1	0.15660
2	0.01559
3	0.07765
4	0.16590
5	0.05603
6	0.01017
7	0.02500
8	0.02662

References

[1] Sinha J K, Friswell M I and Edwards S 2002 Simplified models for the location of cracks in beam structures using measured virbation data *J. Sound Vib.* **251** 13-38

[2] Dimarogonas A D 1996 Vibration of cracked structures: a state of the art review. *Eng. Fract. Mech.* **55** 831-857

[3] Doebling S W, Farrar C R and Prime M B 1998 A summary review of vibration-based damage identification methods *Shock and Vibration Digest.* **30** 91-105

[4] Zaitsev V, Gusev V and Castagnede B 2002 Luxemburg-Gorky effect retooled for elastic waves: a mechanism and experimental evidence *Phys. Rev. Lett.* **89**

[5] Klepka A, Staszewski W J, Jenal RB, Szwedo M, Iwaniec J and Uhl T 2012 Nonlinear acoustics for fatigue crack detection – experimental investigations of vibro – acoustic wave modulations *Structural Health Monitoring* **11** 197-11

[6] Gudmundson P 1983 The dynamic behaviour of slender structures with cross-sectional cracks *J. Mech. Phys. Solids* **31** 329-334

[7] Ruotolo R, Surace C, Crespu P and Storer D 1996 Harmonic analysis of the vibrations of a cantilevered beam with a closing crack *Computers & Structures* **61** pp. 1057-1074

[8] Kisa M and Brandon J 2000 The Effects of closure of cracks on the dynamics of a cracked cantilever beam. *J. Sound Vib.* **238** 1-18

[9] Chondros T G, Dimarogonas A D and Yao J 2001 Vibration of a beam with a breathing brack. *J. Sound Vib.* **239** 57-67

[10] Chati M, Rand R and Mukherjee S 1997 Modal analysis of a cracked beam *J. Sound Vib.* **207** 249-270

[11] Sundermeyer J N and Weaver R L 1995 On crack identification and characterization in a beam by non-linear vibration analysis *J. Sound Vib.* **183** 857-871

[12] Riviola A and White P R 1998 Bispectral analysis of the bilinear oscilaltor with application to the detection of fatigue cracks *J. Sound Vib.* **216** 889-910

[13] Loutridisa S, Douka E and Hadjileontiadis L J 2005 Forced vibration behaviour and crack detection of cracked beams using instantaneous frequency *NDT&E International* **38** 411-419

[14] Donskoy D, Sutin A and Ekimov A 2001 Nonlinear acoustic interaction on contact interfaces and its use for nondestructive testing *NDT&E International* **34** 231-238

[15] Andreaus U, Casini P and Vestroni F 2007 Nonlinear dynamics of a cracked cantilever beam under harmonic excitation *J. Nonlinear Mech.* **42** 566-575

[16] Andreaus U, Baragatti P 2011 Cracked beam identification by numerically analysing the nonlinear behaviour of the harmonically forced response *J. Sound Vib.* **330** 721-742

[17] Hiwarkar V R, Babitsky V I and Silberschmidt V V 2009 Damage assessment of a cracked bar:

effect of material nonlinearity on vibro-impact response *Key Engineering Materials* **413-414** 237-244
[18] Hiwarkar V R, Babitsky V I and Silberschmidt V V 2012 On the modelling of dynamic structures with discontinuities *Nonlinear Dyn.* **67** 2651-2669
[19] Hiwarkar V R, Babitsky V I and Silberschmidt V V 2011 crack as modulator, detector and amplifier in structural health *J Sound Vib.* **331**. 3587-3598

A numerical model on PVB laminated windshield subjected to headform low-speed impact

X Q Xu[1], B H Liu[1], Y Wang[1], Y B Li[1] and J Xu[1,2]

State Key Laboratory of Automotive Safety & Energy, Department of Automotive [1]Engineering, Tsinghua University, Beijing 100084, China

E-mail: xujun06@mails.tsinghua.edu.cn

Abstract. Polyvinyl butyral (PVB) laminated windshield is one of the most important components in automotive to protect vulnerable road users. First, a windshield finite element (FE) model is set up using a piece of interlayer (PVB) sandwiched by two glass layers. Four parameters which have an critical impact on the simulation results, i.e. glass Young's modulus, glass plastic failure strain, PVB stress-strain curve and boundary condition, are suggested to measure the influence on the windshield model. Each windshield model is impacted by a standard headform impactor at the speed of 8m/s based on the LS-DYNA platform and the results are compared with the dynamic experiments of PVB laminated windshield under headform impact to find the most accurate FE model. Furthermore, the most accurate FE windshield model is compacted by the standard headform impactor on various impact velocities (6.6m/s-11.2m/s), angles (60°-90°) compared with the parametric dynamic experiments of PVB laminated windshield to verify the windshield finite element model. This paper provides a useful finite element model of windshield for further systematically numerical studies based on the finite element method to explore the ability of the energy absorption and safety design of PVB laminated windshield.

1. Introduction

Polyvinyl butyral (PVB) laminated windshield is a preferred material for current automotive windshield [1] with a certain extent of impact resistance and energy absorption characteristics. Correlation analysis shows that, the collisions between the head and the windshield is the main cause of the casualties of the vulnerable road users [2].

Several experimental research and numerical studies have been carried out to investigate the behavior of laminated glass under dynamic loading condition. Series of windshield FE models using different modelling techniques have been set up by Pyttel [3], Sun [4] and Peng [5], in order to perform the simulation accurately. Meanwhile, some mono material (e.g. soda-lime glass [6], PVB [7]) property studies have been carried out. Besides, Wingren [8] obtained the mechanical behavior of windshield in the impact of adult head module through impact test. In addition, Xu et al. [9, 10] compared the energy absorption capabilities of two types windshield, PVB and nanoporous energy absorption system (NEAS) interlayers, and carried out a systematic numerical study based on the

[2] Corresponding author: J Xu

extended finite element method (XFEM), to investigate the effects of material and system variables on the crack pattern.

However, little experimental data of the materials and the corresponding impact test verification are available for previous numerical studies. Any attempt to accurately simulate the behavior of the laminated windshield could only be reached via proper modelling techniques with the experiment data embedded verified by corresponding impact test, which is our fundamental motivation for this research. In this research, windshield model is set up using a triple-layered structure. Afterwards, parametric studies are conducted to investigate the parameters influence on the FE model. Finally, windshield FE model is verified via comparison with the results of the PVB laminated windshield subject to headform impact test.

2. FE model and method

2.1. Head model

The FE model of pedestrian headform impactor is built according to the requirement of Global Technical Regulation (GTR) [11]. The internal structure of the impactor is simplified using the mass and centroid position equivalence method during the finite element modeling process and calibrated. Dynamic calibration test has been done using the software LS-DYNA, and the results meet the requirements.

2.2. Windshield model

The windshield FE model is developed with shell elements. The layered structure of the laminated windshield is modelled using three shell elements with shared nodes at their boundaries: the outside two shell elements representing the glass layers and the inside shell element representing the PVB interlayer. The FE model is meshed using 2 mm quadrilateral elements with the triangular elements in the central region with the convergence study.

A series of simulations are conducted based on the LS-DYNA platform for different conditions in order to obtain the parameters' influence on the windshield model. The contact form between head model and windshield apply the surface-to-surface contact with the friction coefficient of 0.3 [10, 12]. Firstly, we vary the glass Young's modulus, glass plastic failure strain and PVB stress-strain curve according to the data listed in Table 1 respectively, meanwhile other conditions are kept at their reference values shown in Table 2. Afterwards, four boundary conditions, i.e. four edges clamped (FC), four edges simply supported (FS), left and right edges clamped while upper and lower edges simply supported (LR), left and right edges simply supported while upper and lower edges clamped (UD), are set up with other simulation conditions at their reference values.

Table 1. The range of the parameters variation.

	1	2	3	4	5
Young's modulus of glass	0.7GPa	7GPa	70GPa	350GPa	700GPa
Plastic failure strain of glass	5×10^{-3}	1×10^{-3}	5×10^{-4}	1×10^{-4}	5×10^{-5}
Strain rate of stress-strain curve of PVB[7, 13]	4×10^{-3}/s	2×10^{-3}/s	4×10^{-2}/s	8×10^{-2}/s	118/s

Table 2. Parameters used in the windshield FE model [10, 14].

layer	Variables	Reference parameters
Glass layer	Density ρ_{glass}	2500kg/m^3
	Young's modulus E_{glass}	70GPa
	Poisson's ratio υ_{glass}	0.227
	Yield stress σ_{glass}	0.143GPa
	Plastic strain to failure ε_{glass}	0.001
	Thickness t_{glass}	0.002m
PVB layer	Density ρ_{PVB}	2000kg/m^3
	Poisson's ratio υ_{PVB}	0.435
	Strain rate of strain-stress curve	118/s
Global settings	Boundary condition	Clamped

2.3. Model verification

Further simulations are conducted to verify the windshield FE model with the dynamic experiments of PVB laminated windshield under headform impact. The head module impact windshield experimental system consists of seven parts, i.e. pedestrian crash test launcher, the windshield gantry system, high-speed photography system, dual photoelectric speedometer, windshield specimen, adult head module, acceleration sensor and buffer system shown in Figure 1.

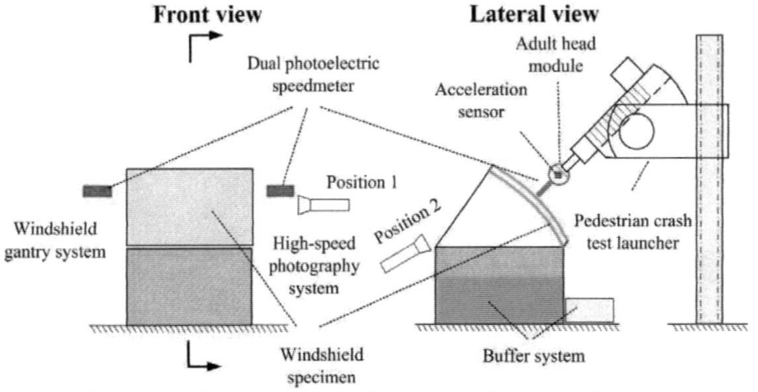

Figure 1. The experimental system schematic diagram.

The headform impact velocities and angles are set to be 6.6m/s-11.2m/s and 60°-90°respectively according to real-world vehicle pedestrian accidents in Tsinghua University traffic accident cases database. A total of 8 experiments are conducted shown in Table 3. The simulations using the most accurate windshield model above-obtained are conducted according to the settings in Table 3 to verify the windshield FE model.

Table 3. The test parameter settings.

NO.	Impact angle (°)	Impact velocity (m/s)
1	90	6.6
2	90	8.1
3	90	8.8
4	90	11.2
5	90	8.1
6	80	8.2
7	71	8.2
8	59	8.2

3. Results and discussions

3.1. Effect of glass Young's modulus

Figure 2 shows the acceleration-time curves of different glass Young's modulus compared with the corresponding test results. When the glass Young's modulus is set to 700MPa and 7GPa, the windshield does not crack and the acceleration trends are different from the experimental results. Once the glass Young's modulus rises beyond 70GPa, the cracks initiated and the crack intensity becomes higher with the Young's modulus. When the glass Young's modulus is set to 70GPa, simulation and experimental results of acceleration curve are closer. Regardless of cracks, head model acceleration increases with increasing glass Young's modulus.

With the glass Young's modulus increasing, the windshield becomes stiffer, so the peak acceleration becomes larger correspondingly. Since the yield stress and plastic failure strain fixed, the failure strain becomes smaller and cracks easier to generate with the increasing glass Yield stress.

3.2. Effect of plastic failure strain

Figure 3 shows the acceleration-time curves of different plastic failure strain compared with the corresponding test results. With the plastic failure strain increasing, the acceleration does not change obviously. As the plastic failure strain decreasing, the cracks become more intensive. When the plastic failure strain is set to 1×10^{-3} and 5×10^{-4}, simulation and experimental results agree better.

Due to the yield stress and the Young's modulus fixed, the failure stress becomes smaller with the plastic failure strain decreasing. The cracks become more intensive for the cracks initiation being more easily.

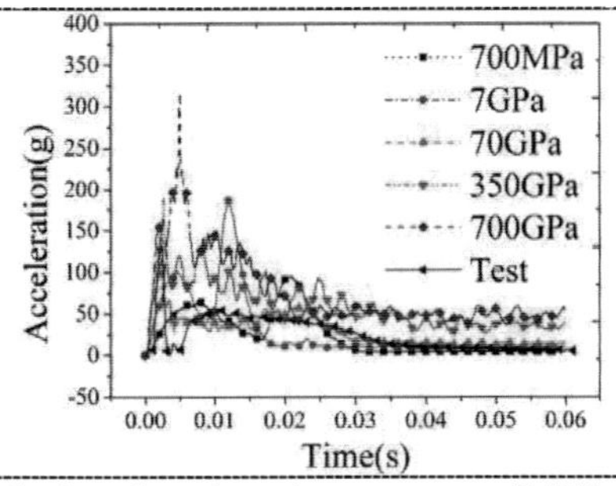

Figure 2. Acceleration-time curves of different glass Young's modulus compared with the corresponding test results.

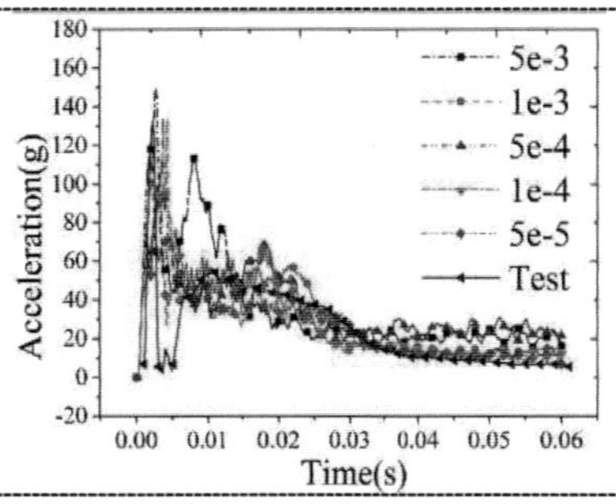

Figure 3. Acceleration-time curves of different plastic failure strain compared with the corresponding test results.

3.3. Effect of PVB stress-strain curve

Figure 4 shows the acceleration-time curves of different PVB stress-strain curves compared with the corresponding test results. Compared with the application of dynamic stress-strain curve, the second acceleration peak is not obvious when the quasi-static stress-strain curves are employed. When we apply quasi-static stress-strain curves with different strain rate, the obtained acceleration curves are not very different. Besides, simulation and experimental results of acceleration curve are closer when we apply the dynamic stress-strain curve.

From the quasi-static and dynamic tensile test results of the specimens, we can see that the PVB materials exhibit different tensile characteristics in dynamic and quasi-static loads [7]. Besides, the impact process itself is a dynamic process. Therefore, the results obtained by the application of dynamic tensile test curve are more in line with the actual situation.

3.4. Effect of boundary condition

Figure 5 shows the acceleration-time curves under different boundary conditions compared with the corresponding test results. With FC and FS boundary conditions, the obtained acceleration curves are almost the same. However, the second acceleration peak is not stable with LR and UD boundary conditions. This shows that, boundary conditions only have little influence. So the FC boundary condition is selected in line with the real situation.

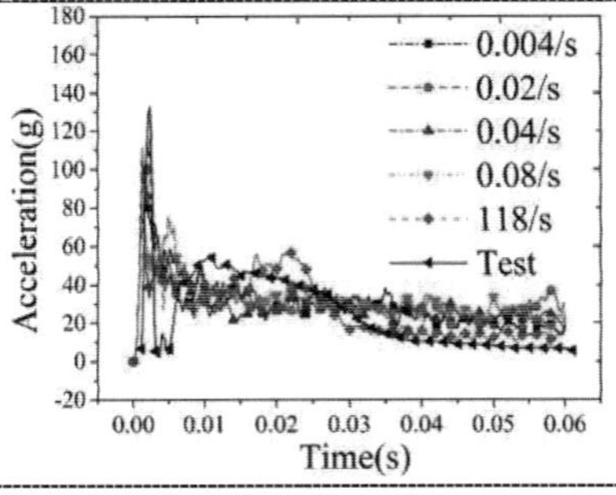

Figure 4. Acceleration-time curves of different PVB stress-strain curves compared with the corresponding test results.

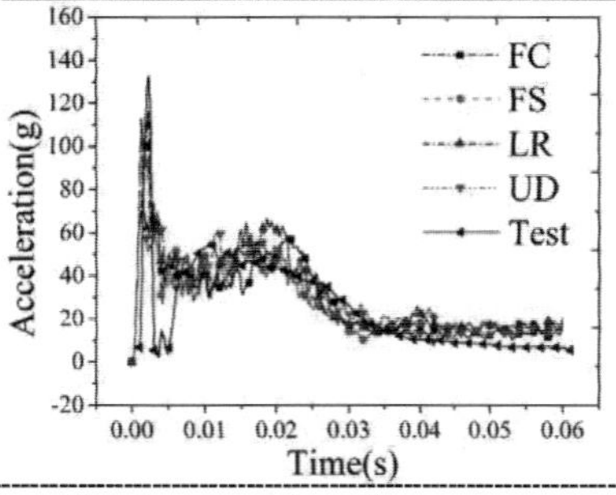

Figure 5. Acceleration-time curves under different boundary conditions compared with the corresponding test results.

3.5. Model verification

Figure 6 shows the obtained acceleration-time curves from the simulations according to table 3 compared with the corresponding results for further verification. We can see that the peak values as well as the entire time course of the simulation fit well with the corresponding impact test results with different impact velocities and impact angles.

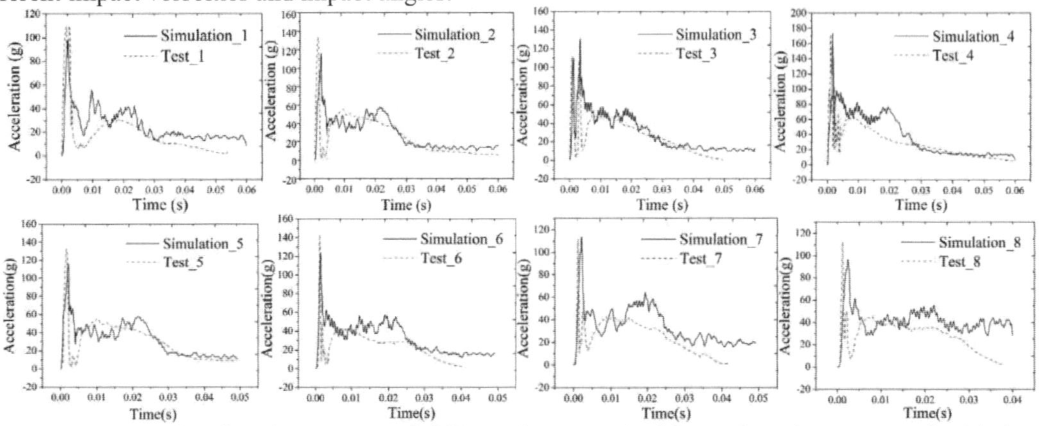

Figure 6. Acceleration-time curves of different impact velocities and angles compared with the corresponding test results for further verification.

Figure 7 shows the comparison of crack propagation patterns for the simulation results and the test results with the velocity of 8m/s at the vertical impact. The crack propagation patterns obtained from the impact test are captured by high-speed cameras [15]. In the beginning of the impact, radial cracks first appear on the windshield panel to release the hoop stress, for the hoop stress is much larger than the radial stress [16]. Circular cracks begin to appear approximately within 1 second and expand from the inside outward consistent with the initial propagation direction of the stress wave concurred with the literature findings [17, 18]. The simulation results are in good agreement with the experimental results.

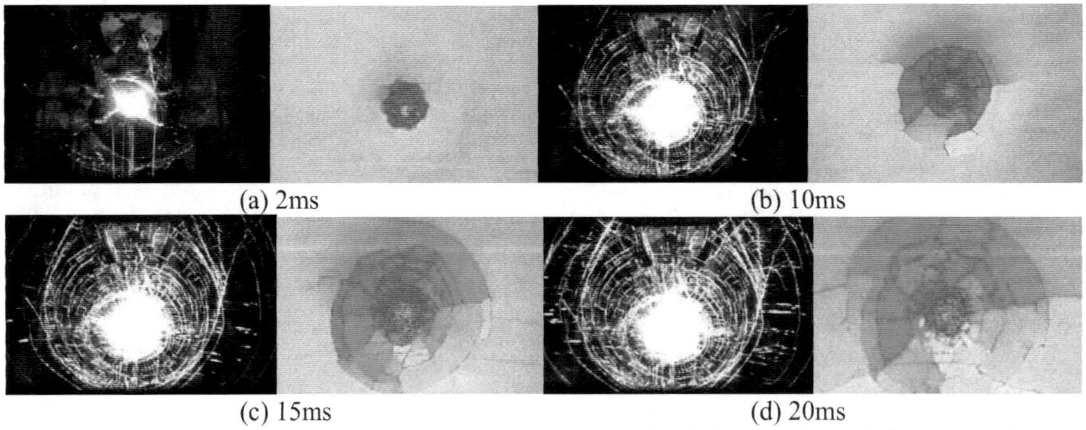

Figure 7. Comparison of crack propagation patterns for the simulation results and the test results with the velocity of 8m/s at the vertical impact.

4. Conclusion

In this paper, windshield with laminated structure is setup and a series of simulations under different conditions are conducted to obtain the parameter's influence on the windshield model such that a finite element model is suggested with proper parameter settings. Values of parameters in Table 2 can best predict the test acceleration curve. This study provides a solid step for systematic numerical studies on windshield dynamics subject to head impact in the future.

Acknowledgements
The work is supported by National Natural Science Foundation of China (11102099), Chinese Post-doctor Foundation and International joint research project sponsored by Tsinghua University (20121080050).

References
[1] Zhao S M, Dharani L R, Chai L and Barbat S D. Analysis of damage in laminated automotive glazing subjected to simulated head impact. *Eng Fail Anal.* 2006;**13(4)**:582-97.
[2] Otte D, Jansch M and Haasper C. Injury protection and accident causation parameters for vulnerable road users based on German In-Depth Accident Study GIDAS. *Accid Anal Prev.* 2012;**44(1)**:149-53.
[3] Pyttel T, Liebertz H and Cai J. Failure criterion for laminated glass under impact loading and its application in finite element simulation. *Int J Impact Eng.* 2011;**38(4)**:252-63.
[4] Sun D Z, Andrieux F and Ockewitz A. Modelling of the failure behaviour of windscreens and component tests. 2005.
[5] Peng Y, Deck C, Yang J K and Willinger R. Finite element modeling of crash test behavior for windshield laminated glass. 2012.
[6] Liu W N, Sun X, Chen W N and Templeton D. Modeling and characterization of dynamic failure of soda-lime glass under high-speed impact. *Int J Damage Mech.* 2012;**21(4)**:577-98.
[7] Liu B H, Sun Y T, Li Y B, Wang Y, Ge D Y and Xu J. Systematic experimental study on mechanical behavior of PVB (polyvinyl butyral) material under various loading conditions. *Polym Eng Sci.* 2012;**52(5)**:1137-47.
[8] Wingren M. Windscreen study using a free moving headform: An investigation of windscreen behaviour when subjected to headform impact. 2011.
[9] Xu J, Li Y B, Chen X, Ge D Y, Liu B H, Zhu M Y and Park T H. Automotive windshield - pedestrian head impact: Energy absorption capability of interlayer material. *International Journal of Automotive Technology.* 2011;**12(5)**:687-95.
[10] Xu J, Li Y B, Chen X, Yan Y A, Ge D Y, Zhu M Y, Liu B H. Numerical study of PVB laminated windshield cracking upon human head impact. *CMC-Comput Mat Contin.* 2010;**18(2)**:183-211.
[11] ECE/TRANS/180/Add.9. Global technical regulation No. 9. 2008.
[12] Xu J, Li Y B, Chen X, Yan Y, Ge D Y, Zhu M Y, Liu B H. Characteristics of windshield cracking upon low-speed impact: numerical simulation based on the extended finite element method. *Comput Mater Sci.* 2010;**48(3)**:582-588588.
[13] Iwasaki R, Sato C and Lataillade J L. Experimental study on the interface fracture toughness of PVB (polyvinyl butyral)/glass at high strain rates. 2007.
[14] Xu J, Bourdet N, Sun Y T, Wang Y, Liu B H, Li Y B and Willinger R. Evaluation of pedestrian's head injury subject to PVB laminated windshield impact. 2011.
[15] Liu B H, Wang Y, Xu X Q, Zhang W H, Xu J and Li Y B. Experimental research on energy absorption characteristic of PVB laminated windshield subjected to headform low-speed impact. Submitted.
[16] Xu J, Li Y B, Chen X, Yan Y, Ge D Y, Zhu M Y and Liu B H. Characteristics of windshield cracking upon low-speed impact: numerical simulation based on the extended finite element method. *Comput Mater Sci.* 2010;**48(3)**:582-588588.
[17] Xu J and Li Y B. Study of damage in windshield glazing subject to impact by a pedestrian's head. *Proc Inst Mech Eng Part D-J Automob Eng.* 2009;**223(D1)**:77-84.
[18] Xu J, Li Y B, Lu G Q and Zhou W. Reconstruction model of vehicle impact speed in pedestrian-vehicle accident. *Int J Impact Eng.* 2009;**36(6)**:783-788788.

Energy and momentum transfer to a 'fully-clamped' elastic plate in an air-blast

Y Yuan and PJ Tan

Department of Mechanical Engineering, University College London,
Torrington Place, London WC1E 7JE, United Kingdom

E-mail: pj.tan@ucl.ac.uk

Abstract. The momentum transfer by a planar wave impinging upon a rigid, free-standing plate in water, a largely incompressible medium, is well understood [1]. Kambouchev et al. [2] extended the results of Taylor [1] to include the nonlinear effects of compressibility whilst Hutchinson [3] has recently addressed the issues of energy and momentum transfer to a rigid, free-standing plate. In this paper, key conclusions from the aforementioned studies are critically re-examined in the context of a 'fully-clamped' elastic plate. The dynamic response of an elastic plate is represented by an equivalent single-degree-of-freedom (SDOF) system. A numerical method based on a Lagrangian formulation of the Euler equations of compressible flow and conventional shock-capturing techn iques, similar to that employed in [2, 3], were employed to solve numerically the interaction between the air blast wave and elastic plate. Particular emphasis is placed on elucidating the energy and momentum transfer to a 'fully-clamped' elastic plate compared to its rigid, free-standing counterpart, and on whether enhancement in the beneficial effects of FSI as a result of fluid compressibility remains and to what extent.

1. Introduction

Over the past decade, there have been renewed interests in understanding how fluid structure interaction (FSI) in a compressible medium may be exploited to design lightweight sandwich panels with better blast amelioration capabilities. To this end, Kambouchev et al. [2] and Hutchinson [3] have carried out detailed numerical studies to quantify the effects of nonlinear compressibility upon the energy and momentum transfer to plates subjected to air blast loading. However, their findings apply only to rigid, free-standing plates. Recent studies in [4–6] used a single-degree-of-freedom (SDOF) representation to study the effects of mass, time ratio (period of the pressure pulse to the natural period of vibration of the structure) and velocity ratio (shock front velocity to the maximum velocity acquired by the structure) upon the maximum deflection of a flexible structure. However, none have addressed the important issue of energy and momentum transfer, and/or the effects of boundary conditions.

In this paper, the numerical method in [2, 3] is employed to study the fluid structure interaction between an air blast wave and a 'fully clamped' elastic plate. The emphasis is on comparing the energy and momentum transfer to a 'fully-clamped' elastic plate with its rigid, free-standing

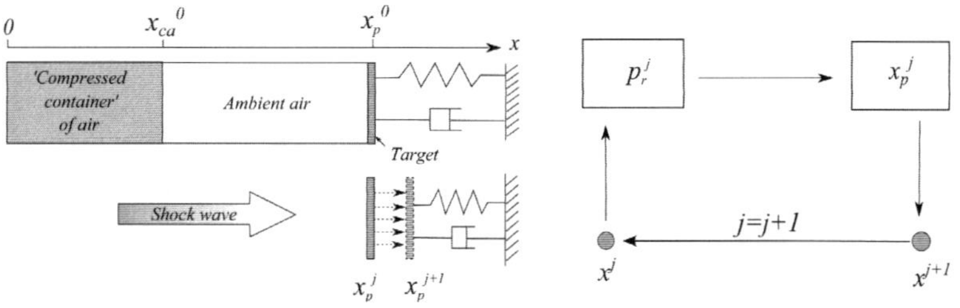

(a) Model setup (b) Fluid-structure coupling

Figure 1. Schematic of model set-up and fluid-structure coupling strategy.

counterpart. Also investigated in this paper is whether enhancement in the beneficial effects of FSI, as a result of fluid compressibility, also applies to a non-free-standing, i.e. supported, structure and to what extent. The results reported here is part of an on-going study to model the influence of boundary conditions upon the momentum and energy transfer to deformable (elasto-plastic) plates through FSI to be published elsewhere.

2. Shock wave generation and its interaction with an elastic plate

Consider a one-dimensional (1D) 'air-column' of two parts, viz. 'compressed container' of isentropic air and quiescent ambient air, and a flexible elastic plate which is represented by an equivalent SDOF system as shown in Fig. 1a. At time $t = 0$ the 'compressed container' of air in the interval $0 \leq x \leq x_{ca}^0$ is prescribed with an initial velocity distribution of

$$v(x) = v_0 \mathrm{e}^{-(x/x_{ca}^0)^2}, \tag{1}$$

where x is the Lagrangian coordinate. It follows immediately that the compressed air has a density distribution $\rho(x)$ and a finite initial energy per unit area ΔE_0 given by [3]

$$\begin{gathered} \left[\frac{\rho(x)}{\rho_a}\right]^\gamma = \left[1 + \frac{\gamma-1}{2}(\frac{v(x)}{c_a})^2\right]^{\gamma/(\gamma-1)} \\ \Delta E_0 = \frac{p_a x_{ca}^0}{\gamma-1}\left[\left(1+\frac{\gamma-1}{2}(\frac{v_0}{c_a})^2\right)^{\gamma/(\gamma-1)} - \left(1+\frac{\gamma-1}{2}(\frac{v_0}{c_a})^2\right)^{1/(\gamma-1)}\right] \end{gathered} \tag{2}$$

where $\gamma(= 1.4)$ is the ratio of specific heats, $p_a(= 104761$ Pa) is pressure, $\rho_a(= 1.225$ kg/m^3) is air density, $c_a(= 346$ m/s) is speed of sound in air and the subscript 'a' denotes ambient conditions. At time $t = 0$, the compressed air is suddenly released into the quiescent ambient air which generates a pressure pulse reminiscent of those in an explosive detonation; this same technique is also employed in [3].

By way of an example, consider a fluid domain of length $d = 10$ m and a 'compressed container' of length $x_{ca}^0 = 0.5$ m. The 1D domain is discretised using a uniform grid of 2000 nodes and the classical Euler equations of compressible flow are solved explicitly by finite difference, see [2]. Artificial viscosity was introduced to smooth the shock discontinuities but in a manner that still preserves energy conservation [3]. The time step was chosen to ensure it satisfies the stability requirement of the iterations. Figure 2a shows the displacement of 100 nodes and the 'shock locus' generated by the travelling wave. Note the dash lines represent the motion of the 'compressed container'. The incident and reflected pressure-time history following wave

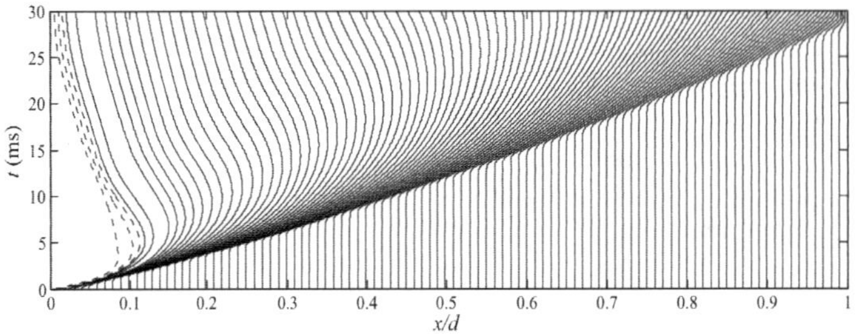

(a) $x - t$ diagram showing a propagating shock wave

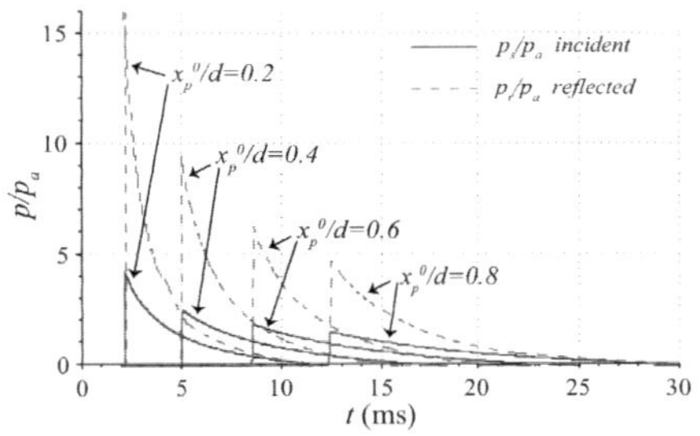

(b) 'Strong' shock ($v_0/c_a = 2.46$)

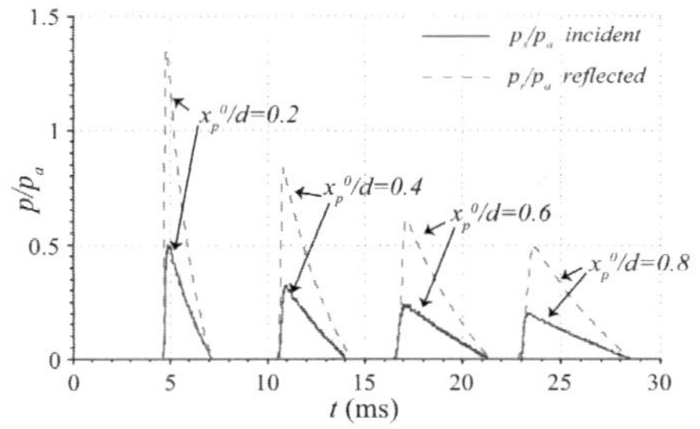

(c) 'Weak' shock ($v_0/c_a = 0.85$)

Figure 2. (a) Shock wave generated by 'compressed container' method and (b,c) incident and reflected pressure-time histories upon interaction with a massive rigid wall at different standoff.

interaction with a massive rigid wall of infinite mass at different standoff are compared in Figs. 2b and 2c for different blast intensities. The intensity of the blast wave is controlled by the non-dimensional parameter v_0/c_a. Some key observations can be made as follows: (1) nonlinear compressibility causes the peak of the evolving pulse to decrease, with a corresponding increase in pulse duration, with distance travelled; (2) the ratio of the peak reflected to incident pressure is between 2 to 8 as predicted by the well-known Rankine-Hugoniot relationships; and (3) the incident and reflected impulse is identical at different standoff since the artificial viscosity was set to preserve energy conservation, i.e. the wave generated is essentially isentropic. In order to ensure that the plate remains elastic, i.e. deforming in Mode I [7], a relatively 'weak' shock of $v_0/c_a = 0.85$ is used in subsequent simulations.

The equation of motion of the elastic plate, represented by an equivalent SDOF system with stiffness k_e and damping coefficient c_e, both per unit area, is given by

$$m_e\ddot{x}_p + c_e\dot{x}_p + k_e x_p = F_e(t) \tag{3}$$

where $(\dot{\ }) \equiv d(\ \ )/dt$, x_p is displacement, m_e is equivalent mass per unit area and $F_e(= p_r K_L)$ is equivalent load per unit area. The damping coefficient c_e is given by $c_e = \alpha(\rho_a c_a)$, where α is chosen to reflect the severity of structural damping. The equivalent values were derived following the procedure outlined by Biggs [8]. Table 1 list the properties of two aluminium plates, of identical mass per unit area, but different flexural rigidity per unit length. Implementation of fluid structure coupling is achieved by solving for the interface pressure (reflected pressure p_r^j) and plate displacement x_p^j which are updated at every time step as shown schematically in Fig. 1b.

Table 1. Properties of square aluminium plates and equivalent parameters of SDOF model.

Parameters	Symbol	Plate-1	Plate-2
Density (kg/m^3)	ρ_p	2760	2760
Young's modulus (GPa)	E	69	69
Poisson's ratio	ν	0.33	0.33
Dimensions (m)	$a \times b \times h$	$4 \times 4 \times 0.0275$	$2 \times 2 \times 0.0275$
Second moment of area (m^4) [8]	$I_a = ah^3/12$	6.93×10^{-6}	3.47×10^{-6}
Flexural rigidity/length (kNm) [8]	$D = EI_a/b$	119.58	119.58
Mass factor [8]	K_M	0.21	0.21
Load factor [8]	K_L	0.33	0.33
Equivalent stiffness/area (kN/m^3) [8]	$k_e = K_L \frac{810D}{a^4}$	124.86	1997.8
Equivalent mass/area (kg/m^2) [8]	$m_e = K_M \rho_p h$	15.94	15.94
Natural period (ms)	T	71.0	17.7

3. Results and Discussions

To check that the SDOF model is correctly implemented and gives sensible predictions, the reflected over-pressure for the 'weak' shock case, shown in Fig. 2c, is prescribed as a stand-alone pressure loading to the fully-validated FE model in [7] and to the SDOF model, both in an uncoupled manner from the fluid domain. Note that damping is initially ignored. Figure 3a shows excellent agreement between the predicted maximum displacement x_p by the uncoupled SDOF model and the maximum central deflection of its equivalent square plate in the uncoupled FE model.

Parametric studies carried out in [4, 6] have established that FSI has the beneficial effects of reducing the maximum displacement of flexible structure. In Fig. 3a, the predictions by both the coupled and uncoupled SDOF model show that increasing standoff reduces the maximum displacement although this effect is more apparent with FSI. This is because the plate experiences a shorter overall loading duration as a result of the coupling which is evident in Fig. 3b. On the other hand, increasing the equivalent plate stiffness per unit area at a given standoff distance does not appear to offer any significant advantage in terms of the maximum plate deflection.

Figure 4 compares the momentum and energy transfer to free-standing (of infinite flexural rigidity) and fully-clamped plates (of two different flexural rigidity given in Table 1). The ratio of transmitted impulse I_p to incident impulse $I_0(=\int_0^{t_d} p\mathrm{d}t)$ is plotted in Fig. 4a. As expected, the rigid free-standing plate acquires a higher momentum (or transmitted impulse) compared to its fully-clamped counterpart. What is surprising, however, is that the plate with a higher flexural rigidity appears to acquire a slightly lower impulse. The total energy E_T transferred to the plate is dissipated as kinetic energy E_K by the free-standing plate and as kinetic E_K plus strain E_S energies by its fully-clamped counterpart shown in Fig. 4b. Since the rigid free-standing plate is free to displace, a greater amount of energy $E_T = \int_0^{x_p} p_r \mathrm{d}x_p$ is transmitted compared to its fully-clamped, flexible counterpart as shown in Fig. 4b. Previous studies on the effects of FSI for free-standing plate have shown that a smaller plate mass displaces further, thus relieving the incident pressure acting on the plate [2, 3, 9]. Figure 4c plots the non-dimensional transmitted impulse vs. FSI index ($\beta_s = \rho_s U_s t_0 / m_e$ where ρ_s is density of gas behind shock front, U_s is shock speed, t_0 is the duration of an equivalent uniform shock and m_e is equivalent mass per unit area). β_s was introduced in [2, 9] to take into account nonlinear compressibility effects. The results in Fig. 4c agrees with those in previous studies where the transmitted impulse is

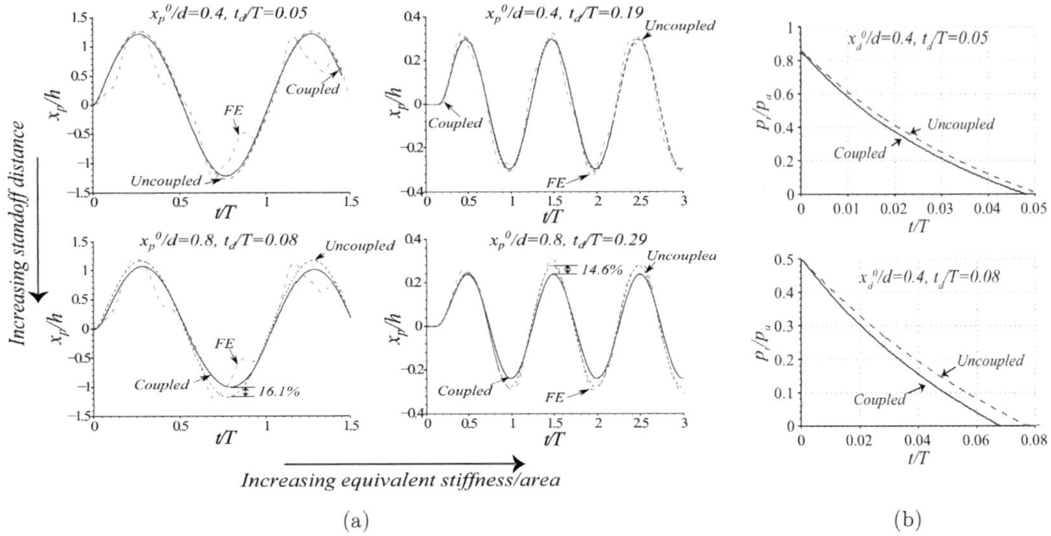

Figure 3. (a) Predicted displacement histories without damping and (b) ratio of the normalised reflected pressure p_r/p_a versus t/T for Plate 1. Coupled SDOF model (—) ; uncoupled SDOF model (- - -) ; uncoupled FE predictions (-.-.-).

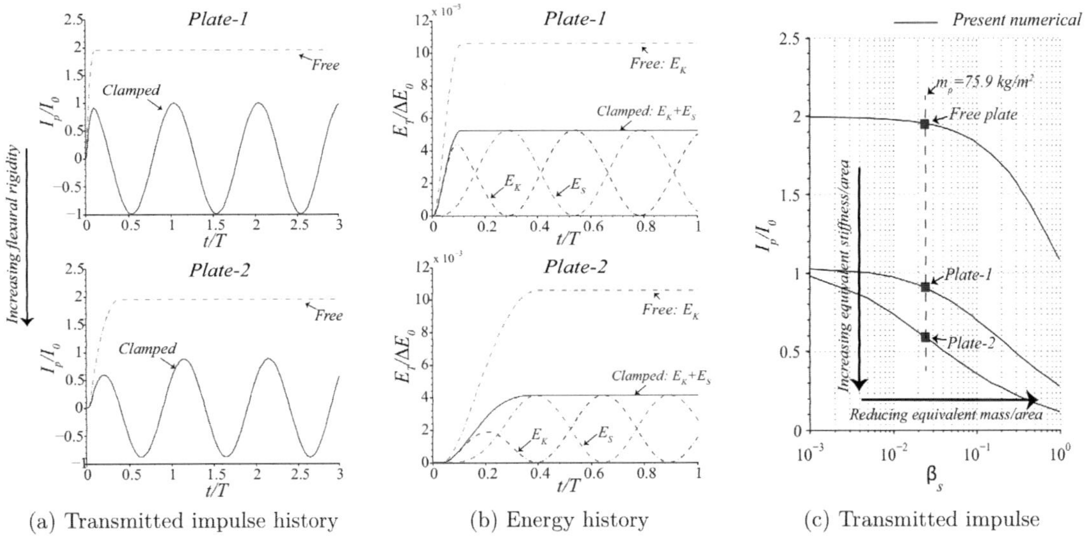

Figure 4. Comparison of momentum and energy transfer to a free-standing and its corresponding fully-clamped plate.

observed to increase with β_s. By contrast, the restrain offered by a clamped boundary leads to a significant reduction in the impulse transmitted to the plate for a given β_s. Therefore, the introduction of a clamped boundary appears to further enhance the beneficial effects of FSI. The implication is that a structure with a higher equivalent stiffness and/or a smaller equivalent mass, both per unit area, acquires less impulse and energy in an air blast.

Lastly, the effects of structural damping is shown in Fig. 5. The results show that greater

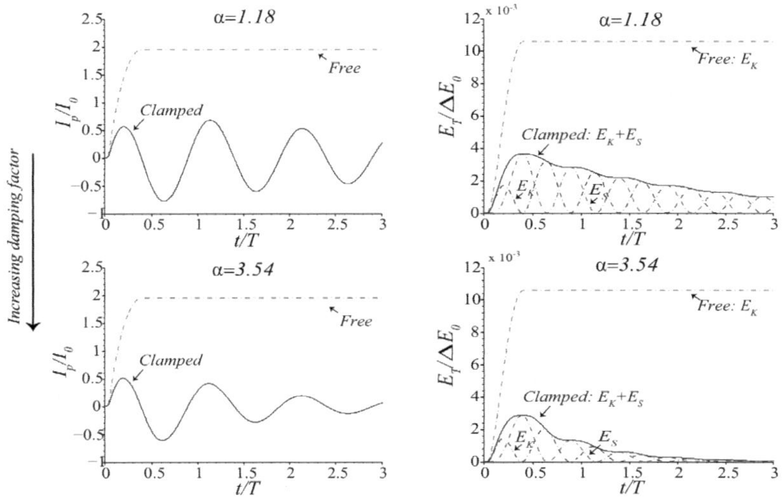

(a) Transmitted impulse history (b) Energy history

Figure 5. The effects of damping upon the momentum and energy transfer for Plate-2.

damping, i.e. a higher α, leads to greater beneficial effects of FSI in terms of the momentum and energy transferred to the structure. This is because the damping force always opposes the direction of motion and this further contributes to the beneficial effects of FSI.

4. Conclusion

Fluid structure interaction between an air blast wave and a fully-clamped elastic plate is studied. Three important conclusions can be drawn:

- The resistance to transverse displacement, offered by a lateral restrain in the form of a support, further enhances the beneficial effects of FSI.
- Higher structural damping also improves the beneficial effects of FSI.
- Stiff and light structures minimise the impulse and energy transmitted in an air blast.

References

[1] Taylor G I 1963 *The Scientific Papers of Sir Geoffrey Ingram Taylor* **3** 287–303
[2] Kambouchev N, Noels L and Radovitzky R 2006 *J App Phys* **100** 063519
[3] Hutchinson J W 2009 *J App Mech* **76** 051307
[4] Subramaniam V K, Nian W and Andreopoulos Y 2009 *Int J Impact Eng* **36** 965–974
[5] Rigby S, Tyas A and Bennett T 2012 *Eng Structures* **45** 396 – 404
[6] Teich M and Gebbeken N 2011 *Eng Structures*
[7] Yuan Y and Tan P J 2013 *Int J Impact Eng* **59** 46 – 59
[8] Biggs J 1964 *Introduction to structural dynamics* (New York: McGraw-hill)
[9] Kambouchev N, Noels L and Radovitzky R 2007 *Computers & Structures* **85** 923–931

Blast response of curved carbon/epoxy composite panels: Experimental study and finite-element analysis

V A Phadnis[1,3], P Kumar[2], A Shukla[2], A Roy[1] and V V Silberschmidt[1]
[1]Wolfson School of Mechanical and Manufacturing Engineering, Loughborough University, Loughborough, LE11 3TU, Leics, UK
[2] Dynamic Photomechanics Laboratory, Department of Mechanical Engineering and Applied Mechanics, University of Rhode Island, Kingston, RI, USA

E-mail: v.a.phadnis@lboro.ac.uk

Abstract. Experimental and numerical studies were conducted to understand the effect of plate curvature on blast response of carbon/epoxy composite panels. A shock-tube system was utilized to impart controlled shock loading to quasi-isotropic composite panels with differing range of radii of curvatures. A 3D Digital Image Correlation (DIC) technique coupled with high-speed photography was used to obtain out-of-plane deflection and velocity, as well as in-plane strain on the back face of the panels. Macroscopic post-mortem analysis was performed to compare yielding and deformation in these panels. A dynamic computational simulation that integrates fluid-structure interaction was conducted to evaluate the panel response in general purpose finite-element software ABAQUS/Explicit. The obtained numerical results were compared to the experimental data and showed a good correlation.

1. Introduction

The ever expanding applications of composite materials expose them to severe loading and environmental conditions, posing new challenges to the designer. Accidental explosions can cause extreme loading on composites-made structures with both flat and curved geometries. Thus, it is important to understand the effect of curvature of the structure exposed to blast, helping manufacture new structures with better blast resistance. Carbon/epoxy composites, in particular, carbon fibre reinforced polymer (CFRPs) composites are widely used in protective structural applications due to their superior mechanical and physical properties such as durability, high strength-to-weight-ratio, and high stiffness.

There are various studies in the literature related to blast loading of structures [1-6]; only the most relevant studies are mentioned here. Rajendran and Lee [1] conducted a detailed review of the phenomena of air and underwater explosions and their effects on plane plates. They found that of the important shock wave parameters significant for an air blast, were the peak overpressure and the impulse. For an underwater explosion, the peak overpressure, time constant, free-field impulse and energy were recognised to be the four vital parameters for the damage process. Tekalur *et al.* [2] studied the effect of various fibres on the blast response of composite panels. They used two different

[3] Author to whom any correspondence should be addressed.

fibre materials: E-glass and carbon, and exposed those composite panels to high strain rates and quasi-static loading. In dynamic loading, the carbon-fibre composites showed catastrophic failure, whereas, E-glass fibre composites exhibited progressive damage behaviour. Arora *et al.* [3] analysed a blast response of glass-fibre sandwich composite panels, fixed at their edges and exposed to real explosives at varying stand-off distances. They used high-speed photography and digital image correlation analysis to characterise the blast response of these panels. Damage was observed to be initiated in the form of a crack in the front skin, giving rise to localised delamination around the cracked region and shear cracking in the core. Interfacial failure between the front skin and the core was also observed. This study also involved a finite-element study to verify experimental observations such as transient boundary conditions. Kumar *et al.* [4] reported the effect of transient boundary conditions in their study where the dynamic response of curved CFRP panels was analysed using high-speed photography, a 3D-DIC technique followed by a post-mortem analysis. They found that curvature had a profound effect on the blast response of the CFRP panels. Ochola *et al.* [5] reported on strain-rate sensitivity of both carbon fibre-reinforced polymer (CFRP) and glass fibre-reinforced polymer (GFRP) composites by testing a single laminate configuration with a strain-rate varying from $10^{-3}\,s^{-1}$ to $450s^{-1}$. The obtained results showed that dynamic material strength for GFRP increased with the increasing strain rate and the strain to failure for both CFRP and GFRP decreased with the increasing strain rate. LeBlanc and Shukla [6] studied an underwater shock loading response of E-glass/vinyl ester curved composite panels. They used a 3D DIC system to measure the transient response during the experiments. They compared experimental results to those obtained in the simulation with the commercial LS-Dyna finite-element code. This comparison showed a high level of correlation based on the Russell error measurement.

In this paper, dynamic response of quasi-isotropic CFRP composite panels with three different radii of curvature exposed to blast loading was studied. A shock tube was employed to impart blast load on these panels to understand the effect of their curvature on the blast mitigation. A real-time analysis was carried out using 3D DIC technique to measure the out-of-plane deflection on the back face of these panels. A finite-element (FE) model of blast response of these panels was developed in ABAQUS 6.11 and was validated using the experimental findings.

2. Experimental details

2.1. Material and specimen

Panels with three different radii of curvature (Fig. 1) were utilized in the experiments: infinite (i.e. flat; Panel A), 304.8 mm (Panel B) and 111.8 mm (Panel C). The specimens were fabricated using ready-to-cure sheets of unidirectional AS4/3501-6 prepreg (fibre volume fraction of 60%) manufactured by the Hercules Corporation of Magna, Utah. The stacking sequence of the composite laminate was selected to have quasi-isotropic properties ($[0^0/90^0/+45^0/-45^0]_{4s}$). The specimens were 203 mm × 203 mm × 2 mm in size, made out of 32 layers of carbon fibres. For the curved panels, arc lengths of curved edges correspond to the plate length, i.e. 203 mm. The material properties of this laminate are listed in Table 1.

2.2. Experimental setup

Experimentally, blast loading onto a structure can be imparted using two different methods: using explosives or shock tubes. The use of real explosives is dangerous and produces spherical wave fronts and pressure signatures, which are spatially complex and difficult to measure. On the contrary, a shock tube offers the advantage of planar wave fronts so that the wave parameters can be reasonably easily controlled. Furthermore, the loading conditions are easy to replicate. Therefore, the shock tube was the preferred choice to apply blast loading in our experiments. The shock tube apparatus used in this study and location of pressure transducers is shown in Fig. 2a and b respectively. A complete description of the shock tube and its calibration can be found in [7]. The specimens were clamped on all their four edges.

Table 1. Mechanical properties of AS4/3501-6 UD composite laminate [8]

Elastic modulus	$E_{xx} = 147$ GPa, $E_{yy} = 11$ GPa, $E_{xy} = 5$ GPa
Tensile strength in fiber direction	2004 MPa
Compressive strength in fiber direction	1197 MPa
Tensile strength in transverse direction	53 MPa
Compressive strength in transverse direction	200 MPa
Shear strength	137 MPa

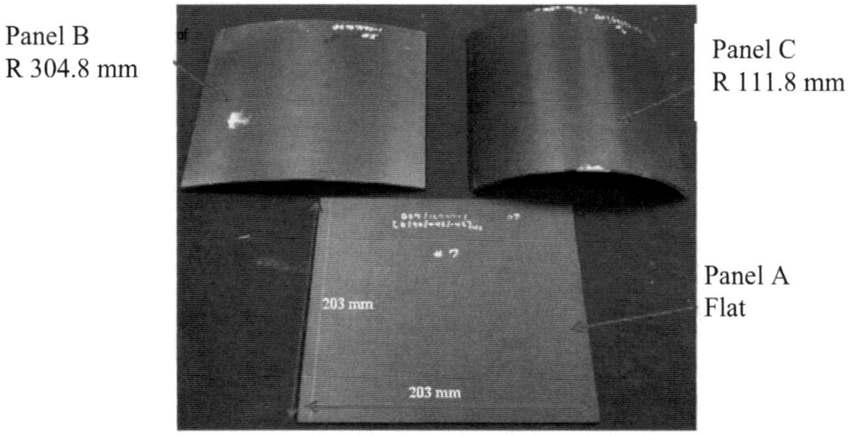

Figure 1. Specimens used in blast study

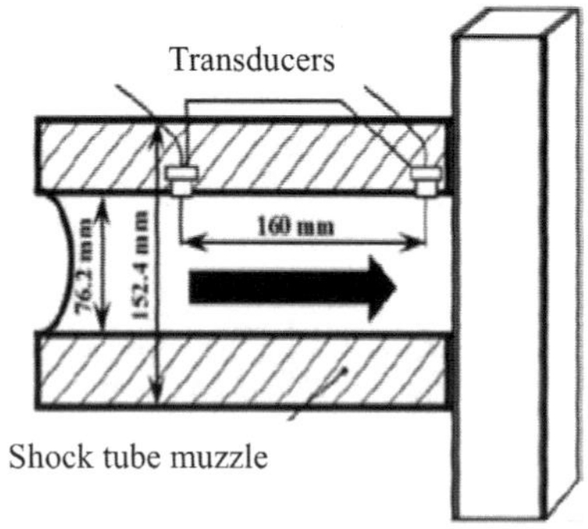

(a) (b)

Figure 2 a. Shock tube appartus used in blast study **b.** Shock tube nozzle and location of transducers

The shock tube consisted of a long rigid cylinder divided into a high-pressure driver section and a low pressure driven section separated by a diaphragm. By pressurizing the driver section, a pressure difference was created across the diaphragm. When this pressure differential reached a critical value, the diaphragm ruptured. The subsequent rapid release of gas created a shock wave which travelled down the shock tube to impart shock loading on the specimen at the muzzle end.

The shock tube utilized in the present study had an overall length of 8 m, consisting of driver, driven, converging, and muzzle sections. The diameter of the driver and driven section was 0.15 m. The final muzzle diameter was 0.07 m. Two pressure transducers (Fig. 2b), mounted at the end of the muzzle section measured the incident shock pressure and the reflected shock pressure during the experiment. The pressure transducers were oriented along the horizontal line of symmetry of the specimen. The specimens were clamped on all four edges. Appropriate fixtures were designed and manufactured to hold the specimens. The specimens were blast loaded at three different pressures varying from 3 MPa to 8 MPa. The experiments were repeated three times under same conditions. The pressure profiles obtained at the transducer location closer to the specimen A and B are shown (Fig. 3).

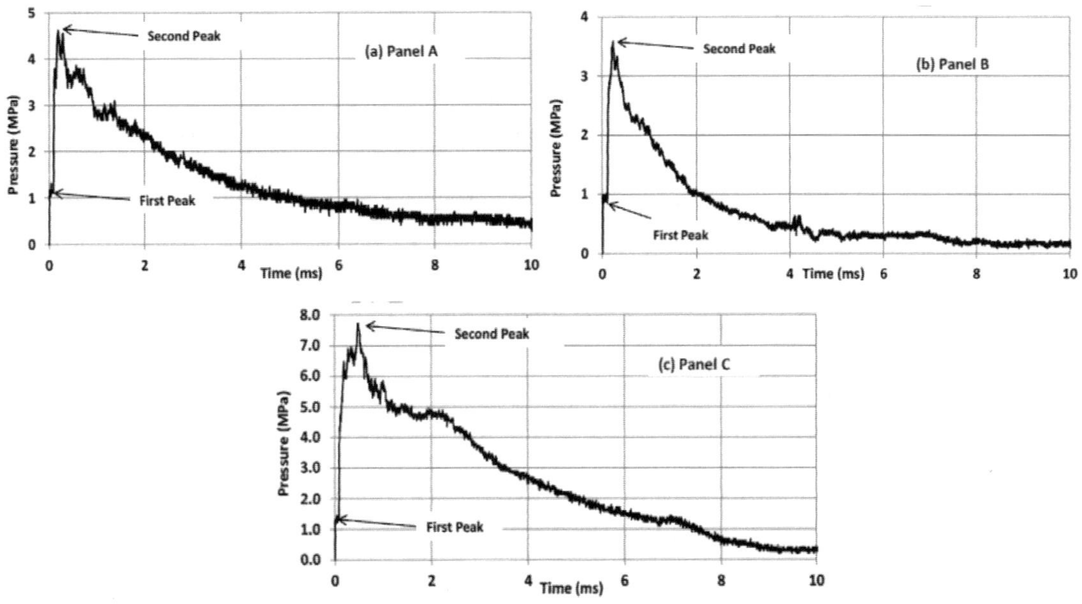

Figure 3. Pressure profiles for studied carbon/epoxy composite panels [9]

2.3. Digital image correlation

Digital Image Correlation (DIC) technique is a non-contact optical method for analysing full-field shape and deformation. Here, DIC technique was employed to measure in-plane strains and out-of-plane deflection at the centre of the back face of all studied specimens. Two high speed digital cameras, Photron SA1s, were positioned behind the shock tube apparatus to capture the real-time deformation and displacement of the panel. The high speed cameras were set to capture synchronized images at 20,000 frames per second (inter-frame time of 50 μs). During the blast loading event, as the panel responds, the cameras record the speckles on the painted back face. The high speed images were analysed using DIC software to correlate the images from the two cameras and generate real-time in-plane strain and out-of-plane deflection/ velocity histories. The further details regarding DIC analysis can be found in [9].

3. Finite-element model

Blast experiments are generally laborious, difficult to repeat under the same conditions and demand plethora of safety precautions. In this regard, a validated FE model can prove to be a valuable tool to estimate the blast performance of underlying materials. Here, a dynamic FE model imitating the mechanical response of curved CFRP panels exposed to the blast was developed in commercial software ABAQUS/EXPLICIT [10]. The details about material modelling, fluid-structure interaction, mesh and boundary conditions are discussed in the next sections.

3.1. Material model

A user-defined 3D damage model (VUMAT) with solid elements was developed and implemented into the finite-element code ABAQUS/EXPLICIT to predict the character and extent of damage through the laminate thickness when exposed to the blast load. Interface cohesive elements were inserted between the plies of the modelled laminate to simulate delamination. The general contact algorithm in ABAQUS/EXPLICIT was used to simulate contact conditions between the shock wave and the composite laminate, and between the layers by defining appropriate contact-pair properties. An element deletion approach [13] was used to represent the laminate degradation based on initiation and evolution of damage in the meshed CFRP elements. The results of numerical simulations were evaluated using comparison with the experimental data.

3.2. Element deletion

The blast load induces large deformation in the CFRP laminate. During the FE modelling, these excessively deformed elements can cause unreasonable increase in the simulation time or premature-termination of the analysis. Hence these elements were removed from the analysis as soon as they satisfied a stress based damage criteria based on the work of Hashin [11] and Puck [12] implemented through VUMAT in Abaqus/explicit. The details can be found in our previous work [13].

3.3. Fluid-structure coupling and shock wave loading

The fluid model consisted of the air inside and outside the shock tube as well as the air surrounding the plate as shown in Fig. 8. The air outside the tube was modelled in an Eulerian domain as a cuboid with a domain size of 400 mm × 400 mm in X-Y plane and 2000 mm along the tube axis. The model had 100 mm of air along the tube axis behind the plate to ensure that the plate remained in air during deformation caused by shock wave loading. The air inside the tube was also modelled in the Eulerian domain with the element size of 3 mm. All the fluid elements were meshed with the Eulerian eight-node, one-integration point hexahedral elements EC3D8R. The acoustic structural coupling between the fluid-mesh acoustic pressure and the CFRP panel structural displacements was accomplished with a surface-based tie constraint at their common surface. The master-slave type of contact was established between the annular surface of the shock tube in contact with the CFRP panel and the top surface of the panel. The surface of the external fluid at the interface was designated as the master surface. This pairing created an internal coupling of the acoustic pressure and structural displacements at 14 the surface nodes of CFRP panel (slave) and coupled the acoustic pressure exerted through the shock tube to the fluid mesh acoustic pressure at the interface. The incident wavefront was assumed to be planar. For a planar wave, which does not decay, two reference points, namely, the standoff point and the source point, were defined, the relative positions of which were used to determine the travel direction of the incident shock wave. The pressure history at the standoff point was used to drive the incident wave. The 'amplitude' definition [9] was used specifying the top surface of the CFRP panel to which the incident-wave loading was applied using a pressure-history curve. The entire analysis was divided into two steps pertaining to the wave incidence and reflection where appropriate average shock wave velocity and density was used. Linear fluid mechanics was used. The observed total pressure in the fluid was decomposed into two components: the incident wave itself, which was known, and the wave field excited in the fluid due to reflections at the fluid boundaries and interactions with the solid. The setup of the FE model of the shock loading in the CFRP panel with coupled fluid-structure domain is shown in Fig. 4.

The entire model consisted of about 6.5 million meshed elements. A mesh convergence study was carried out for both fluid and structure domains. The mesh adopted in this FE model represented the refinement level achieved after our mesh-convergence study. The computational time required was approximately 42 hours on 12 Intel Xeon processors (64 bit) using the high performance computing Hydra cluster, available at Loughborough University, UK.

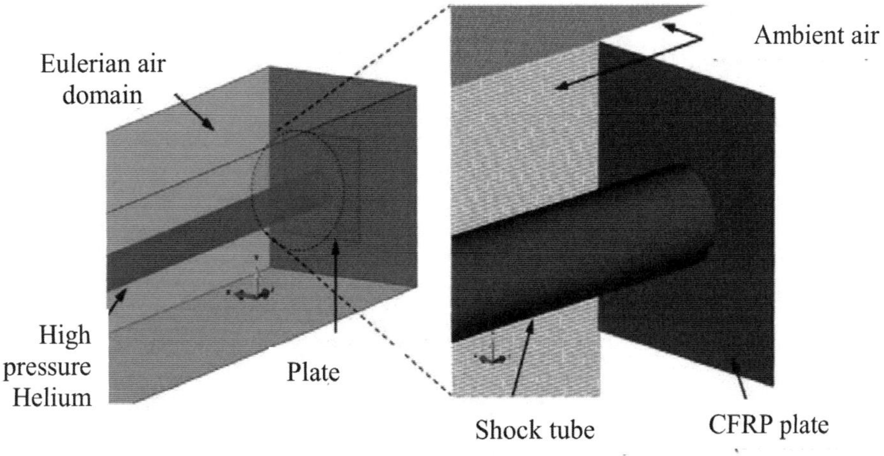

Figure 4. Fluid-structure coupling in FE model

4. Results and discussions

The results of FE model are discussed below. The FE analysis was focussed to determine the out-of plane deflection response of CFRP plates and their energy absorption capacity under blast loading. The local damage at the impacted location was also studied but is not discussed here for the brevity of space.

4.1. Modes of deflection in CFRP panels

From the FE analysis, it was observed that the deflection in the studied CFRP panels was the combined output of two deflection modes, namely, the indentation mode and the flexural mode. It was seen that all the panels started deflecting in the indentation mode. In flat panel (Panel A), the global flexural mode quickly took over and dominated the deflection process. This was also evidenced by the continuous nature of displacement contour that showed monotonic increase in deflection from the edge to the centre of the specimen after t = 200 μs. Deflection of Panel B continued in the indentation mode till about 400 μs after which it snapped into the global flexural deflection mode. These deflection contours showed continuous increase in deflection from the edge to the centre of panel and 16 the transition from elliptical contours back to the circular shape.

In Panel C, the deflection remained in the indentation mode since only the central loading region was affected. In addition to the out-of-plane deflection, velocities and in-plane strain data were also extracted from FE model at the centre point of the back of the studied CFRP panels. Fig. 4 shows that the deflection rate (35 m/s), for the initial 200 μs, was almost same in all the three panels, though Panels A and B attained a higher deflection as compared to Panel C. This showed that the Panel C was stiffer than the other two panels since it sustained higher pressure and had a lower deflection. Panel A and B had similar deflection trend till 1000 μs. At this time, damage was initiated in Panel B which explains the rapid rise in its deflection. The in-plane strain e_{xx} as well as e_{yy} were nearly same for studied panels at the centre point. The lower in-plane strain and out-of-plane deflection in Panel C showed that this panel had higher flexural rigidity. Panel B had higher in-plane shear strain which explained the catastrophic failure in the panel.

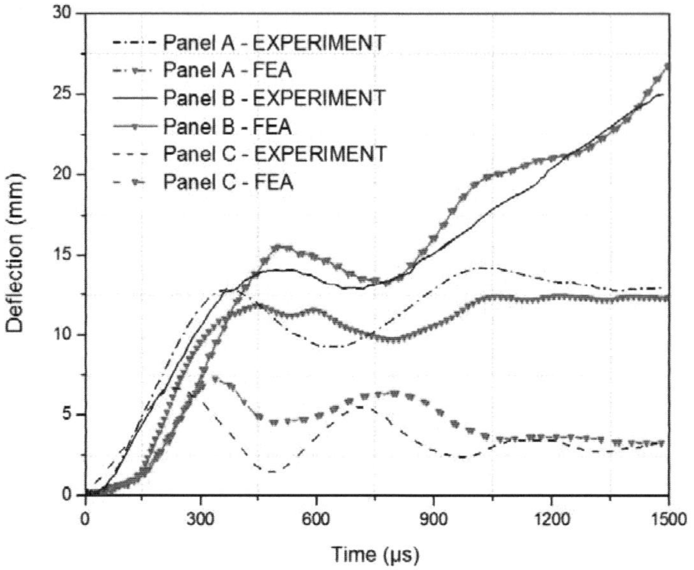

Figure 4. Out-of-plane deflection in CFRP plate: experimental and FE comparison

4.2. Damage dissipation during blast

The incident and remaining energies associated with the shock loading intensities were analysed with the FE model. The damage dissipation energy was obtained numerically. Panel C was subjected to the highest intensity of shock loading and so the incident energy was the highest for this panel, though the damage dissipation energy was the lowest, while it was highest for the Panel B, the Panel A showed intermediate response. The magnitude of the energy dissipated towards damage for studied panels is listed in Table 2.

Table 2. FE results showing magnitude damage dissipation energy for studied CFRP panels

Specimen	Energy dissipated towards damage in specimens (kJ)
Specimen A (Flat)	50
Specimen B (R = 305 mm)	62
Specimen C (R = 112 mm)	27

5. Conclusions

The effect of curvature of CFRP panel on their blast mitigation property was studied using a shock tube. The blast mitigation performance of these panels was characterised in terms of their out-of-plane deflection, in-plain stress. The FE model of blast loading on curved CFRP panels was developed and validated using 3D DIC data coupled with high-speed photography. This model accurately accounted for the interaction between the curved panels, shock tube and the surrounding air. The strain-rate sensitive material model enabled to capture the appropriate mechanical response of these panels under high load rates. The following conclusions were made:

1. There were two preliminary deformation modes with which the studied panels deflected under shock loading: flexural mode and indentation mode. Flexural deformation decreased and indentation deformation increased as the radius of curvature was reduced. Indention mode was found to be more severe since it led to the damage initiation in the panels.
2. The experimental analysis (macroscopic post-mortem analysis and DIC deflection, velocity, and in-plane strain analysis) showed that Panel C was capable of sustaining the highest threshold

failure load, though further optimisation study using validated FE model revealed the fact that Panel D (76 mm radius of curvature) performed better than Panel C reflecting about 90% of the incident shock energy.

References

[1] Rajendran R and Lee J M 2009 *Mar. Struct.* **22** 99-127
[2] Tekalur S A, Shivakumar K and Shukla A 2008 *Compos. Part B.* **39** 57-65
[3] Arora H, Hooper P A and Dear J P 2011 *Compos. Part A* **42** 1651-1662
[4] Kumar P, LeBlanc J, Stargel D S and Shukla A 2012 *Int. J. Impact Eng.* **46** 74-85
[5] Ochola R O, Marcus K, Nurick G N and Franz T 2004 *Compos. Struct.* **63** 455-467
[6] LeBlanc J and Shukla A 2011 *Compos. Struct.* **93** 3072-3081
[7] Wright J 1961 *Shock tubes.* New York: John Wiley and Sons Inc
[8] LeBlanc J, Shukla A, Rousseau C and Bogdanovich A 2007 *Compos. Struct.* **79** 344–55
[9] Kumar P, Stargel D S and Shukla A 2013 *Compos. Struct.* **99** 19-30
[10] Abaqus 6.11 2011 *User manual* Dassault System Rhode Island US
[11] Hashin Z 1986 *Eng. Fract. Mech.* **5** 771-778
[12] Puck A and Schürmann H 1998 *Compos. Sci. Tech.* 7 1045-67
[13] Phadnis V, Makhdum F, Roy A and Silberschmidt V V 2013 *Compos. Part A.* **47** 41-51

Ballistic impact behaviour of woven fabric composite: Finite element analysis and experiments

V A Phadnis[1,3], K S Pandya[2], N K Naik[2], A Roy[1] and V V Silberschmidt[1]
[1]Wolfson School of Mechanical and Manufacturing Engineering, Loughborough University, Loughborough, LE11 3TU, Leics, UK
[2]Aerospace Engineering Department, Indian Institute of Technology, Powai, Mumbai - 400076, India

E-mail: v.a.phadnis@lboro.ac.uk

Abstract. A mechanical behaviour of plain-weave E-glass fabric/epoxy laminate composite plate exposed to ballistic impact is studied using a finite-element (FE) code Abaqus/Explicit. A ply-level FE model is developed, where a fabric-reinforced ply is modelled as a homogeneous orthotropic elastic material with potential to sustain progressive stiffness degradation due to fiber/matrix cracking, and plastic deformation under shear loading. The model is implemented as a VUMAT user subroutine. Ballistic experiments were carried out to validate the FE model. A parametric study for varying panel thickness is performed to compare impact resistance of the studied composite.

1. Introduction

In recent years, woven-fabric polymer-matrix composites (PMCs) have been used increasingly in defence-related applications due to their higher energy absorption, high through-thickness stiffness and strength properties. Additionally, their shape and properties can be tailored to meet the needs of a specific application. High-performance fibers such as carbon, boron and Kevlar are of interest for military and aerospace applications of composites primarily due to their resilience at high temperatures and corrosion resistance in comparison to traditional structural materials.

In ballistic impact, PMCs retard a projectile by absorbing its kinetic energy through different mechanisms such as deformation, delamination, and shear between layers. The condition for perforation, also called the *ballistic limit velocity* (V_{50}) is the most important factor for design of a suitable protective structure [1]. In this regard, significant research has been carried out on the behavior of composite materials under ballistic impact loading. Few studies can be mentioned: Zhu *et al.* [2] investigated the response of woven Kevlar/polyester laminates of varying thicknesses to quasi-static and dynamic penetration by cylindroconical projectiles. Ballistic limits were also determined and terminal velocities measured. It was reported that deliberately introduced delamination and changes in the volume fraction of fibres did not result in significant changes in the impact resistance. Some studies revealed that a damage pattern for dynamic loading was, significantly different from that in the corresponding quasi-static penetration conditions. Cheng *et al.* [3] developed a model for high-

[3] Author to whom any correspondence should be addressed.

velocity impact of thick composites based on a continuum orthotropic constitutive behavior with stress-based failure criteria and a simplified degradation model of failure. The model was implemented into a hydrodynamic finite element code. Punching, fiber breakage, and delamination were the major energy-absorbing mechanisms of the penetration processes. Silva [4] carried out experiments to study effect of ballistic impact on Kevlar-29 impacted with simulated fragments. Numerical modelling was used to obtain an estimate for the limit perforation velocity (V_{50}) and simulate failure modes and damage. Good correlation between results of computational simulation and experimental data was reported, both in terms of deformation and damage of the laminates. Naik and Doshi [5] presented the ballistic impact behavior of typical woven fabric E-glass/epoxy thick composites analytically. It was reported that shear plugging was the major energy-absorbing mechanism.

Numerical models for predicting the performance of woven fabric composites have been the subject of interest for many years. However, due to the complexity involved in describing the response of fabric panels to ballistic impact, most models attempt to provide the most acceptable trade-off in performance analysis. To model a woven fabric down to a level of individual yarn crossovers would certainly be preferred in order to study frictional and crimping effects, but such studies are computationally impracticable. The complexity of the problem has forced researchers to accept certain simplifying assumptions to make it computable.

The work presented here focuses on the finite-element analysis of a ballistic impact response of plain-weave E-glass fabric/epoxy composite using the general-purpose finite-element software ABAQUS/Explicit [6]. The experimental details are discussed first, followed by a physics-based finite -element model that accounts for both delamination and in-plain failure of fabric-reinforced composite plies. Next, the obtained numerical results are compared with the experimental data and followed by conclusions.

2. Experimental studies

Experimental studies were carried out using a single-stage gas-gun-operated ballistic-impact test facility shown in figure 1. The material studied was commercially available plain-weave E-glass/epoxy fabric composite. The specimens were prepared in a way to obtain symmetric balanced cross-ply laminates of dimension 125 mm × 125 mm with thicknesses of 2.5, 3, 4.5 and 5 mm. The mass of the projectile of diameter 6.36 mm made of hardened steel was 6.42 g. Experimental studies were carried out on at least 6 specimens for each impact condition. Further experimental details are provided in [1].

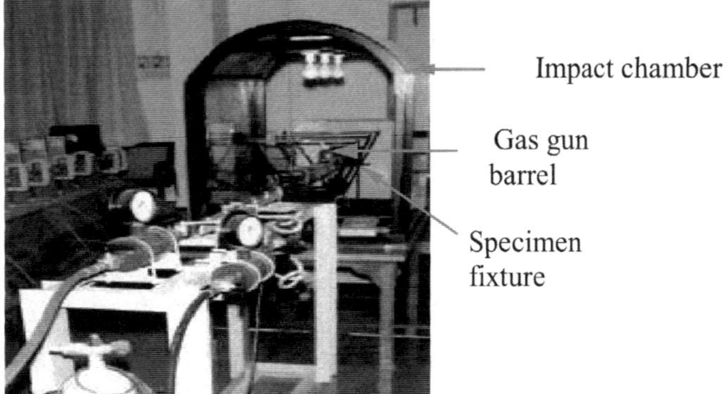

Figure 1. Ballisic impact test facility **[1]**

3. Constitutive model for fabric-reinforced composites

A constitutive material model representing the mechanical response of the plain-weave glass fabric/epoxy composite at ply level was developed. Each ply was modelled as a homogeneous orthotropic elastic material with the potential to sustain progressive stiffness degradation due to fiber/matrix cracking and plastic deformation under shear loading. The model was implemented as a material subroutine, VUMAT in ABAQUS [6]. Its main details are discussed below.

3.1. Elastic continuum-damage-mechanics model

The fibre directions of woven fabric reinforcement are assumed to be orthogonal. The constitutive stress-strain relations are formulated in a local co-ordinate system with its base vectors aligned with fibre directions. The in-plane elastic stress-strain relations are given according to orthotropic elasticity with damage [7].

$$\begin{Bmatrix} \varepsilon_{11} \\ \varepsilon_{22} \\ \varepsilon_{12}^{el} \end{Bmatrix} = \begin{pmatrix} \dfrac{1}{(1-d_1)E_1} & -\dfrac{\nu_{12}}{E_1} & 0 \\ -\dfrac{\nu_{21}}{E_2} & \dfrac{1}{(1-d_2)E_2} & 0 \\ 0 & 0 & \dfrac{1}{(1-d_{12})2G_{12}} \end{pmatrix} \begin{Bmatrix} \sigma_{11} \\ \sigma_{22} \\ \sigma_{12} \end{Bmatrix}, \qquad (1)$$

where $\varepsilon = \left(\varepsilon_{11}, \varepsilon_{22}, \varepsilon_{12}^{el}\right)^T$ is the elastic stain vector; $\sigma = \left(\sigma_{11}, \sigma_{22}, \sigma_{12}^{el}\right)^T$ is the stress vector; E_1 and E_2 are the Young's moduli in the principal orthotropic directions; G_{12} is the in-plane shear modulus; ν_{12} and ν_{21} are the principal Poisson's ratios; d_1 and d_2 are the damage parameters associated with the fibre fracture along the principal orthotropic directions; and d_{12} is the damage parameter associated with the matrix micro-cracking due to in-plane shear deformation. The damage parameters vary between 0 and 1 and represent the stiffness degradation caused by micro-damage in the material. The material properties were taken from [1] owing to the similarity of specimens.

3.2. Elastic-plastic shear model

The in-plane shear response is dominated by a non-linear behavior of the matrix that exhibits both stiffness degradation due to matrix micro-cracking and plasticity [8]. The elastic part of the shear response was calculated using (1). Here, the plastic shear response of the material is considered. The matrix response may be inelastic due to extensive cracking or plasticity. This leads to permanent deformations in the ply upon unloading. To account for these effects, a classical plasticity model with an elastic domain function and a hardening law, which is applied to the effective stresses in the damaged material, is used. The elastic domain function F is given by

$$F = \left|\tilde{\sigma}_{12}\right| - \tilde{\sigma}_0(\bar{\varepsilon}^{pl}) \leq 0 . \qquad (2)$$

The hardening law is of the form:

$$\tilde{\sigma}_0(\bar{\varepsilon}^{pl}) = \tilde{\sigma}_{y0} + C(\bar{\varepsilon}^{pl})^p , \qquad (3)$$

where $\tilde{\sigma}_{y0}$ is the initial effective shear yield stress; C and p are coefficients; and $\bar{\varepsilon}^{pl}$ is the equivalent plastic stain due to shear deformation. The vales of $\tilde{\sigma}_{y0}$, C and p are listed in Table 2.

4. Description of the finite element model

4.1. Intralaminar response of the plate

The in-plane response of the fabric plies was modelled using the constitutive model described. The elasticity constants include the Young's moduli in the fiber directions, the principal Poisson's ratio, and the in-plane shear modulus. The damage initiation coefficients account for tensile and compressive strengths along the fiber directions and shear strength at the onset of shear damage. The values of the user material constants and damage initiation coefficients are given in Table 1. The damage-evolution coefficients are represented by tensile and compressive fracture energies per unit area along the fiber directions and by parameters α_{12} and $d_{12\max}$ .We assume α_{12} = 0.15 and $d_{12\max}$ = 1, in our simulations. The fracture energies can be determined through the testing procedure described in [9].

Table 1. Material properties of plain-weave E-glass fabric/epoxy composite [1, 9, 11]

Parameter	Variable	Value
Longitudinal modulus (GPa)	E_{11}	12.5
Transverse moduli (GPa)	$E_{22} = E_{33}$	11.8
Principal Poisson's ratio	ν_{12}	0.169
Shear moduli (GPa)	$G_{12} = G_{23} = G_{31}$	3.5
Longitudinal tensile strength (MPa)	X_{1+}	322
Longitudinal compressive strength (MPa)	X_{1-}	204
Transverse tensile strength (MPa)	Y_{1+}	290
Transverse compressive strength (MPa)	Y_{1-}	200
In-plane shear strength (MPa)	S	39
Mode-I fracture toughness (J/m^2)	G_{IC}	504
Mode-II fracture toughness (J/m^2)	G_{IIC}	1566
Ultimate failure strain	$\varepsilon_{\max}$	4.2%

The critical values of the fracture energies along the fiber direction 2 were calculated so that the ratios G_f^{2+} / G_f^{1+} and G_f^{2-} / G_f^{1-} were approximately equal to the ratios Y_{2+} / X_{1+} and Y_{2-} / X_{1-}, respectively.

Table 2. Shear plasticity coefficients [11]

Parameter	Variable	Value
Initial effective shear yield stress (MPa)	$\tilde{\sigma}_{y0}$	150
Hardening coefficient in equation (3)	C	1125
Exponent in equation (3)	p	1.05

4.2. Interlaminar response of plate

In order to account for adhesive bonds between adjacent plies and model the delamination phenomenon, a generalized traction-separation law was used. Damage initiation at the interfaces was defined using a quadratic nominal-stress criterion, according to which damage was assumed to initiate when the quadratic interaction function involving the stress ratios reaches a value of one [10]:

$$\left[\frac{t_n}{t_n^0}\right]^2 + \left[\frac{t_s}{t_s^0}\right]^2 + \left[\frac{t_t}{t_t^0}\right]^2 = 1 , \qquad (4)$$

where $t = \{t_n, t_s, t_t\}$ is the nominal traction vector. The interface damage-initiation properties used in the present analysis are listed in Table 3. These properties are usually difficult to determine experimentally with good accuracy; therefore, they can be used as calibration parameters, if required. In the present work, the variation of the damage initiation parameters within 15% from the typical values given in Table 3 did not influence significantly the overall analysis results. It should be noted that experimental determination of material properties typically yields a variation of ~15%. Progression of damage at the interfaces was modelled using a critical mixed-mode energy behaviour based on the Benzeggagh–Kenane (BK) law [7]:

$$G_n^c + (G_s^c - G_n^c)\left\{\frac{G_S}{G_T}\right\}^{\eta} = G^c , \qquad (5)$$

where $G_S = G_s + G_t$ and $G_T = G_n + G_S$; G_n, G_s and G_t refer to the work done by the normal and the first and the second shear forces acting in the interface, respectively; G_n^c and G_s^c are critical fracture energies required to cause failure in normal and shear directions, respectively. G^c is the total critical mixed-mode fracture energy and η is the power coefficient taken to be 2.284. After damage initiation, its evolution until failure was modelled using exponential softening law [7].

Table 3. Interface damage-initiation properties

Parameter	Variable	V alue
Maximum nominal stress in normal mode (MPa)	t_n^0	35
Maximum nominal stress in shear directions (MPa)	$t_s^o = t_t^o$	65

4.3. Element deletion

In simulations, failed continuum shell elements were removed from the model to prevent excessive distortion and premature termination of the analysis. In the present work, damage-based element deletion is combined with a deformation-based deletion criterion. The damage-based element deletion was activated when any damage variable along the fiber directions or the equivalent plastic strain due to shear deformation reached a maximum specified value. The deformation-based deletion criterion was used when either the tensile or compressive principal logarithmic strain reached its maximum or minimum specified value, respectively. Additionally, detached composite fragments were deleted when they moved far away from the impact zone to prevent non-physical numerical distortions.

4.4 Loading and boundary conditions
A schematic of the developed FE model is shown in figure 2. Both woven fabric plate and bullet were modelled as 3D continuum deformable solids. The dynamic explicit solver was used in simulations to account for the time-dependent loading and the complex interaction between the bullet and the fabric composite plate. A 3 mm thick symmetric cross-ply laminate in this FE model consisted of 12 plies with an individual ply thickness of 0.25 mm. The local co-ordinate systems were defied to account for orientations of individual plies and to model the laminate and material behaviour precisely. The cylindrical bullet of mass 6.42 g and length 25.3 mm impacts the centre of the workpiece in the axial direction using a velocity boundary condition. The details are listed in Table 4.

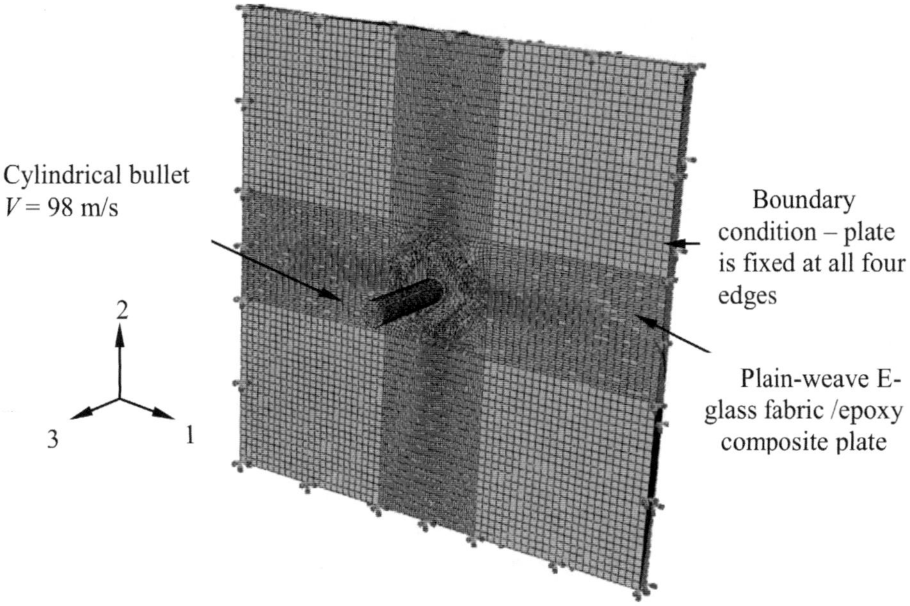

Figure 2. FE model: bullet impact on plain weave E-glass fabric/epoxy composite plate

A parametric study was carried out to analyse the effect of plate thickness on the ballistic limit velocity. The mechanical properties and geometry of the bullet used earlier simulations were kept constant. The plate thicknesses studied are listed in Table 4.

4.5 Finite elements and mesh sensitivity
In the rectangular composite plate, each ply was discretised with eight-node linear brick elements with reduced integration (C3D8R) through its thickness in the vicinity of the impact zone. Mesh-sensitivity analysis is important in simulations involving high deformations with a non-linear material response. Thus, a rigorous mesh-sensitivity study was carried out to obtain a computationally accurate finite-element mesh. In the current study all results are presented based on simulations performed with an optimised mesh. A computing clock time was reduced by introducing different mesh sizes in distinct regions of the FE model. A planar mesh size of 0.25 mm × 0.25 mm (in 1–2 plane) in the vicinity of

the impacted area was used; while a coarser mesh of 1.25 mm × 1.25 mm was employed in the area away from the zone of interest.

Table 4. Loading and boundary conditions

Material studied	Plain-weave E-glass fabric composite (symmetric balanced cross-ply), thickness : 3mm
Parametric study	Plate thickness : 2.5 mm, 4.5 mm and 5 mm
Impactor	Cylindrical bullet (mass : 6.42 g, length : 25.3 mm)
Impactor velocity, V_{50} (m/s)	98 m/s
Boundary condition	All four edges of plate fixed

Localised stiffness reduction due to internal damage can cause excessive element distortion that could lead to difficulties in numerical convergence. To resolve this numerical issue, 'distortion control' was used in ABAQUS, and the damage variables were limited to a maximum value of 0.999. At each ply interface, cohesive elements of type COH3D8 with a thickness of 10 μm were embedded and used to model delamination initiation and growth with the failure criterion discussed in Section 4.2. The degradation parameters were set to 0.99, and failed cohesive elements were removed from the FE model, once the failure criteria were satisfied.

It was observed in the wave-stability study that cohesive zone elements govern stability of the solution (due to near-zero element size) with a very low stable time increment of the order of 10^{-9} s which affected the overall solution run time and, hence, a selective mass-scaling technique was used [10]. The density of cohesive-zone elements was artificially increased by a factor of 25 so that the total mass of the laminate remained practically unaffected. The bullet was meshed with one-integration-point, hexahedral elements of type C3D8R.

The model consisted of 575000 finite elements and analysis was completed in two and half hours using eight parallel processors on Intel Core i7 CPU with 8 GB RAM.

5. Simulation results and discussion

The effect of increase in the thickness of composite plate on the ballistic limit velocity (V_{50}) is shown in figure 3 and the magnitudes of V_{50} are listed in Table 5.

Table 5. Magnitude of ballistic limit velocity (V_{50}) at varying composite plate thickness

Thickness of composite plate	Ballistic limit velocity (V_{50})	
(mm)	FE Analysis	Experiment
2.5	73.5	72
3	98	98
4.5	118.5	118
5	130.5	130

It can be observed form Table 5 that, the increase in plate thickness by 66% (considering t = 3 mm as the base plate thickness) resulted in a higher ballistic limit increase by 33%, while decrease in plate thickness by 16% caused reduction in V_{50} by 25%.

It was found that the kinetic energy absorbed by the composite plate during impact did not increase linearly with increase in the plate thickness, thus the primary reason of this increase in the ballistic limit can be attributed to the prolonged contact. It can be safely assumed that at high loading rates, as normally observed in ballistics, the glass fabric undergoes considerable hardening, before failure – thus providing extended resistance.

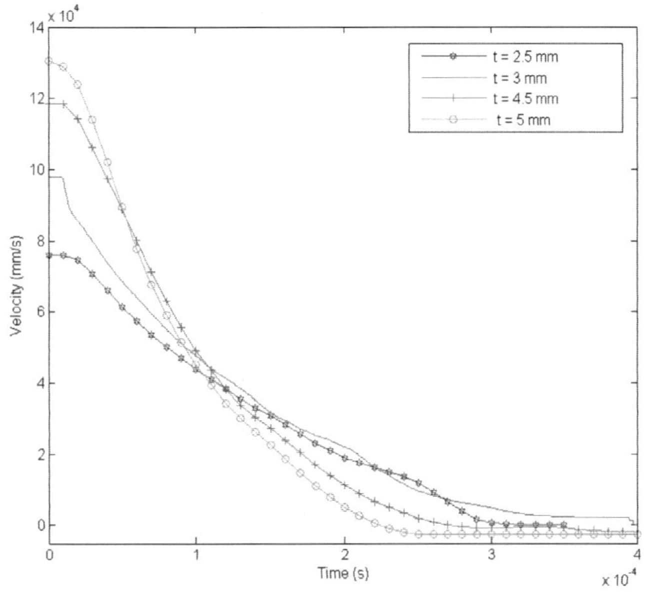

Figure 3. Effect of composite plate thickness on ballistic limit velocity (V_{50})

6. Conclusions

The ballistic impact behaviour of plain-weave E-glass fabric/epoxy composite was studied experimentally and numerically. A ply-level 3D finite element model was developed to analyse the response of the plate of this composite impacted by a cylindrical bullet. This FE model was validated with the experimental results and later used to predict the ballistic limit velocity for varying plate thicknesses. A good correlation was obtained with the experimental data emphasising the effectiveness of this model to predict the mechanical response of woven fabric composites under high-velocity impact loading.

Acknowledgement

The authors are grateful to the Royal Society, UK and Department of Science and Technology, Government of India for the partial support of this study.

References

[1] Pandya KS, Pothnis JR, Ravikumar G and Naik NK 2013 *Mater. Design.* **44** 128-135
[2] Zhu G, Goldsmith W and Dharan CKH 1992 *Int. J. Solids Struct.* **29** 399–420
[3] Cheng WL, Langlie S and Itoh S 2003 *Int. J. Impact Eng.* **29** 167–84
[4] Silva M 2005 *Int. J. Impact Eng.* **31** 289–306
[5] Naik N and Doshi A 2008 *Compos. Struct.* **82** 447–64
[6] Abaqus 6.11 2011 *User manual* Dassault System Rhode Island US
[7] Maimí P, Camanho PP, Mayugo JA and Dávila CG 2006 *A thermodynamically consistent damage model for advanced composites.* Technical Report NASA/TM-214282 Langley US
[8] Sokolinsky VS, Indermuehle KC and Hurtado JA 2011 *Compos. Part A.* **42**(9) 1119-1126
[9] Pinho ST, Robinson P and Iannucci L 2006 *Compos. Sci. Tech.* **66** 13
[10] Phadnis VA, Makhdum F, Roy A and Silberschmidt VV 2013 *Compos. Part A.* **47** 41-51
[11] Ullah H, Harland AR and Silberschmidt VV 2012 *Comput. Mater Sci.* **64** 130-135

Damage and fracture in fabric-reinforced composites under quasi-static and dynamic bending

H Ullah, A R Harland and V V Silberschmidt[1]

Wolfson School of Mechanical and Manufacturing Engineering, Loughborough University, Leicestershire, LE11 3TU, UK

E-mail: V.Silberschmidt@lboro.ac.uk

Abstract. Fabric-reinforced polymer composites used in sports products can be exposed to different in-service conditions such as large deformations caused by quasi-static and dynamic loading. Composite materials subjected to such bending loads can demonstrate various damage modes - matrix cracking, delamination and, ultimately, fabric fracture. Damage evolution in composites affects both their in-service properties and performance that can deteriorate with time. Such behaviour needs adequate means of analysis and investigation, the main approaches being experimental characterisation and non-destructive examination of internal damage in composite laminates. This research deals with a deformation behaviour and damage in carbon fabric-reinforced polymer (CFRP) laminates caused by quasi-static and dynamic bending. Experimental tests were carried out to characterise the behaviour of a CFRP material under large-deflection bending, first in quasi-static and then in dynamic conditions. Izod-type impact bending tests were performed on un-notched specimens of CFRP using a Resil impactor to assess the transient response and energy absorbing capability of the material. X-ray micro computed tomography (micro-CT) was used to analyse various damage modes in the tested specimens. X-ray tomographs revealed that through-thickness matrix cracking, inter-ply and intra-ply delamination such as tow debonding, and fabric fracture were the prominent damage modes both in quasi-static and dynamic test specimens. However, the inter-ply damage was localised at impact location in dynamically tested specimens, whereas in the quasi-static specimens, it spread almost over the entire interface.

1. Introduction

Composite materials have found many applications in aerospace, automotive, medical and construction components and structures due to their better specific strength and stiffness, excellent fatigue strength, good corrosion behaviour and low thermal conductivity. One specific class of composites - woven fabric-reinforced laminates - offer a good resistance to fracture and transverse rupture and high impact strength compared to their unidirectional and cross-ply tape counterparts [1]. These properties have attracted the sports industry to incorporate woven CFRP laminates in the design of sports products. Expanding the application away from the traditional aerospace structures results in

[1] Author to whom any correspondence should be addressed.

new types of loading regimes such as large-deflection bending experienced in sports products during their service, such regimes were rarely studied. New types of quasi-static and dynamic loads generate high local stresses, leading to complex damage modes, due to heterogeneity and anisotropy of composite laminates. The damage mechanisms typically caused by out-of-plane quasi-static and dynamic bending loads are matrix cracking, fibre breakage and delamination at interfaces within the composite structure [2]. Evolution of these intralaminar and interlaminar damage modes results in significant reduction of in-service mechanical properties and leads to a loss of structural integrity of composite sports products with time [3]. Such internal damage mechanisms, which often cannot be detected by visual inspection, degrade the load-bearing capacity of the structures. Therefore, it is important to study damage suffered by composites under quasi-static and dynamic loading conditions through X-ray micro-computed tomography (micro-CT), which can provide 3D images of internal realisation of deformation mechanisms and their interaction in composites with sufficient resolution, to investigate these internal damage mechanisms. This relatively recent technique has been used to investigate damage mechanisms in composites at micron length scale [4-6]. Therefore, in this study, damage imposed in composites by the new types of loading regimes such as large-deflection quasi-static and dynamic bending experienced by sports products in service is investigated using micro-CT.

The large-deflection behaviour of woven laminates under quasi-static bending have been previously studied by the authors [3, 4, 7, 8]. Most of these studies were based on material characterization using experimental tests and numerical simulations. Similarly, the low velocity impact response of woven-fabric composite laminates has been extensively studied in the literature [9-11]. However, most of the studies were dedicated to the impact behaviour of composites in drop-weight tests, which usually cause localised damage such as penetration and perforation. A large-deflection dynamic bending behaviour of composites caused by a pendulum-type impactor was rarely investigated. Although such an instrumented impactor was used by Silberschmidt *et al.* [12] and Casas-Rodriguez *et al.* [13] to study damage in adhesively bonded CFRP joints under repeated impacts, the loading mode was tensile. Recently, the large-deflection dynamic bending behavior of CFRP laminates was studied in [14, 15], using experimental tests and finite-element analyses. One of the latest study conducted by the authors revealed that the damage formation under such dynamic bending conditions was rather different from that observed under drop-weight impact tests [16]. In the present work, which is an extension of the previous studies, formation of various damage mechanisms under quasi-static and dynamic loading regimes is investigated at microstructural level using X-ray micro-CT technique. Quasi-static tests were performed under three-point bending conditions, whereas dynamic tests were carried out using a pendulum type impact tester at various energy levels. The type and location of damage mechanisms in woven CFRP laminates under both types of loadings was identified and compared.

Table 1. Material properties of twill 2/2 woven CFRP

E_{11} (GPa)	E_{22} (GPa)	E_{33} (GPa)	G_{12} (GPa)	$G_{13}=G_{23}$ (GPa)	ν_{12}	Density ρ (kg/m^3)
55.0	52.0	8.0	3.8	3.0	0.05	1470

2. Experimental methods

2.1. Material

The materials studied were laminates of woven fabric made of carbon fibres reinforcing thermo-plastic polyurethane (TPU) polymer matrix. The material was manufactured from 0°/90° prepregs in the form of four plies designated as $[0°,90°]_{2s}$, where 0° and 90° represent yarns in the warp and weft directions, respectively. The woven laminate had a 2/2 twill balanced weaving pattern; the fabric had the same number of yarns in the warp and weft directions. Quasi-static bending tests were performed at various speeds on rectangular specimens of 80 mm length and 25 mm width with a span-to-thickness ratio of 40 according to the ASTM D790 standard [17]. Un-notched rectangular specimens of 40 mm length and 25 mm width were prepared for dynamic tests according to the ASTM D4812 standard [18]. The elastic constants of CFRP determined from the quasi-static tests are presented in Table 1; details of the tests for material characterisation are presented elsewhere [7].

2.2. Experimental testing

Quasi-static tests were performed on CFRP specimens at indenter speeds of 100, 200, 300 and 400 mm/min under conditions of three-point large-deflection bending till their ultimate fracture. The flexural stress-strain diagrams of CFRP specimens are shown in Figure 1. Dynamic impact tests were carried out according to the ASTM D4812 standard on an instrumented pendulum type CEAST Resil impactor. In the impact tests, one (lower) part of the specimen was fixed firmly in the machine vice as a cantilever beam. The upper part with a length of 30 mm of the specimen was hit by the striking nose of the pendulum hammer with a controlled level of initial energy, resulting in dynamic large-deflection bending. The distance between the fixed support and the line of contact of the hammer's striking nose was 22 mm, according to the standard. In this work, a calibrated impact hammer with a mass of 0.6746 kg and a length of 0.3268 m was used. CFRP specimens were tested at energy levels of 0.4J, 0.5 J and 0.6 J to determine the energy inducing ultimate fracture of the specimens. It was found that the specimen fractured at 0.6 J, corresponding to the initial angle of 64°. Typical records of force *vs.* time for un-fractured and fractured CFRP specimens are presented in Figure 2.

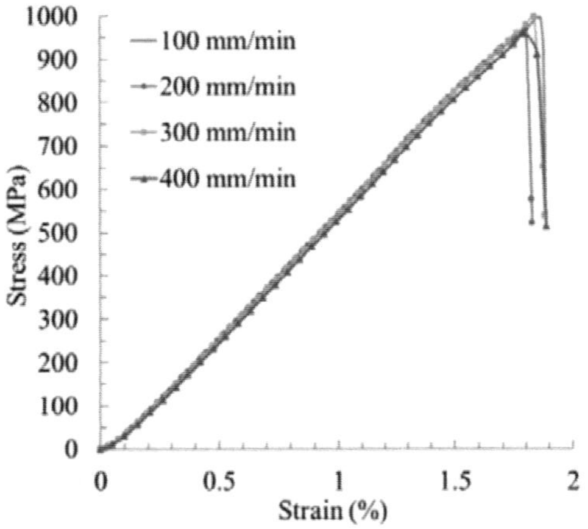

Figure 1. Flexural stress-strain response of CFRP laminates in quasi-static tests at various displacement rates

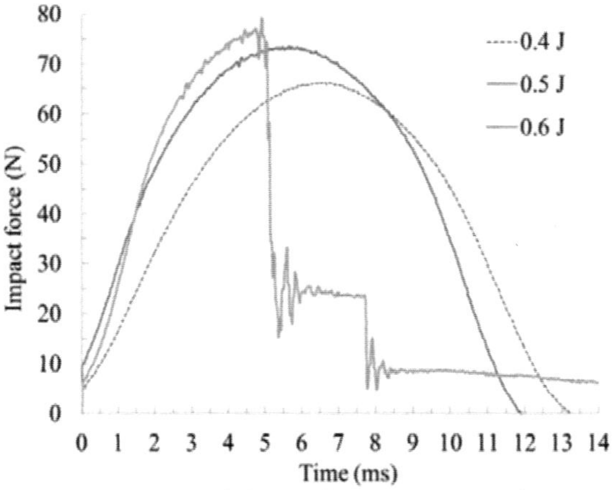

Figure 2. Typical force-time response of CFRP laminates in dynamic bending tests at various energy levels

3. Damage characterisation

Damage such as delamination and matrix cracking usually occurs inside composites laminates and is barely visible. A non-destructive evaluation method, micro computed tomography (micro-CT), was used to visualise the 3D microstructure and internal damage of the tested composite laminates. In this study, micro-CT measurements were performed using an XT H 225 X-ray scanner. The system consists of an X-ray detector and an electronic X-ray source, creating 2D cross-sections of the object. The source is a sealed X-ray tube operating at 25–225 kV with a 3 μm spot size. Following acquisition of tomographic data of the specimen, a software program builds a precise 3D map from 2D radiograph images by 'stacking' the individual slices one on top of the other; this process is known as *reconstruction*. As denser materials absorb more X-rays than voids and air, this attenuation contrast allows detection and characterisation of cracks and flaws in tomographic images. A small sample with dimensions of 30 mm x 7 mm was prepared from the damaged region of the CFRP specimen tested under quasi-static bending. Additionally, two samples were prepared - one from the fracture region and another from the impact region of the fractured CFRP laminate tested at 0.6 J in dynamic bending. Dimensions of the fractured and impact samples were 16.4 mm (length) x 8.8 mm (width) and 20.6 mm (length) x 10.2 mm (width), respectively; all the samples had thickness of 1 mm. The data for those samples was collected at 75 kV and 80 μA. Those settings resulted in tomographs with a resolution of 12.7 μm and 14.7 μm for the fractured and impact specimens, respectively.

4. Discussion of results

Results of experimental tests and micro-CT scanning of woven CFRP laminates subjected to quasi-static and dynamic bending are presented in this section. In quasi-static tests as shown in Figure 1, the on-axis specimens exhibited a quasi-brittle response because the applied load was carried by the fibres, which were stronger but brittle than the TPU matrix. Hence, the stress-strain curves were almost linear till ultimate failure, represented by a sudden drop in the stress level. The material also showed strain-rate insensitive behaviour, since carbon fabric-reinforced composites are essentially strain-rate independent in the fibre-dominated modes as elaborated in [16]. The dynamic response of the material at various energy levels is presented in Figure 2. At the energy level of 0.4 J, the loading and unloading parts of the curve have a symmetric parabolic shape suggesting that the respective stages during the contact duration were almost the same, and no significant damage occurred. As the energy level was increased to 0.5 J, the response became unsymmetrical. The oscillations in the load-time response before the peak load represent delamination damage. The impact force at energy of 0.6 J, causing the specimen's ultimate fracture, is higher. Here, higher impact energy induced larger deformation and therefore, a larger impact force. The load *vs.* time plot in Figure 2 shows oscillations due to significant damage in the specimen before its ultimate bending fracture. The latter is represented by a sudden drop in the contact force, implying a momentary loss of contact between the impactor and specimen.

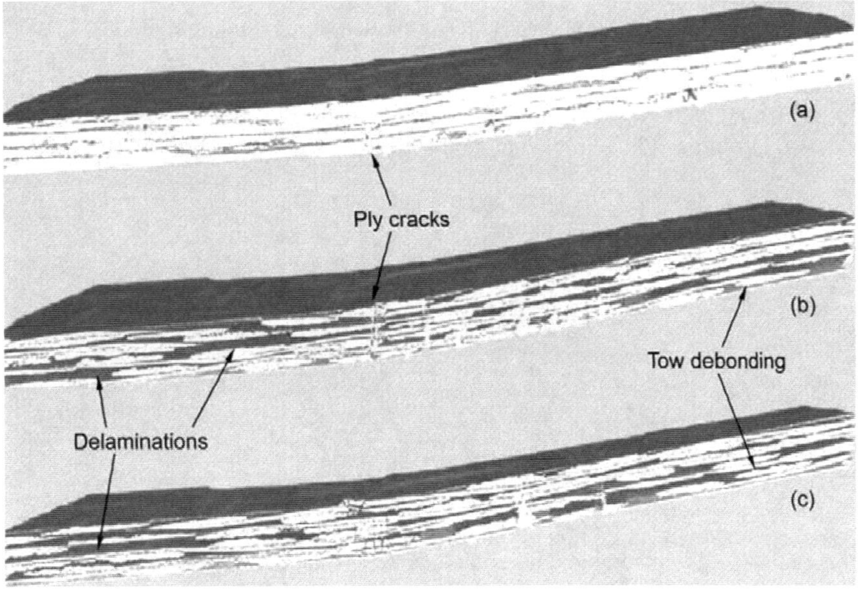

Figure 3. Damage mechanisms at various locations across width of CFRP specimen tested under quasi-static bending (resolution 11 μm): (a) edge; (b) 50% of width; (c) 75% of width

Realization of damage mechanisms captured with micro-CT analysis of the CFRP specimen subjected to quasi-static loading is shown in Figure 3, at various locations across the sample's width. Dark grey regions in the images represent cracks and damage whereas light grey regions represent a higher density material i.e. carbon-fibre yarns. Apparently, before ultimate fracture, the laminate exhibited matrix cracking, and then delamination in the form of tow debondings. Matrix cracks developed in the weak resin-rich pockets around the tows. Inter-ply and intra-ply delamination (tow debonding) can also be observed. Examination of the internal structure showed that almost every ply was delaminated at the time of fabric fracture. The prominent failure modes were tow debonding, delaminations and, subsequently, transverse ply fracture. The reconstructed 3D images at the bending (fractured) location of the transversely fractured CFRP specimen under dynamic loading are shown in Figure 4. Realisation of damage mechanisms at the outer edge, 50% and 75% of the sample width is shown in Figures 4a, b, and c, respectively. Similarly, tomographs presented in Figure 5 show matrix cracking and delamination at the hammer impact location along the specimen's height. The X-ray CT images showed that before its ultimate fracture, the laminate exhibited matrix cracking and then delaminations and tow debondings. At the time of fabric fracture, as the analysis of the internal structure revealed, almost every ply was delaminated. All the tomographs showed that matrix cracking

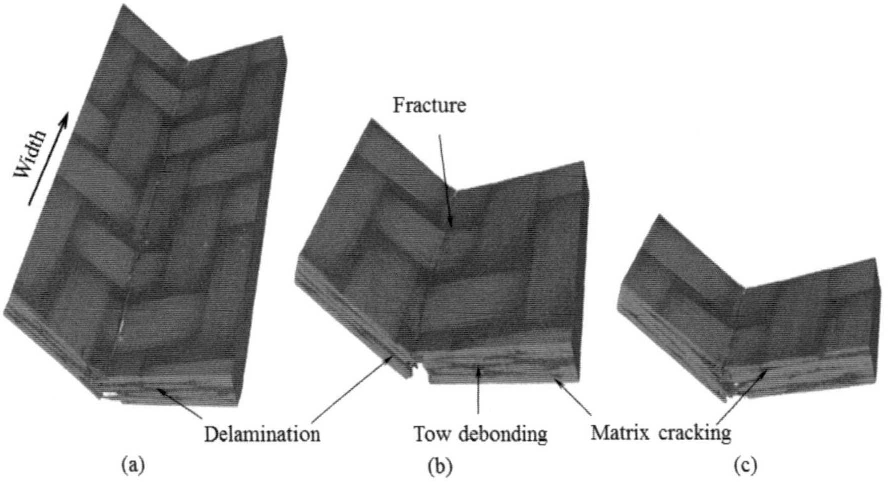

Figure 4. Reconstructed 3D images of dynamically tested CFRP specimen at failure location across width of sample (resolution 12.7 μm): (a) edge; (b) 50% of width; and (c) 75% of width

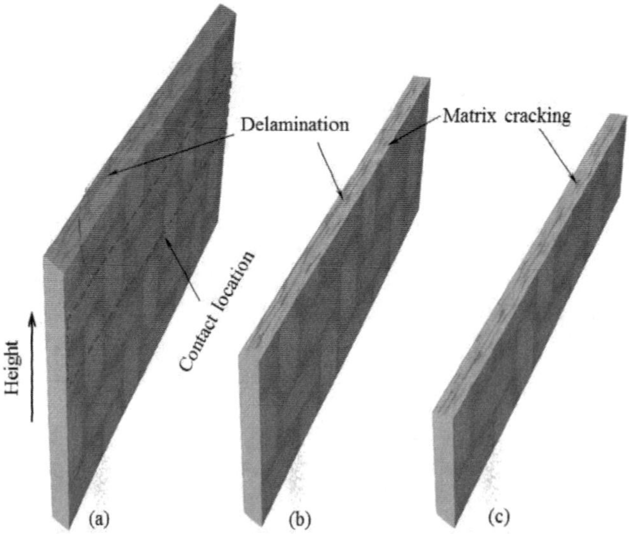

Figure 5. Reconstructed 3D images of twill 2/2 CFRP specimen at impact location across height of sample (resolution 14.7 μm): (a) edge; (b) 50% of height; and (c) 75% of height

and inter-ply delamination are the prominent damage modes at the specimen's impact location, whereas at the bending (fractured) location, these modes are coupled with tow debonding and, subsequently, transverse ply fracture. As evident from Figures 3 and 5, the major modes of damage are the same under quasi-static as well as dynamic loading conditions. However, the interlaminar damage is more localized at the impact region under dynamic loading, whereas in quasi-static loading, these damage modes spread along the length of the specimen.

5. Conclusions

The large-deflection bending behaviour of woven CFRP laminates under quasi-static and dynamic bending was studied using experimental tests and micro-CT scanning. Quasi-static bending tests were carried out to characterise the material's flexural behavior. Large-deflection dynamic bending tests were performed to investigate the dynamic behaviour and energy absorbing capability of the material. The microstructural analysis was used to observe different damage modes in the impacted specimens with micro-CT, providing a detailed picture of the damage modes and interaction among different failure mechanisms at various locations. It was observed that damage modes such as matrix cracking and delamination were prominent at the impact location, whereas these modes were coupled with tow debonding and fabric fracture at the bending location. Both types of tests resulted in almost similar types of dmage mechanisms. However, the spread of damage in specimens subjected to quasi-static loading was more than the localised one in the impact tested samples. The microstructural examination formed a basis for identification of interlaminar and intralaminar damage locations for computational models of woven composite laminates.

References

[1] Aktaş M, Atas C, İçten B M and Karakuzu R 2009 *Compos. Structures* **87** 307-313
[2] Menna C, Asprone D, Caprino G, Lopresto V and Prota A 2011 Int. J. Impact Engng **38** 677-685
[3] Ullah H, Harland A R, Lucas T, Price D and Silberschmidt V V 2012 *Comput. Mater. Sci.* **52** 147-156
[4] Ullah H, Harland A R and Silberschmidt V V Applied Composite Materials, 19 (2011) 769-783.
[5] Scott A, Mavrogordato M, Wright P, Sinclair I and Spearing S, 2011 *Compos. Sci. Technol.* **71** 1471-1477
[6] Tsigkourakos G, Silberschmidt V V and Ashcroft I A 2012 *Shock Vibrat.* **19** 573-584
[7] Ullah H, Harland A R and Silberschmidt V V 2012 *Comput. Mater. Sci.* **64** 130-135
[8] Ullah H, Harland A R, Lucas T, Price D and Silberschmidt V V 2011 *J. Phys. Conf. Ser.* **305** 12045
[9] Naik N, Chandra Sekher Y and Meduri S 2000 *Compos. Sci. Technol.* **60** 731-744
[10] Evci C and Gülgeç M 2012 *Int. J. Impact Engng* **43** 40-51
[11] Sutcliffe M P F, Monroy Aceves C, Stronge W J, Choudhry R S and Scott A E, 2012 *Compos. Structures* **94** 1781-1792
[12] Silberschmidt V V, Casas-Rodriguez J P and Ashcroft I A 2009 *Key Engng Mater.* **399** 71-78
[13] Casas-Rodriguez J P, Ashcroft I A and Silberschmidt V V 2008 *Compos. Sci. Technol.* **68** 2663
[14] Ullah H, Abdel-Wahab A A, Harland A R and Silberschmidt V V *J. Phys. Conf. Ser.* **382** 012015
[15] Ullah H, Harland A R and Silberschmidt V V 2012 *DYMAT 2012 - 10th Int. Conf. on the Mechanical and Physical Behaviour of Materials under Dynamic Loading*, EPJ Web of Conferences, Freiburg, Germany, pp. 04004
[16] Ullah H, Harland A R and Silberschmidt V V 2013 *Compos. Structures* **101** 144–156
[17] *ASTM D790 – 10: Standard Test Methods for Flexural Properties of Unreinforced and Reinforced Plastics and Electrical Insulating Materials* (2010)
[18] *ASTM D4812 – 06: Standard Test Method for Unnotched Cantilever Beam Impact Resistance of Plastics* (2006)

Numerical analysis of dynamic out-of-plane loading of nonwovens

E Demirci[1], F Farukh[1,3], M Acar[1], B Pourdeyhimi[2] and V V Silberschmidt[1]

[1]Wolfson School of Mechanical and Manufacturing Engineering
Loughborough University, UK
[2]Nonwovens Cooperative Research Center
North Carolina State University, Raleigh, NC, USA

E-mail: F.Farukh@lboro.ac.uk

Abstract. This paper presents finite element (FE) modelling of deformation behaviour of thermally bonded bicomponent fibre nonwovens under out-of-plane dynamic loading. Nonwoven fabric was treated as an assembly of two regions with distinct mechanical properties. Bond points were treated as composite material having a matrix of the sheath material reinforced with fibres of the core material. Elastic-plastic and viscous properties of the constituent fibres, obtained with tensile and relaxation tests were implemented into the FE model. The mechanical behaviour of the material under out-of-plane dynamic loading was observed with visual techniques. The deformation behaviour of nonwoven under out-of-plane dynamic loading computed with the numerical model was compared with that observed in the tests.

1. Introduction

Nonwovens are used in wide range of consumer and industrial products and applications ranging from disposable items such as wipes and diapers to durable products, such as, geotextiles [1]. Being engineered fabrics, properties of nonwovens can be tailored to serve specific purposes. Regardless of the application areas, structural integrity of nonwovens is the most basic requirement so that they can perform their intended functions under service conditions.

This study focuses on thermally bonded nonwoven materials with polymer-based bicomponent fibres. These bicomponent fibres have a core/sheath structure, with the core material having higher melting point than that of the sheath. During thermal bonding of such fibres, the temperature of the hot calender is adjusted in such a way that only the sheath material melts and forms the bond spots whereas core part of fibres remains intact. On the other hand, the regions where the fibres are not in contact with the calendar remain unaffected and form the fibrous matrix [2]. The structure of resulting thermally bonded nonwoven obtained with scanning electron microscopy (SEM) is shown in Fig. 1. The two regions — bond points and fibre matrix — have distinct characteristics which define unique but complex deformation behaviour of nonwovens. Additionally, this complex behaviour is affected by a specific orientation distribution of fibres and their visco-elasticplastic properties of constituent fibres. Several studies were performed to study the response of the material under mechanical loading [2-10]. The developed finite-element models predicted the deformation behaviour of the fabric

[3] Author to whom any correspondence should be addressed

effectively. Still, these studies focused only on in-plane quasi-static loading neglecting dynamic effects related to time-dependent deformation mechanisms. None of these studies deal with the deformation behaviour of the fabric under dynamic out-of-plane loading.

The susceptibility to deform under out-of-plane dynamic loading is one of the major concerns for various applications of nonwovens such as filters and scaffolds. This paper's aim is to introduce a practical way to simulate a 3D behaviour of thermally bonded nonwoven fabrics under dynamic out-of-plane loading in order to reduce costly and time-consuming trial-and-error stages in product development.

2. Assessment of properties

2.1. Network structure

A thermally bonded nonwoven with planar density of 50 g/m^2 manufactured with PP/PE 75/25 bicomponent fibres is the object of our case study. Information about the bond point size, shape and pattern was obtained from SEM images (Fig. 1).

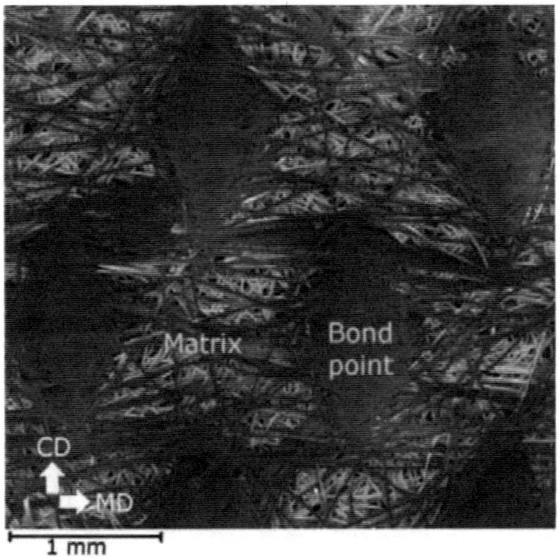

Figure 1. SEM image of 50 g/m^2 thermally bonded nonwoven fabric (PP/PE 75/25)

The fabric was composed of two regions — bond points and fibre matrix— that have different microstructures. Bond points were solid and continuous whereas the fibre matrix was porous (Fig. 1). Due to such microstructure, mechanical properties of these two regions were different. Moreover, it was observed from SEM images that fibres were not fully randomly oriented but preferentially aligned along the machine direction (MD) (Fig. 1), making the mechanical behaviour of bond points and fibre matrix anisotropic. In order to assess the level of anisotropy in mechanical properties of the nonwoven regions, quantification of non-uniformity in orientation distribution of fibres was performed with the use of in-house software — Nonwoven Anisotropy V1— described in [5]. The orientation distribution function (ODF) of fibres obtained from the software is given in Fig. 2.

2.2. Mechanical properties

The mechanical properties of regions of nonwoven were obtained in terms of orthotropic parameters defining direction-dependency of their behaviour. These orthotropic parameters, having three symmetry planes, were sufficient to define the level of anisotropy for thermally bonded nonwovens having three principal directions, named *machine direction* (MD), *cross direction* (CD) and *thickness direction* (TD). Orthogonal parameters for MD and CD were calculated based on ODF using the following equations:

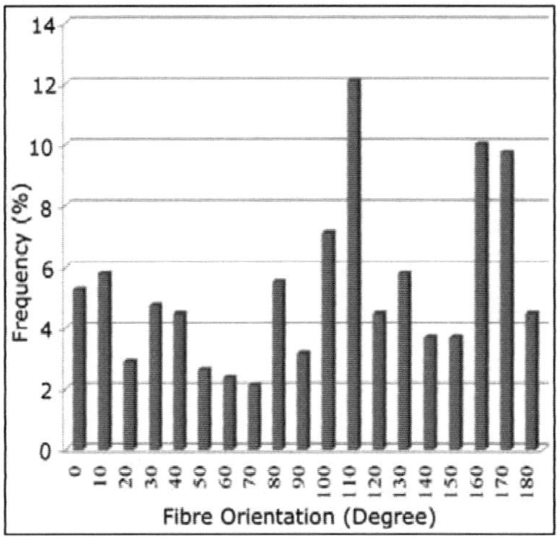

Figure 2. ODF of 50 g/m^2 thermally bonded nonwoven fabric (PP/PE 75/25)

$$C_{\mathrm{MD}} = \frac{\sum_{i=1}^{N} |\sin \alpha_i|}{\sum_{i=1}^{N} |\sin \alpha_i| + \sum_{i=1}^{N} |\cos \alpha_i|}, \tag{1}$$

$$C_{\mathrm{CD}} = \frac{\sum_{i=1}^{N} |\cos \alpha_i|}{\sum_{i=1}^{N} |\sin \alpha_i| + \sum_{i=1}^{N} |\cos \alpha_i|}, \tag{2}$$

where C_{MD} and C_{CD} are the parameters defining the level of orthotropy in MD and CD; α_i is the angle between the axis of the ith fibre and CD; N is the total number of fibres accounted by the ODF algorithm. After computing the orthotropic constants, elastic, plastic and viscous properties of bond points and matrix were assessed using mechanical properties of constituent fibres, obtained experimentally, and manufacturing parameters of nonwoven fabric obtained from manufacturer's datasheet [11]. Elastic-plastic properties of bond points and fibre matrix were computed using the Rule of Mixture (RoM) for MD and CD whereas the Halpin-Tsai equations were employed to calculate the mechanical properties along TD. Dynamic properties were determined by normalization of time-dependent behaviour of single fibre used to manufacture the fabric. The details of single-fibre tests and computation of mechanical properties are given elsewhere [2]. The computed flow curves and relaxation moduli of the constituting regions of nonwovens obtained from the algorithm are shown in Fig. 3.

3. Fabric's behaviour

The nonwoven fabric under study, being composed of polymer fibres with viscous properties, demonstrated a time-dependent behaviour. In order to assess the dynamic effects related to its time-dependent deformation behaviour under out-of-plane dynamic loading, drop-weight testing was performed (Fig. 4). The metal sphere ball with a mass of 43.4 g and diameter 22 mm was released from a 100 mm height to fall freely under gravitational acceleration onto the fabric. The ball was consecutively bounced back several times with a gradual decrease in the bouncing height after each impact due to energy-dissipation mechanisms related to viscous properties of the nonwoven. As a result of the impact of ball, the nonwoven material was exposed to dynamic out-of-plane loading. The maximum heights of the ball after each bounce with respective times are shown as experimental data in Fig. 6. These data for each bounce was recorded with a camera and observed visually.

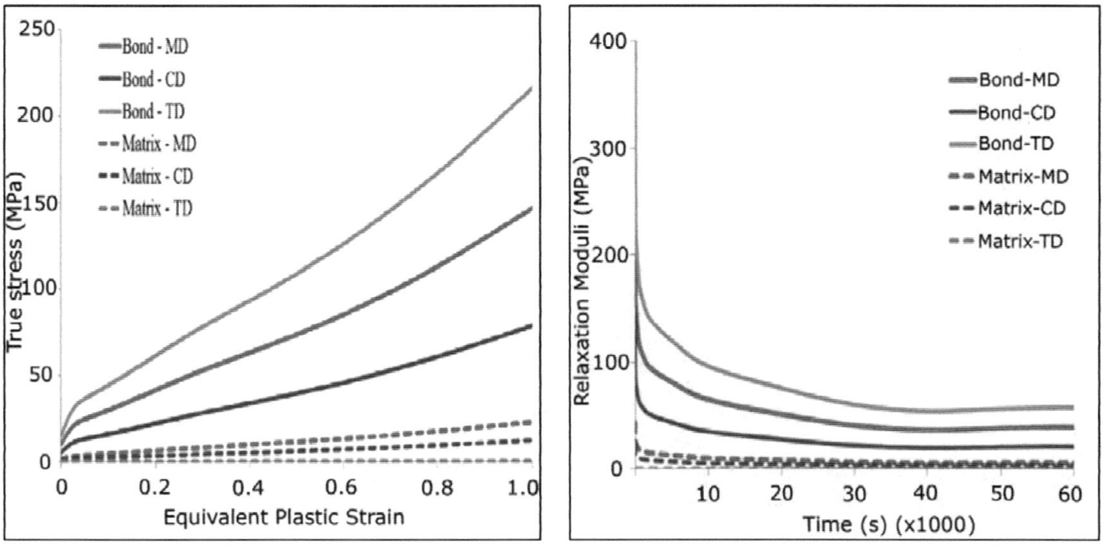

(a) **(b)**

Figure 3. Flow curves and relaxation moduli of 50 g/m^2 (PP/PE 75/25) nonwoven regions

4. Finite-element model

The FE model for the nonwoven under study was composed of two distinct regions—bond points and matrix. The material properties of these regions obtained with the discussed algorithm were used to characterise their orthotropic viscoelastic-plastic behaviour in the FE model. The simulations were carried out using commercial software MSC.Marc®. The model with dimensions 200 mm × 200 mm, equal to that of the fabric in our experiments, was prepared (Fig. 5) using 88500 bi-linear four-node thin shell elements. The bond-point shape, size and pattern of the model were similar to that of the fabric. Since the stiffness of the metallic sphere was much higher than that of the fabric, it was modelled as a rigid body with weight. To allow comparability of simulation results with experimental ones, the boundary conditions implemented in the FE model were similar to those used in experiments. The size and dropping distance of the rigid sphere were the same as in the experiments. A constant gravitational acceleration of 9.812 m/s^2 was applied to the rigid-body sphere, which was released from its initial steady-state position. The edges of the modelled fabric were fully constrained as in our tests (Fig. 4).

5. Results and discussions

A drop-weight simulation was performed with exact specimen size, dimensions, and boundary conditions to replicate the experiments. Stress distributions in the fabric's model as a result of four successive impacts obtained for the moments of maximum deformation are given in Fig. 6.

In order to compare the response of the model with experiments alone, evolution of height of the sphere obtained in the simulation and its maximum height, recorded at each bounce in the experiments, were compared (Fig. 5). The heights in the experiment and simulation correspond to the distances between the initial fabric plane and the metal sphere. The results obtained with the FE model were in good agreement with the experimental ones confirming the success of the model to predict the time-dependent behaviour of the fabric under dynamic out-of–plane loading. The average difference in heights obtained from simulation was less than 20% as compared to experimental data. This time-dependent behaviour is manifested by the changing bouncing height of the metal sphere after each successive impact. An interesting observation here is that the stress distributions in the fabric under loading extended more along MD than CD due to fabric's microstructure. Besides, the stress magnitudes decreased with each successive bounce showing energy dissipation by the fabric due to its time-dependent properties. The values of energy dissipation by the fabric obtained from experiment and simulation as a percentage of initial energy are given in Table 1.

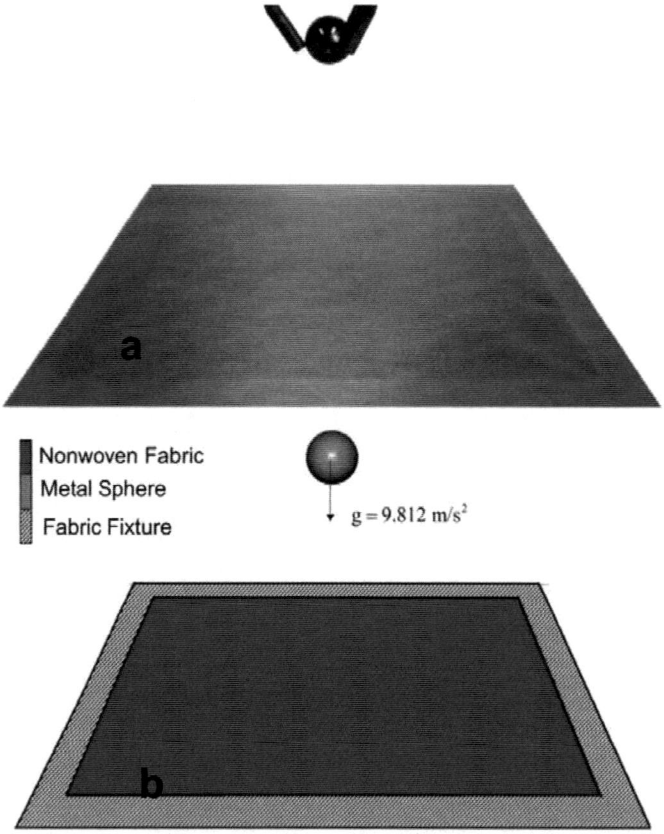

Figure 4. (a) Test setup; (b) corresponding FE model

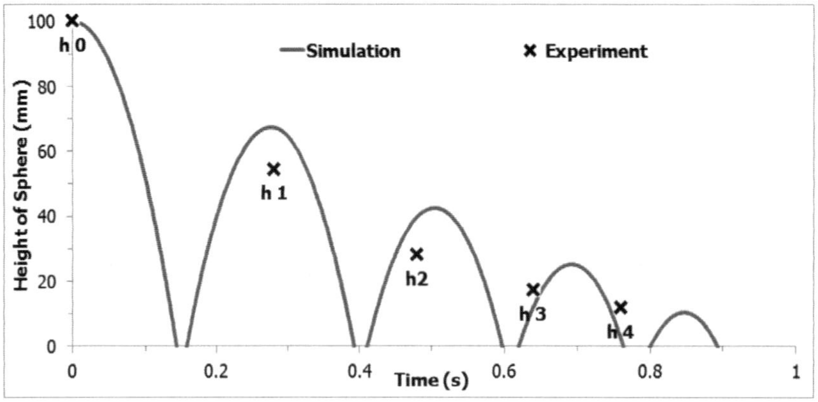

Figure. 5. Height of sphere in experiment and FE simulations after each impact

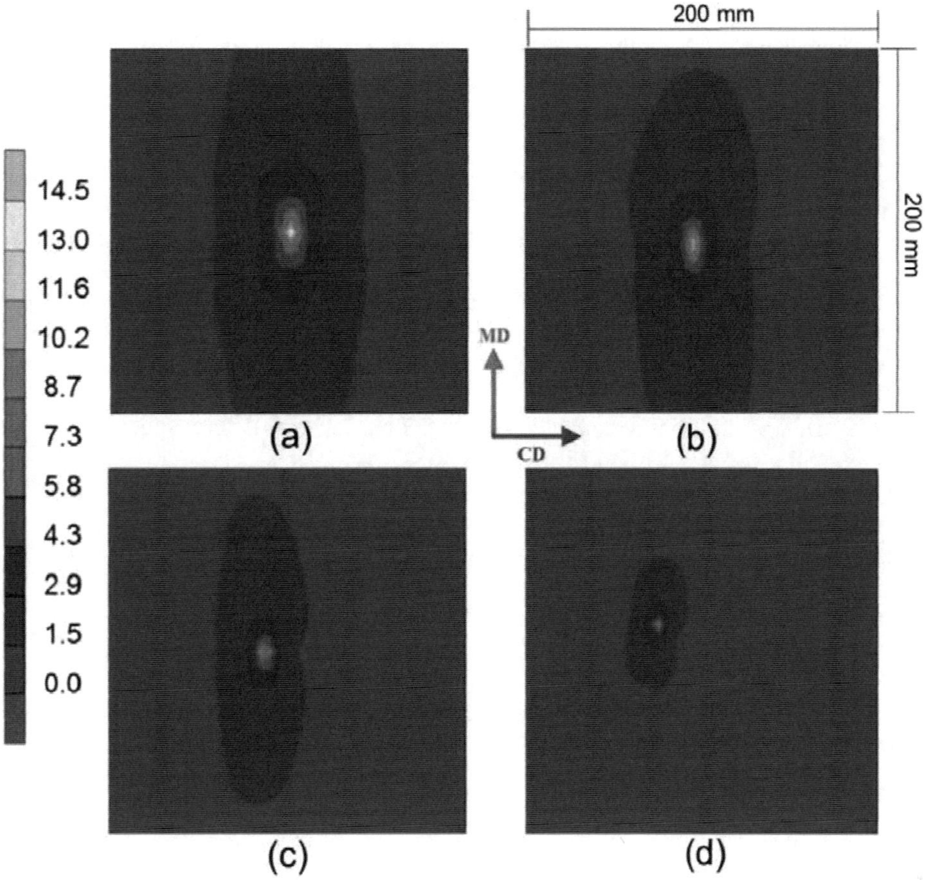

Figure 6. Distribution of equivalent von Mises stress (in MPa) of deformed nonwoven in model at first (a), second (b), third (c) and fourth (d) impacts

Table 1. Percentage of initial energy dissipated in each bounce

	1st bounce	**2nd bounce**	**3rd bounce**	**4th bounce**
Experiment	45.63	26.32	10.79	5.24
Simulation	32.67	24.53	17.2	15.8

6. Conclusions

In this paper, a practical way of simulating the real-life behaviour of a thermally bonded bicomponent fibre nonwoven under dynamic loading is introduced. A preferential orientation distribution of fibres and polymer-based constituents led to a complex mechanical behaviour of the fabric, which was modelled by employing two distinct regions with different mechanical properties. A straight-forward method for developing the model of nonwoven was introduced, based on two steps: assessment of fabric's properties and FE modelling. Anisotropic material properties for the bond points and fibre matrix implemented into the model were computed using the developed algorithm. The novel numerical model proposed in this study is effective in predicting the mechanical behaviour of thermally bonded nonwoven fabrics under dynamic out-of-plane loading including time effects, as was verified with experiments. Since the model accounts for anisotropy and constituent material's properties, it is expected that determination and implementation of damage criteria into it can increase its capability to predict damage behaviour of the fabric as well.

Acknowledgement
We greatly acknowledge the support by the Nonwovens Cooperative Research Centre at North Carolina State University, Raleigh, USA.

References
[1] Farukh F, Demirci E, Acar M, Pourdeyhimi B and Silberschmidt V V 2012 *J. Phys.:Conf. Ser.* **382** 012018
[2] Demirci E, Acar M, Pourdeyhimi B and Silberschmidt V V 2011 *Comput. Mater. Sci.* **50** 1286-1291
[3] Demirci E, Acar M, Pourdeyhimi B and Silberschmidt V V 2010 *ASME Conf. Proc; ESDA* **2010(49187)** 117-122
[4] Demirci E, Hou X, Acar M, Pourdeyhimi B and Silberschmidt V V 2012 *Solid State Phen.* **188** 164-169
[5] Farukh F, Demirci E, Sabuncuoglu B, Acar M, Pourdeyhimi B and Silberschmidt V V 2012 *Comput. Mater. Sci.* **64** 112-115
[6] Farukh F, Demirci E, Sabuncuoglu B, Acar M, Pourdeyhimi B and Silberschmidt V V 2013 *J. Mater. Sci.* **48**(6) 2334-2345
[7] Farukh F, Demirci E, Sabuncuoglu B, Acar M, Pourdeyhimi B and Silberschmidt V V 2013 *Comput. Mater. Sci.* **71** 165-171
[8] Hou X, Acar M and Silberschmidt V V 2009 *Comput. Mater. Sci.* **46** 700-707
[9] Hou X, Acar, M and Silberschmidt V V 2011 *Comput. Mater. Sci.* **50** 1292-1298
[10] Hou X, Acar M and Silberschmidt V V 2011 *J. Mater. Sci.* **46** 307-315
[11] Demirci E, Acar M, Pourdeyhimi B and Silberschmidt V V 2012 *Comput. Mater. Sci.* **52** 157-163

A multiscale constitutive model for intergranular stress corrosion cracking in type 304 austenitic stainless steel

A Siddiq[1,2] and S Rahimi[1]

[1] Advanced Forming Research Centre (AFRC), 85 Inchinnan Drive, Glasgow, UK
[2] Department of Design, Manufacture and Engineering Management, University of Strathclyde, Glasgow, UK
Email: muhammad.amir@strath.ac.uk

Abstract. Intergranular stress corrosion cracking (IGSCC) is a fracture mechanism in sensitised austenitic stainless steels exposed to critical environments where the intergranular cracks extends along the network of connected susceptible grain boundaries. A constitutive model is presented to estimate the maximum intergranular crack growth by taking into consideration the materials mechanical properties and microstructure characters distribution. This constitutive model is constructed based on the assumption that each grain is a two phase material comprising of grain interior and grain boundary zone. The inherent micro-mechanisms active in the grain interior during IGSCC is based on crystal plasticity theory, while the grain boundary zone has been modelled by proposing a phenomenological constitutive model motivated from cohesive zone modelling approach. Overall, response of the representative volume is calculated by volume averaging of individual grain behaviour. Model is assessed by performing rigorous parametric studies, followed by validation and verification of the proposed constitutive model using representative volume element based FE simulations reported in the literature. In the last section, model application is demonstrated using intergranular stress corrosion cracking experiments which shows a good agreement.

1. Introduction

Austenitic stainless steels due to their superior corrosion resistance, excellent mechanical properties and good weldability have consistently been used in power generation plants and oil and gas industries [1–3]. Intergranular stress corrosion cracking (IGSCC) is a dominant failure mechanism in sensitized austenitic stainless steels subjected to critical environments where intergranular cracks extend along susceptible grain boundaries. Sensitization refers to the precipitation of chromium carbides at grain boundaries which is accompanied by the depletion of chromium in the vicinity of these regions and therefore susceptibility to IGSCC [4–7]. The susceptibility to sensitization and the kinetics of sensitization are strongly dependent on the geometry and crystallographic structure of the boundaries [8]. Grain boundary structure has commonly been classified in the framework of coincidence site lattice (CSL) model, which attributes a Σ (sigma) notation to each grain boundary. This describes the orientation relationship between the crystal lattices of the adjoining grains in the grain boundary and can be obtained using two-dimensional electron back scattered diffraction (EBSD) technique [9]. Based on the CSL model grain boundaries are often classified into two groups of low sigma ($\Sigma \leq 29$) and random ($\Sigma > 29$) or high angle grain boundaries. High angle grain boundaries ($\Sigma > 29$) that are frequently been referred to as random boundaries have higher susceptibility to sensitization than those with low sigma [10,11]. It has been observed that under similar mechanical loading and environment conditions, intergranular cracks have tendency to propagate more slowly in microstructure with higher fraction of low Σ ($\Sigma \leq 29$) boundaries [12]. From the correlation between intergranular crack path and boundary characters it can be deduced that random grain boundaries are more prone to chromium carbide precipitation while low Σ boundaries show more resistance [8,12,13]. Generally, coherent twin boundaries with a {111} habit pane have been found to be the only resistant feature to sensitisation phenomenon [8,13].

Several predictive methods have been developed to define the relations between intergranular crack propagation and the grain boundary properties. These predictive approaches determine the probability if crack arrest and distribution of arrested crack lengths for a given microstructure. These models can be categorized into percolation based models [10,11,14–30] and mechanics based modeling approaches [31–35]. Most of these predictive models were either analytical or FE based which require high amount of computational power and time.

In this study, a multiscale constitutive model has been presented for intergranular stress corrosion crack propagation in metals with emphasis on type 304 austenitic stainless steel. The presented model is computationally robust as it is based on homogenization technique and will save ample amount of computational time and resources. The proposed constitutive model takes into account the material's microstructure character distribution (GBCD) such as grain size and fraction of boundaries with different characteristics obtained by EBSD, and predicts the maximum intergranular crack length. The constitutive model assumes each grain to be a two phase material comprising of grain interior and grain boundary zone. Grain interior is modeled using crystal plasticity theory while grain boundary zone is modeled by proposing a phenomenological constitutive model motivated from cohesive zone modeling approach. Overall, the response of representative volume is calculated by volume averaging the individual grain behavior. In the first part of this paper, a brief outline of the constitutive model is discussed. In the second section, the validations and verification of the proposed constitutive model is presented by comparing with representative volume element based FE simulations reported in the literature. In the last section, model application is demonstrated using post-mortem stress corrosion experiments.

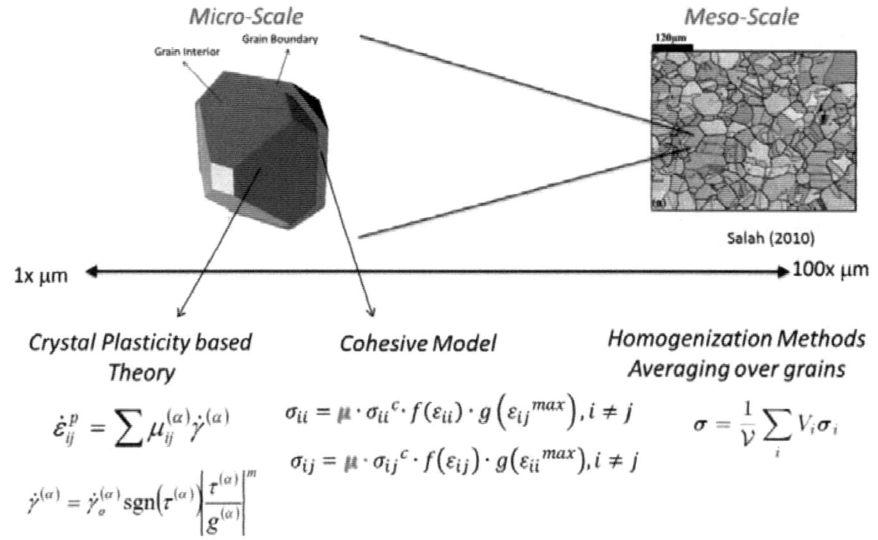

Figure 1. Schematic of multiscale constitutive model showing how scales are bridged.

2. Constitutive model

A brief outline of the constitutive model is discussed in the following. A detailed description of the model and implementation will be reported later in a full journal article. The grain structure is assumed to be truncated octahedron or Kelvin cell with 14 faces comprising of 6 hexagons and 6 tetragons. Each individual grain is considered as a two-phase material consisting of grain interior and the grain boundary zone as a thin shell (4 μm) encompassing the grain interior. Grain interior is modeled using crystal plasticity theory [36–42] while the grain boundary zone is modeled using a phenomenological constitutive model motivated from cohesive zone modeling approaches (please see references there in [36–42]). Schematic of the constitutive model is shown in Figure 1.

In the following we only present details of the calculation of overall response of the grain and a brief description of the models for both grain interior and grain boundary zone.

The stress conjugate to strain before fracture initiation is given by

$$\boldsymbol{\sigma} = \xi \boldsymbol{\sigma}_{gi} + (1 - \xi)\boldsymbol{\sigma}_{gb}$$

where ξ is the volume fraction of the grain interior region and is given by

$$\xi = \frac{V_{gi}}{V_{gb}}$$

where σ_{gl} and σ_{gb} are the stresses in the grain interior and grain boundary zone. V_{gl} and V_{gb} are volume fractions of grain interior and grain boundary zone, respectively.

Each phase, i.e. grain interior and grain boundary zone are assumed to have same deformation gradient

$$\boldsymbol{F} = \boldsymbol{F}_{gb} = \boldsymbol{F}_{gi}$$

Once the crack is initiated stresses are computed using the following relation

$$\boldsymbol{\sigma} = \xi \cdot \frac{e^{1-\varsigma \cdot tanh\left(\frac{\varepsilon_{ij} - \mu\varepsilon_{ij,1}}{\mu(\varepsilon_0 - \varepsilon_{ij,1})}\right)^2}}{2.718} \cdot \boldsymbol{\sigma}_{gi} + (1 - \xi)\boldsymbol{\sigma}_{gb}$$

The first term of the above equation accounts for unloading of the grain interior instantly after the initiation of a crack, i.e. when the crack propagation starts. Where μ and ζ can be estimated from experiments. ζ controls the damage evolution and μ links directly the rise in normalised impurity concentration with drop in strength and strain to failure.

Polycrystalline response is computed using classical Taylor type averaging method. The homogenized stress response is computed through arithmetic averaging of overall grains.

$$\bar{\boldsymbol{F}} = \boldsymbol{F}_1 = \ldots = \boldsymbol{F}_n \qquad \bar{\boldsymbol{P}} = \frac{1}{\mathcal{V}} \sum_{i=1}^{N} V_i \boldsymbol{P}_i$$

Constitutive model has the capability to use microstructure character distribution data obtained by EBSD such as mean grain size and the grain orientation relationships in the form of Euler angles.

Grain interior is modelled using the following hardening model [36–42]

$$h_{\alpha\alpha} = \left\{ (h_o - h_s) \sec h^2 \left[\frac{(h_o - h_s)\gamma^\alpha}{\tau_s - \tau_o} \right] + h_s \right\}$$

$$h_{\alpha\beta} = q h_{\alpha\alpha} \qquad (\alpha \neq \beta)$$

As can be seen the above equation requires 4 parameters of h_o, h_s, τ_o, τ_s to be identified from experiments.

The constitutive relation for grain boundary zone is given in Figure 1. In the following, we present the list of expressions for the constitutive model without going into the detail of derivations.

The relationships for $f(\varepsilon_{ij})$ and $g(\varepsilon_{ij})$ are given below.

Before grain boundary crack initiation the following equation is dominant

$$f(\varepsilon_{ij}) = 2\left(\frac{\varepsilon_{ij}}{\mu \cdot \varepsilon_{ij,1}}\right) - \left(\frac{\varepsilon_{ij}}{\mu \cdot \varepsilon_{ij,1}}\right)^2$$

while instantly after grain boundary crack initiation the following relationships are dominant.

$$f(\varepsilon_{ij}) = \frac{e^{1-\varsigma \cdot tanh\left(\frac{\varepsilon_{ij}-\mu \cdot \varepsilon_{ij,1}}{\mu \cdot (\varepsilon_0-\varepsilon_{ij,1})}\right)^2}}{2.718} \qquad \varepsilon_{ij,1} < \varepsilon_{ij} < \varepsilon_0$$

$$g(\varepsilon_{ij}) = 2\left(\frac{\varepsilon_{ij}}{\mu \cdot \varepsilon_0}\right)^3 - 3\left(\frac{\varepsilon_{ij}}{\mu \cdot \varepsilon_0}\right)^2 + 1$$

where

$\varepsilon_{ij,1} = Strain\ at\ \boldsymbol{\sigma_c}$ $\qquad \varepsilon_0 = \varepsilon_{N0} = Normal\ strain\ at\ complete\ failure$

$\varepsilon_0 = \varepsilon_{T0} = Shear\ strain\ at\ complete\ failure$

All simulations presented in the following are based on the code developed in the research group using FORTRAN language and all the simulations are based on local model. Total number of grains used in the model are 200. Grain misorientation was randomly distributed using experimental EBSD data reported in [46] and is not repeated have for brevity.

3. Parameter identification for type 304 austenitic stainless steel

The material data used in this analysis is a mill annealed type 304 austenitic stainless steel (UNS30400) plate with a chemical composition of (wt.%) 18.15Cr–8.60Ni–0.45Si–1.38Mn–0.055C–0.032P–0.038N–0.005S. The constitutive model requires several parameters to be determined experimentally. These are uniaxial tensile test data along with the microstructure character distribution data such as mean grain size and grain orientations. The tensile properties were obtained according to the ASTM standard for sheet, strip, flat wire and plate [43].

The HKL-EBSD system used was interfaced to a Philips XL-30 FEG-SEM. Channel 5 software was used to characterize the data, applying Brandon's criterion in the CSL model framework [44,45] to define grain boundary character. A 15° threshold was used to identify high angle grain boundaries (HAGB). A minimum misorientation angle of 2° was implemented to distinguish low angle boundaries. Grain boundaries having $\sum \leq 29$ were considered as resistant boundaries, with the low angle grain boundaries (LAGBs) ($\sum 1$) included. Information on the details of the tensile experiments and microstructure characterizations are published elsewhere [46,47].

The materials mechanical and microstructure data reported in [44] and [47] are for fully sensitized austenitic steel at room temperature. Since the experiments were performed in air the μ parameter is set to 1, i.e. no degradation of properties in the absence of corrosive environment. The identified set of parameters is listed in Table 1 and a comparison of stress-strain data is shown in Figure 2.

Table 1. Parameters identified through uniaxial tension test data on Austenitic steel at room temperature.

1(a) Grain Interior

h_0 (MPa)	τ_0 (MPa)	τ_s (MPa)	h_s (MPa)
10.	10.5	24.8	0.1

1(b) Grain Boundary

d_{gb} (mm)	N_{grain}	σ_c (MPa)	ε_0	$\varepsilon_{ij,1}$	ζ
0.005	200	1100	0.3	0.05	2.

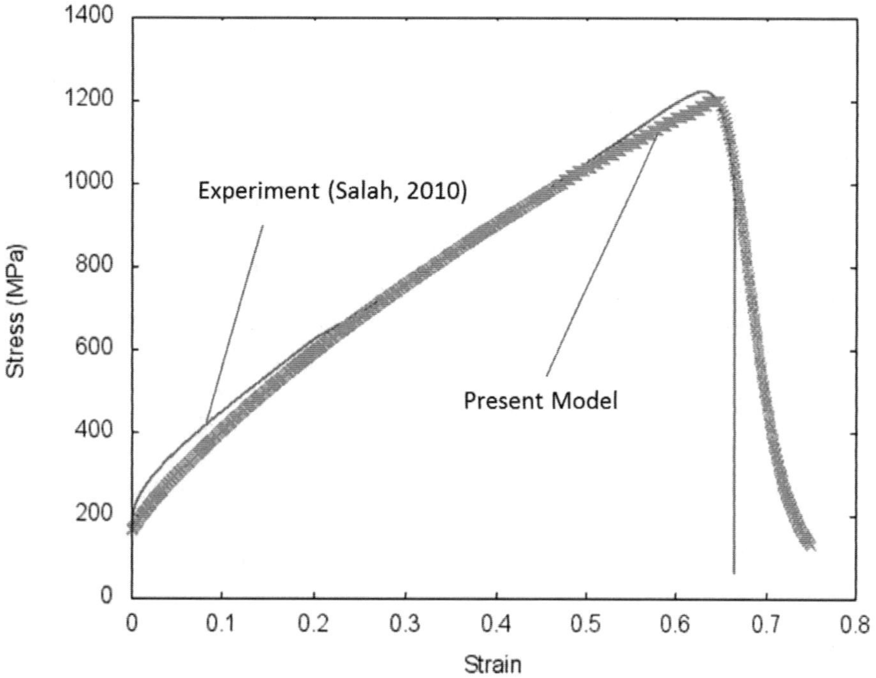

Figure 2. Comparison between uniaxial experimental [46] and simulated response of austenitic steel at room temperature

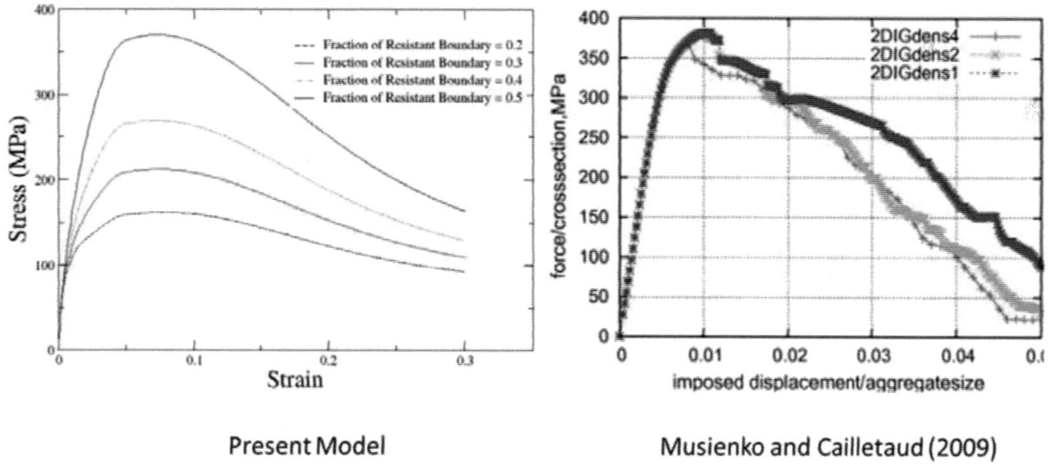

Figure 3. Effect of resistant grain boundaries number fraction (left; present model) and FE based RVE study reported in the literature (Musienko and Cailletaud [35]).

4. Model capability and validation

One of the features defined in the constitutive model is the incorporation of grain boundary characters based on their $\sum$ value, i.e. resistant and susceptible grain boundaries as has been discussed in section 3. Using these fractions of resistant and susceptible grain boundaries, constitutive model randomly distributes the grain boundaries in the representative volume of the microstructure being considered in the constitutive model. For the results discussed below parameters used are the same as Table 1 while for μ a constant value of 0.1 is used.

To show the model capability of capturing the effect of fraction of resistant and susceptible grain boundaries, uniaxial tests were performed in the corrosive environment by assuming the value of μ equals to 0.1 for susceptible grain boundaries and 1 for resistant grain boundaries. It can be inferred from Figure 3 (left) that as the fraction of resistant grain boundary increases the failure curves shift up, i.e. material is more resistant to failure. To qualitatively compare the overall stress-strain response, finite element based RVE results on a similar material are plotted from Musienko and Cailletaud [35] which shows a very similar stress-strain response. To analyze the effect of fraction of resistant grain boundaries on crack growth rate, crack fraction as a function of time has been plotted in Figure 4 (left). It can be inferred from Figure 4 (left) that as the fraction of resistant grain boundary increases crack growth rate decreases. To qualitatively compare the overall crack growth trend, FE based RVE study reported in literature (Musienko and Cailletaud [35]) is shown in Figure 4 (right) showing overall trend in crack growth is similar.

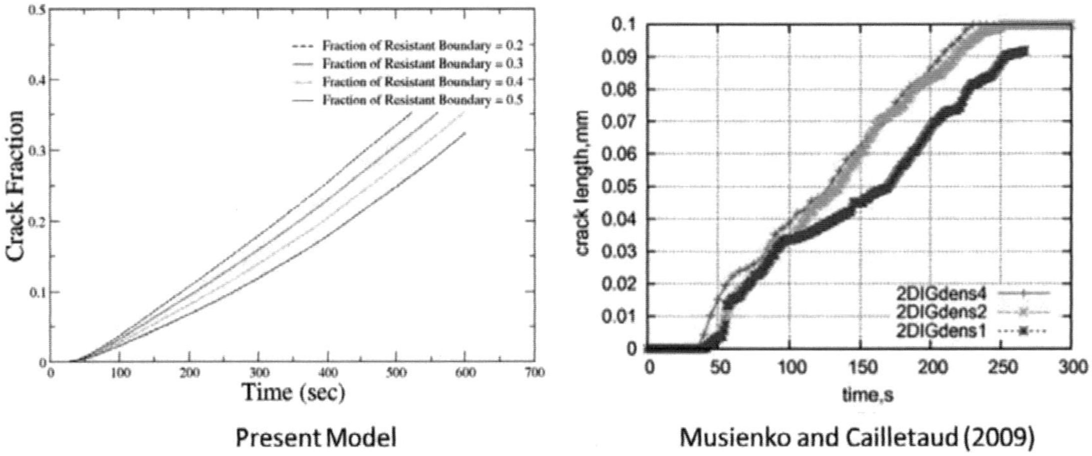

Figure 4. Effect of resistant grain boundaries number fraction on crack fraction (left; present model) and FE based RVE study reported in the literature (Musienko and Cailletaud [35]).

In order to show the effect of exposure time and load on crack growth rate, experimental double bent beam (DBB) test [46] performed on fully sensitized microstructure using constant loads of 100 and 260 MPa are simulated. Three different exposure times of 144, 288 and 432 hours were used during simulation. Based on experimental data [46] it is assumed that ζ and μ are a function of exposure time and are given by

$$\varsigma(t) = \left(\left(1 - sech^2(a_1 t)\right).b_1\right)$$
$$\mu(t) = (sech^2(a_1 t)).b_1$$

where a_1 and b_1 are material parameters to be identified from experiments.

Figure 5 shows the comparison between experimental data and simulated response. Results show a good agreement between experimental and simulated data. Parameters identified from experiments are $a_1 = 0.08$/hr and $b_1 = 8$.

5. Conclusion

We presented a multiscale constitutive model for intergranular stress corrosion cracking. Model takes into account the grain size, and grain orientation relationship using a crystal plasticity theory. Grain boundary zone is modeled using a cohesive based constitutive model. Constitutive model has the capability of using EBSD data to define grain size, orientation and distribution of grain boundary types (resistant and susceptible). Conventional uniaxial tests in air and corrosive environment were simulated and showed a good qualitative agreement with finite element based simulations. In the last part, double bent beam experiment in corrosive environment was simulated and showed a good quantitative agreement with experimental data.

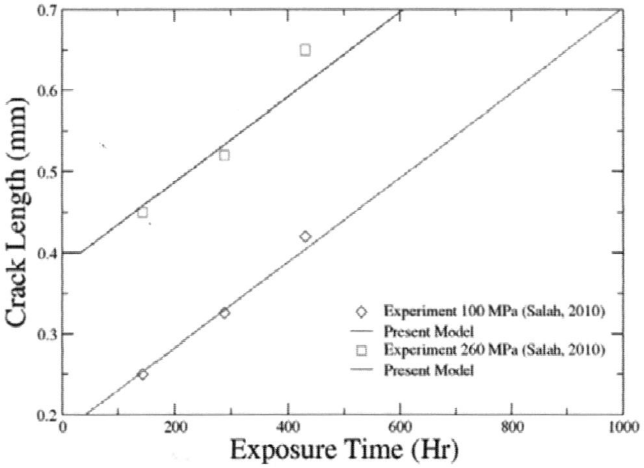

Figure 5. Crack length v/s exposure time for two different constant loads during IGSCC

References

[1] Llewellyn D and Hudd R 1998 *Steels: Metallurgy and Applications* (Boston: Butterworth Heinemann)

[2] Davis J 1994 *ASM Speciality Handbook: Stainless Steels,* (Materials Park, Ohio: ASM International)

[3] Sourmail T 2001 Review: Precipitation in Creep-Resistant Austenitic Stainless Steels *Materials Science and Technology* **17** 1–14

[4] Marshal P 1984 *Austenitic Stainless Steels, Microstructure and Mechanical Properties* (Elsevier Applied Science)

[5] Sedriks A J 1996 *Corrosion of Stainless Steels* (New York: Wiley Interscience)

[6] Parvathavarthini N 2002 Sensitisation and Testing for Intergranular Corrosion *Corrosion of Austenitic Stainless Steels. Mechanism, Mitigation and Monitoring* ed H. S. Khatak B. Raj (New Delhi: Norasa Publishing House) pp 117–38

[7] Dayal R K N, Parvathavarthini N and Raj B 2005 Influence of Metallurgical Variables on Sensitisation Kinetics in Austenitic Stainless Steels *Int. Mater. Rev.* **50** 129–55

[8] Bruemmer S M and Was G S 1994 Microstructural and microchemical mechanisms controlling intergranular stress corrosion cracking in light-water-reactor systems *J. Nucl. Mater.* **216** 348–63

[9] E8-04 A 2004 *Standard Test Methods for Tension Testing of Metallic Materials*

[10] Lehockey E M, Brennenstuhl A M and Thompson I 2004 On the relationship betweengrain boundary connectivity, coincidence site lattice boundaries and intergranular stress corrosion cracking *Corrosion Sci.* **46** 2383–404

[11] Palumbo G, King P J, Aust K T, Erb U and Lichtenberger P C 1991 Grain Boundary Design and Control for Intergranular Stress-Corrosion Resistance *Scripta Metallurgica et Materialia* **25** 1775–80

[12] Rahimi S Engelberg D L Duff J A Marrow T J 2009 In situ observation of intergranular crack nucleation in a grain boundary controlled austenitic stainless steel *J. Micros.* **233** 423–31

[13] ertsman V Y and Bruemmer S M 2001 Study of grain boundary character along intergranular stress corrosion crack paths in austenitic alloys *Acta Materialia* **49** 1589–98

[14] Stauffer D and Aharony A 1994 *Introduction to percolation theory* (London: Taylor & Francis Publication)
[15] Wells D B, Stewart J, Herbert A W, Scott P M and Williams D E 1989 The use of percolation theory to predict the probability of failure of sensitized, austenitic stainless steels by intergranular stress corrosion cracking *Corrosion* **45** 649–60
[16] Gaudett M A and Scully J R 1994 Applicability of Bond Percolation Theory to Intergranular Stress-Corrosion Cracking of Sensitised AISI 304 Stainless Steel *Metallurgical and Materials Transactions A* **25A** 775–87
[17] Pan Y, Olson T and Adams B L 1995 Application of Orientation Imaging Analysis to Microstructural Control of Intergranular Stress Corrosion Cracking *Canadian Metallurgical Quarterly* **34** 147–54
[18] Gertsman V Y and Tangri K 1997 Modelling of Intergranular Damage Propagation *Acta Materialia* **45** 4107–16
[19] Kumar M, King W E and Schwartz K J 2000 Modification to the Microstructural Topology in F.C.C Materials Through Thermomechanical Processing *Acta Materialia* **48** 2081–91
[20] Schuh C A, Kumar M and King K E 2003 Analysis of grain boundary networks and their evolution during grain boundary engineering *Acta Materialia* **51** 687–700
[21] Palumbo G, Lehockey E M Lin P 1998 Applications for grain boundary engineered materials *J. Minerals, Metals and Mater. Society* **50** 40–3
[22] Aust K T, Erb U and Palumbo G 1994 Interface control for resistance to intergranular cracking *Mater. Sci. Engng* A **176A** 329–34
[23] Lin H Pope D P 1993 The influence of grain boundary geometry on intergranular crack propagation in Ni3A1 *Acta Metallurgica et Materialia* **41** 553–62
[24] Babout L Marrow T J Engelberg D L Withers P J 2006 X-ray microtomographic observation of intergranular stress corrosion cracking in sensitised austenitic stainless steel *Mater. Sci. Technol.* **22** 1068–75
[25] Marrow T J, Babout L, Connolly B J, Engelberg D L, Johnson, G Buffiere J Y and Newman R C 2004 High Resolution in-situ, Tomographic Observations of stress corrosiobn cracking *EICM-2* vol Banff (Canada)
[26] Thomas L E Bruemmer S M 2000 High-resolution characterization of intergranular attack and stress corrosion cracking of Alloy 600 in high-temperature primary water *Corrosion* **56** 572–87
[27] McLean D 1957 *Grain boundaries in Metals* (London: Clarendon Press)
[28] Was G S Thaveeprungsriporn V Crawford D C 1998 Grain boundary misorientation effects on creep and cracking in Ni-based alloys *J. Minerals, Metals Mater. Society* **50** 44–9
[29] Palumbo G 1998 *Metal alloys having improved resistance to intergranular stress corrosion cracking* (No. 5,817,193: US patent)
[30] Gertsman V Y Janecek M Tangri K 1996 Grain boundary ensembles in polycrystals *Acta Materilia* **44** 2869–82
[31] Jivkov A P, Stevens N P C and Marrow T J 2006 A two dimensional mesoscale model for intergranular stress corrosion crack propagation *Acta Materialia* **54** 3493–501
[32] Jivkov A P, Stevens N P C and Marrow T J 2006 A three-dimensional computational model for intergranular cracking *Compiut. Mater. Sci.* **38** 442–53
[33] Jivkov A P Stevens N P C Marrow T J 2007 *Modelling of Microstructure Effects on Environment Assisted Cracking (EAC), Final Report (Modelling)* (Report Number MPC/R/051, Materials Performance Centre, School of Materials, University of Manchester)
[34] Jivkov A P and Marrow T J 2007 Rates of intergranular environment assisted cracking in three-dimensional model microstructures *Theoretical and Applied Fracture Mechanics* **48** 187–202
[35] Musienko A and Cailletaud G 2009 Simulation of inter- and transgranular crack propagation in polycrystalline aggregates due to stress corrosion cracking *Acta Materialia* **57** 3840–55

[36] Dattaguru B 2010 *IUTAM Symposium on Multi-Functional Material Structures and Systems: Proceedings of the the IUTAM Symposium on Multi-Functional Material Structures and Systems, Bangalore, India, December 10-12, 2008* vol 19 (Springer Verlag)

[37] Siddiq A and Sayed T E 2012 A thermomechanical crystal plasticity constitutive model for ultrasonic consolidation *Comput. Mater. Sci.* **51** 241–51

[38] Siddiq A and Schmauder S 2006 Interface fracture analyses of a bicrystal niobium/alumina specimen using a cohesive modelling approach *Modelling Simulation Mater. Sci. Engng* **14** 1015

[39] Siddiq A, Schmauder S and Huang Y 2007 Fracture of bicrystal metal/ceramic interfaces: a study via the mechanism-based strain gradient crystal plasticity theory *Int. J. Plasticity* **23** 665–89

[40] Siddiq A, Schmauder S and Ruehle M 2008 Niobium/alumina bicrystal interface fracture: A theoretical interlink between local adhesion capacity and macroscopic fracture energies *Engng Fracture Mech.* **75** 2320–32

[41] Siddiq A and Schmauder S 2005 Simulation of hardening in high purity niobium single crystals during deformation *Steel Grips, J. Steel Related Mater.* **3** 281–6

[42] Siddiq A and Schmauder S 2006 Crystal plasticity parameter identification procedure for single crystalline material during deformation *Journal of Computational and Applied Mechanics* **7** 1–15v

[43] Randle V 1997 No. 127-The role of the grain boundary plane in cubic polycrystals *Acta Materialia* **46** 1459–80

[44] Brandon D G 1966 The structure of high angle grain boundaries *Acta Metallurgica* **14** 1479–84

[45] Grimmer H, Bollmann W and Warrington D H 1974 Coincidence site lattice and complete pattern shift lattices in cubic crystas *Acta Crystallographica* **A30** 197–207

[46] Rahimi S 2009 *Behaviour of Short Intergranular Stress Corrosion Crack in Type 304 Austenitic Stainless Steel* (University of Manchester)

[47] Rahimi S and Marrow T J 2011 Effects of orientation, stress and exposure time on short intergranular stress corrosion crack behaviour in sensitised type 304 austenitic stainless steel *Fatigue & Fracture Engng Mater. & Struct. J.* **35** 359–73

Stability of transverse vibration of rod under longitudinal step-wise loading

A K Belyaev[1], **D N Ilin**[2] **and N F Morozov**[1,2]

[1] Institute for Problems in Mechanical Engineering, Russian Academy of Sciences, 199178, V.O. Bolshoy pr. 61, St. Petersburg, Russia
[2] St. Petersburg State University, 199034, Universitetskaya nab. 7-9, St. Petersburg, Russia

E-mail: vice.ipme@gmail.com

Abstract. The problem of dynamic stability of a simply supported rod subjected to axial jump loading is considered. A systematic application of the method of expansion in terms of the normal axial and bending vibration modes is utilised. Longitudinal vibrations give rise to periodic longitudinal forces which in turn causes unstable bending vibrations. Application of the Galerkin approach results in a system of ordinary differential equations with periodic coefficients which are reduced to Mathieu equation. The instability regions whose form depends on the spectral properties of the longitudinal and flexural vibrations, damping values and longitudinal force are obtained. An example of unusual shapes of the instability regions is shown: the twelfth transverse mode caused by the first longitudinal mode turns out to be unstable for some parameters of the rod. The critical value of the jump load leading to instability of the considered transverse vibration modes is derived.

1. Introduction. The case history

The problem of the static buckling of the simply supported beam of length l loaded by the longitudinal force P was first set by L. Euler. This problem is now referred to as Euler elastica and the boundary-value problem is as follows

$$EI\frac{d^2w}{dx^2}+Pw=0,\ w(0)=w(l)=0\ ,$$

where EI denoted the bending stiffness and w stands for the beam deflection. The solution is given by $w(x)=A\sin(\lambda x),\ \lambda=\sqrt{P/EI}$ and the bifurcation takes place at $P_E=\pi^2 EI/l^2$ known as the Euler (buckling) force. The straight form of the beam ($A=0$) remains stable if $P<P_E$. More realistic is the non-linear formulation of the problem in terms of the angle of rotation of the beam cross-section $\psi=dw/dx$. In this case the boundary-value problem is given by

$$EI\frac{d^2\psi}{dx^2}+P\sin\psi=0,\ \frac{d\psi}{dx}(0)=\frac{d\psi}{dx}(l)=0\ ,$$

whose solution gives the laterally deformed forms of equilibrium for each value of the longitudinal force $P > P_E$ in addition to the straight form. It should be noted that this formulation of the problem has nothing in common with the problem of transition from unstable equilibrium to a stable equilibrium because this approach operates only with the concept of bifurcation of solutions and the existence of the alternative forms of static equilibrium. From a perspective of the theory of motion stability the transition from the straight form to a laterally deformed form corresponds to the divergent loss of stability, also called the static instability, cf. [1].

For the first time the problem of dynamic buckling of a simply supported beam subjected to the axial jump loading was addressed in the article by Lavrentiev and Ishlinsky [2]. It was assumed that the beam has some initial curvature and the external compressive force applied to one end immediately covers the entire beam and then remains constant. This particular problem statement is inherently contradictory since the beam is assumed to behave statically in axial direction however stability analysis in transverse direction is carried out by means of dynamic approach. It is interesting that this specific dynamic formulation also predicts a divergent loss of stability of the original pre-curved form. It is shown that there is no one-to-one correspondence between the number of the Euler force and the number of the buckling mode with the highest rate of increase in the deflection. For example, if the rod is suddenly loaded by the third Euler force ($9P_E$) then the second buckling shape will have the highest rate of instability. When the fourth Euler force ($16P_E$) is applied, the fastest instability will develop in the third form, etc. It is easy to see that these assumptions are quite restrictive:

1. The result is sensitive to the shape of the initial deflection. The highest rate of increase in a particular buckling form only implies that this form will have a larger deflection at $t \to \infty$ regardless of its contribution to the initial deflection shape. For example, if the initial shape is orthogonal, say, to the second buckling form, i.e. this form does not represented in the initial deflection, then the conclusion of paper [2] that this buckling form possesses the highest rate of instability is useless for stability analysis as this particular form will not be excited at all. Another shortcoming is as follows.
2. The conclusions are valid only for the higher buckling modes,
3. Axial dynamics of the rod is not considered.

Let us also notice paper [3] where the authors carried out a study of the lateral stability of the rod taking into account the longitudinal waves generated by a jump of axial force.

In the present paper we propose an alternative approach to the problem by Lavrentiev and Ishlinsky [2]. In contrast to [2] and [3], we carry out a systematic application of the method of series expansion in terms of both axial and bending vibration modes. The longitudinal vibrations give rise to periodic longitudinal forces, which in turn can cause unstable bending vibrations. This phenomenon is known as the parametric resonance, cf. [4].

2. Regions of dynamical instability

Let us consider a simply supported beam with a constant cross-sectional area loaded at the end $x=l$ by an axial force $P(t) = P_0 H(t)$ where $H(t)$ is the Heaviside step function. Elastic axial vibrations are assumed, that is, we are allowed to expand the axial displacement in series in terms of the normal modes of axial vibration $u_k(x)$

$$u(x,t) = \sum_{k=1}^{\infty} u_k(x) q_k(t),\ u_k(x) = \sin\frac{\omega_k}{a}x,\ \omega_k = \frac{\pi a}{2l}(2k-1), a = \sqrt{\frac{E}{\rho}},\ k = 1,2,.. \ . \tag{1}$$

where ρ is the mass density of the rod and ω_k denotes the natural frequency of the axial vibration. Accounting for zero initial conditions one obtains the following expression for the generalized coordinate

$$q_k(t) = (-1)^{k+1} \frac{2a^2 P_0}{lEF\omega_k^2}(1 - \cos\omega_k t) \tag{2}$$

and the equation or the axial force in the form of a series

$$P(x,t) = EF\frac{\partial u}{\partial x} = \frac{4}{\pi}P_0\sum_{k=1}^{\infty}\frac{(-1)^{k+1}}{2k-1}(1-\cos\omega_k t)\cos\pi\left(k-\frac{1}{2}\right)\frac{x}{l}, \tag{3}$$

where F stands for the cross-sectional area of the rod.

The inner damping is introduced by means of the Kelvin-Voigt rheological visco-elastic model [5], then the governing equation for the transverse vibration of the model accounting for the varying axial force $P(x,t)$ is given by

$$E\left(1+\gamma\frac{\partial}{\partial t}\right)I\frac{\partial^4 w}{\partial x^4} + \frac{\partial}{\partial x}\left(P(x,t)\frac{\partial w}{\partial x}\right) + \rho F\frac{\partial^2 w}{\partial t^2} = 0 \tag{4}$$

where γ is a non-dimensional viscosity factor. The necessity of introducing the viscosity is due to the fact that the damping has a crucial influence on the shapes of the regions of dynamic instability at small magnitudes of the axial force, cf. [4]. The transverse displacement $w(x,t)$ is represented as a series in terms of the normal modes of bending vibration

$$w(x,t) = \sum_{m=1}^{\infty}\sin\frac{m\pi x}{l}Q_m(t)$$

The normal modes of bending vibration $\sin m\pi x/l,\ m = 1,2,3...$ are also the buckling form of the simply supported beam. It means that both the beam vibration and beam buckling are accurately modeled.

Substituting the expression (3) for the axial force into equation (4) and applying Galerkin method we arrive at the infinite system of coupled ordinary differential equations for the generalised coordinates $Q_m(t)$:

$$\rho F\frac{l}{2}\ddot{Q}_j(t) + E\gamma I\frac{j^4\pi^4}{l^4}\frac{l}{2}\dot{Q}_j(t) + EI\frac{j^4\pi^4}{l^4}\frac{l}{2}Q_j(t) - \\ - EF\frac{\pi^2}{8l^2}\left[\sum_m m\sum_k(2k-1)\alpha_{mk}q_k(t)Q_m(t)\right] = 0 \tag{5}$$

where

$$\alpha_{mk} = \left(k-\frac{1}{2}+m\right)\frac{2j}{(k-\frac{1}{2}+m)^2-j^2}(-1)^{k+m+j+1} + \left(k-\frac{1}{2}-m\right)\frac{2j}{(k-\frac{1}{2}-m)^2-j^2}(-1)^{k-m+j+1},$$

j and m are the numbers of the bending modes whereas k denotes the number of the axial mode. One can see that the generalized coordinates of bending vibration $Q_j(t)$ depend on the coefficients of the generalized coordinates of axial vibration $q_k(t)$ as well as on the other generalized coordinates of bending vibration $Q_m(t)$. This coupling reflects the interaction of the generalized coordinates of bending vibration.

Let us insert expression (2) for $q_k(t)$ into equation (5) and neglect the mutual interaction of the different modes of bending vibration, then only the term with $m=j$ remains in the double series (5). This assumption is widely used in the vibration analysis and is known to deliver a good approximation. The result is a system of uncoupled equations, each describing the influence of the axial vibration mode (with number k) on the bending mode under consideration (with number m):

$$\ddot{Q}_{mk} + \frac{E\gamma I\pi^4 m^4}{\rho F l^4}\dot{Q}_{mk} + \left(\frac{EI\pi^4}{\rho F l^4}m^4 + \frac{(-1)^k 2m\alpha_{mk}}{2k-1}\frac{P_0}{\rho F l^2}(1-\cos\omega_k t)\right)Q_{mk} = 0 \tag{6}$$

Here the subscripts at Q_{mk} $(k = 1,2,3,...m = 1,2,3,...)$ correspond to number m of the bending vibration and number k of the axial vibration. This equation provides one with the answer to the question of the axial load magnitude P_0 leading to the instability of *m-th* bending vibration caused by *k-th* axial vibration mode.

Let us rewrite equation (6) in the standard form of the problem of parametric instability, cf. [5,6]

$$\ddot{Q}_{mk} + 2\delta_{mk}\Omega_{mk}\dot{Q}_{mk} + \Omega_{mk}^2\left(1 - 2\mu_{mk}\cos\theta t\right)Q_{mk} = 0 \tag{7}$$

where the coefficients are given by

$$\Omega_{mk}^2 = \frac{EI\pi^4}{\rho F l^4}m^4 + \frac{(-1)^k 2m\alpha}{2k-1}\frac{P_0}{\rho F l^2}, \quad \mu_{mk} = \frac{\dfrac{(-1)^{k+1}\alpha}{2k-1}P_0}{\dfrac{EI\pi^4}{l^2}m^3 + \dfrac{(-1)^k 2\alpha}{2k-1}P_0}$$

$$\delta_{mk} = \frac{E\gamma I\pi^4 m^4}{2\rho F l^4\Omega}, \quad \theta = \omega_k = \frac{\pi a}{2l}(2k-1) \tag{8}$$

Before we begin to determine the instability regions let us state that these regions exist only under the condition $\Omega_{mk}^2 > 0$. If $\Omega_{mk}^2 \le 0$ then the solution of equation (7) will be unbounded even for $\mu_{mk} = 0$, i.e. the corresponding bending mode will be unstable. Therefore the critical value of the axial force affected by *k-th* axial vibration mode can be determined from the condition

$$\frac{EI\pi^4}{\rho F l^4}m^4 + \frac{(-1)^k 2m\alpha}{2k-1}\frac{P_0}{\rho F l^2} \le 0 . \tag{9}$$

Provided that $\Omega_{mk}^2 > 0$, equation (7) is nothing else than the Mathieu equation with damping. It is known, cf. [4], that the damping does not considerably affect the main instability region and makes the side regions not feasible in practical engineering. Based on this remark, we restrict ourselves to the main region of dynamic instability.

In the above notation, the main region of instability of transverse vibration modes is determined by the systems of inequalities, cf. [4],

$$2\Omega\sqrt{1 - \sqrt{\mu^2 - 4\delta^2}} \le \theta \le 2\Omega\sqrt{1 + \sqrt{\mu^2 - 4\delta^2}} , \tag{10}$$

where in what follows the subscripts are omitted for simplicity. These inequalities are formally identical with the inequalities of the classical Mathieu equation however it should be borne in mind that the parameters Ω, μ, δ are functions of P_0, i.e. the axial vibration changes fundamentally the formula for each parameter, see equation (8), thus drastically changing the dynamic instability regions. This system of inequalities describes implicitly the critical value of the load at which *m-th* transverse vibration mode will be unstable under the impact of *k-th* axial mode.

By constructing the dynamic instability regions defined by the inequalities (10) for all possible axial – bending pairs we can obtain the critical load values which are smaller than the values obtained from equation (9).

3. An example of calculation of the critical force

Let us determine the critical value of the force by constructing the regions of dynamic instability for the following parameters of the rod: $E = 2.1 \cdot 10^{11}\, N/m^2$, $\rho = 7.8 \cdot 10^3\, kg/m^3$, $l = 2m$, $h = 0.005m$, $b = 0.01m$, $\gamma = 10^{-4}$, where h and b are the height and width of the cross section respectively.

Figure 1 displays the effect of the first axial mode on the main instability region for the seventh - twelfth bending modes. In the standard problem of dynamic buckling of beams [4] the frequency of the axial force θ can be varied continuously. In the problem under consideration the frequency of the axial force θ belongs to the discrete spectrum of the natural frequencies of axial vibrations. The natural frequency of the first longitudinal mode of vibration is given by $\theta = \omega_1 = \pi a / 2l$. Figure 1 shows that the horizontal line $\theta = \omega_1$ does not cross the main instability regions for the bending modes with the number below 12, that is, the twelfth bending vibration mode becomes unstable and is allows us to determine the minimum value of the load leading to instability twelfth transverse vibration mode which is denoted by P_{cr}. For the lower bending vibration modes the critical value of the force can be found from equation (9).

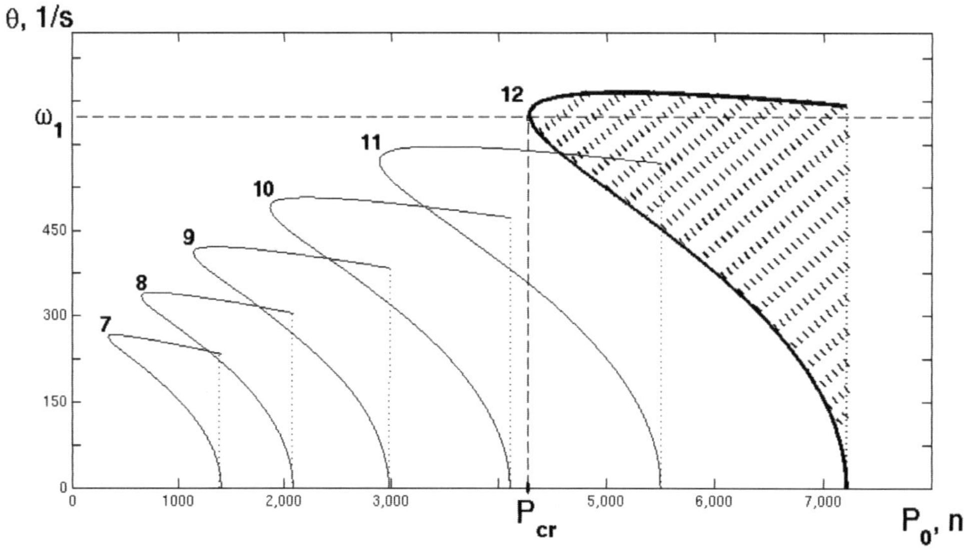

Figure 1. The boundaries of the main region of instability for the seventh - twelfth transverse vibration modes caused by the first axial mode and the critical force P_{cr}.

The estimates below are obtained with the help of the above criteria of dynamic instability.
Euler's force:

$$P_E = \frac{EI\pi^2}{l^2} \approx 53.97\,N.$$

The condition of instability of the *first* bending mode:

$$P_0 \geq 1{,}32 P_E \approx 71.3\,N.$$

The condition of instability of the *second* bending mode:

$$P_0 \geq 5P_E \approx 270.6\,N.$$

The condition of instability of the *third* bending mode:

$$P_0 \geq 11.2 P_E \approx 603.44\,N \text{ etc.}$$

The condition of instability of the *seventh* bending mode:

$$P_0 \geq 3266.6\,N.$$

For values of the load $P_0 \geq 3924.5\,N$ the twelfth bending mode can be theoretically unstable (in accordance with the results obtained from the consideration of equation (10)).

The condition of instability of the *eighth* bending mode:

$$P_0 \geq 4265.3\,N \text{ etc.}$$

It should be noted that the location of the instability regions in the plane (P_0, θ) depends on the rod parameters (Young's modulus, coefficient of viscous damping etc.). Depending upon the rod geometry and the mechanical characteristics of the material the axial force can lead to dynamic instability of another bending mode rather than that of the twelfth bending mode.

4. Conclusions

Construction and analysis of the regions of dynamic instability lead to the following conclusions:

1. The standard approaches of static stability analysis predict the static (divergent) loss of stability. In contrast to them, the present study demonstrates that the dynamic (oscillatory) character of instability is also possible depending on the magnitude of the external load.
2. For the lower bending vibration modes the value of the critical load can be obtained from equation (9) whereas that for the higher modes - from conditions (10).
3. The dependence between the number of buckling mode and the axial load value is not directly proportional.
4. The rod buckling due to the highest modes is possible for values of the compressive force which are less than the values calculated by Euler's formula.
5. The smallest value of the critical load corresponds to the pair first axial mode – first bending mode. This value is greater than Euler's force and is obtained from equation (9) to give $P_{cr} = 1.3P_E$.
6. The formulae concerning the critical value of the axial force and the conclusion about instability due to the twelfth bending form are valid only for the particular geometric and mechanical parameters taken for the numerical work. If one takes other characteristics of the beam or frequency-dependent internal damping or other boundary conditions then one should expect that another bending form becomes unstable. Nevertheless, the general conclusions predicting both static and dynamic character of instability remain valid.
7. The result of the present analysis could be related to the practical and industrial applications, first of all, they are important from a perspective of the health of steel structural elements of buildings. The well-established standard methods of stability analysis rely on the static (linear or nonlinear) approaches. The present analysis has demonstrated that, in general, account for the dynamic effects of parametric resonance is required as well. This conclusion is of special importance for the structures subjected to intensive dynamic loading, such as those due to earthquake, explosion etc.

Acknowledgment

Financial support by Area 2 of the Austrian Center of Competence in Mechatronics (ACCM, Linz) and the Russian Foundation for Basic Research, grant 12-01-92000 HHC_a is acknowledged.

References

[1] Bolotin V V 1963 *Nonconservative Problems of the Theory of Elastic Stability* (Oxford: Pergamon Press).
[2] Lavrentiev M A and Ishlinsky A Yu 1949 Dynamic forms of loss of stability of elastic systems (in Russian). *Doklady (Transactions) of the USSR Academy of Sciences* **64** 776-82.
[3] Morozov N F and Tovstik P E 2009 Dynamics of rod under axial impact (in Russian). *Vestnik of St. Petersburg State University, Series 1* 105-11.
[4] Bolotin V V 1964 *The Dynamic Stability of Elastic Systems* (San-Francisco: Holden Day).
[5] Palmov V A 1998 *Vibration of Elasto-Plastic Bodies* (Berlin-Heidelberg: Springer).

Structured interfaces for flexural waves - trapped modes and transmission resonances

S G Haslinger[1], R C McPhedran[1,2], N V Movchan[1] and A B Movchan[1]

[1] Department of Mathematical Sciences, Mathematical Sciences Building, Peach Street, Liverpool L69 3BX, United Kingdom
[2] CUDOS, School of Physics, University of Sydney 2006, NSW, Australia

E-mail: sgh@liv.ac.uk

Abstract. The article combines the analytical models of scattering and Bloch waves for a stack of periodic gratings in an infinite elastic plate. The waves represent flexural deflections of the plate governed by a fourth-order partial differential equation. The emphasis is on the analysis of trapped modes and transmission resonances for different configurations of the grating stack and physical parameters of the flexural waves. Special attention is given to the phenomenon of Elasto-Dynamically Inhibited Transmission (EDIT). The analytical model is supplemented with comprehensive numerical examples.

1. Introduction

An important branch of structural mechanics is the study of the vibration of plates. The analysis of such models gives us an insight into the predicted response of elastic structures to earthquakes and other dynamic loads. In this paper we consider structured plates that contain an interface consisting of a finite number of periodic gratings, and our particular interest is in the localisation of flexural waves within the grating structure. Many elastic systems such as aircraft, long bridges and reinforced roads encompass periodic arrangements of inclusions, voids and masses. These structures are frequently subjected to stress concentrations and therefore the study of the resonant action of incident flexural waves within our structured plates has applications in engineering.

Our model allows for the gratings to consist of inclusions of any shape or size. Here we will concentrate on rigid, cylindrical voids with a finite radius, in the limiting case of the radius tending to zero, which can be thought of as giving a pinned point. We demonstrate that the periodically structured interface supports sharp transmission resonances for low-frequency flexural vibrations arising from the interaction of a plane wave, characterised by its angle of incidence θ_i and spectral parameter β, with the plate.

We show an illustrative example in Fig. 1. The outer pair of this triplet consists of pinned points, whilst the central grating contains rigid voids with a finite radius a. The other parameters are the angle of incidence θ_i, the period of the gratings d and the grating separation η. We also indicate the scattered field. An important additional parameter is the relative shift of the central grating denoted by ξ, which is crucial for supporting a filtering effect similar to Electromagnetically Induced Transparency (EIT). We term this Elasto-Dynamically Inhibited

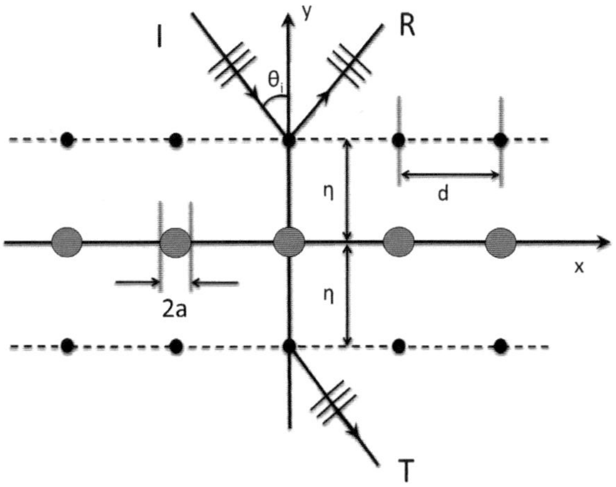

Figure 1. Stack of gratings consisting of an outer pair of rigid pins and a central grating of finite nonzero inclusions with radius a and period d. The relative grating separation between consecutive gratings is denoted by η.

Transmission (EDIT) and this novel phenomenon for elasticity problems was first observed by Haslinger*et al.* [1-2]. It is characterised by a resonant peak in transmission being cut in two by a resonant dip with an extremely high quality factor.

We discuss how the radius of the inclusions affects EDIT, which is dependent on the coincidence of symmetric and anti-symmetric modes. For a triplet consisting only of rigid pins, the anti-symmetric mode's frequency is invariant with ξ [1]. This makes it easier to tune the system's symmetric mode to coincide with the odd mode. This is no longer true for inclusions of nonzero radius.

The increased radius makes it necessary to take into account higher-order multipole terms characterising the scattered field, and the periodicity of the grating will also lead to the use of higher order lattice sums. It follows that more care is required to locate the frequency for EDIT for structures incorporating finite-sized voids.

Furthermore, the grating stack can be considered as a structured waveguide for trapped modes. For a simple case of a stack of rigid pins, a quasi-periodic Green's function is employed to derive the dispersion equation for Bloch waves in such a waveguide. A connection is established with the transmission problem by identifying parameters of the grating stack and of the incident wave, to generate a transmission resonance linked to a trapped Bloch wave within the structured waveguide.

2. Transmission resonances for gratings stacks

A structured interface consisting of a periodic array of cells of chosen geometry separates two half-planes in an infinite flexural plate. A plane wave representing flexural displacements is incident on such an interface, and it is shown that a nearly 100 % transmission is achievable for a certain combination of parameters within a narrow frequency range. A new phenomenon of Elasto-Dynamically Inhibited Transmission is outlined for a stack of three gratings.

2.1. Governing equations and method of solution

Let the amplitude W be a solution of the scattering problem for the biharmonic operator (see Movchan *et al.* [3]):

$$\Delta^2 W(\mathbf{x}) - \beta^4 W(\mathbf{x}) = 0, \tag{1}$$

where $\beta^2 = \omega\sqrt{\rho h/D}$. Within this expression, $D = Eh^3/(12(1-\nu^2))$ is the flexural rigidity of the plate, h denotes the plate thickness, E is the Young modulus and ν is the Poisson ratio. In addition, ρ is the mass density and ω is the angular frequency. Here we have assumed time-harmonic vibrations with angular frequency ω. Consequently, equation (1) is derived by substituting the out-of-plane displacement $w(\mathbf{x};t) = W(\mathbf{x})\sin(\omega t)$ into the standard Kirchhoff plate equation of motion:

$$D\Delta^2 w(\mathbf{x};t) + \rho h \frac{\partial^2 \omega}{\partial t^2}(\mathbf{x};t) = 0. \tag{2}$$

This yields an equation satisfied by the amplitude $W(x,y)$ or $W(r,\theta)$, where (r,θ) are the polar coordinates applicable to the case of cylindrical inclusions.

The solution of equation (1) can be divided into two parts W_H and W_M, which satisfy the Helmholtz equation and its counterpart form, the modified Helmholtz equation:

$$(\Delta + \beta^2)W_H = 0, \qquad (\Delta - \beta^2)W_M = 0. \tag{3}$$

An important aspect of the physics of the problem is that W_H contains both propagating and evanescent waves, while W_M consists entirely of evanescent waves.

The field W satisfies the Bloch quasi-periodicity condition along the horizontal x-axis:

$$W(\mathbf{x} + md\mathbf{e}^{(1)}) = W(\mathbf{x})e^{i\alpha_0 md}, \tag{4}$$

where the diffraction order m is an integer, d is the period, and α_0 is the Bloch parameter $\alpha_0 = \beta\sin\theta_i$, where θ_i is the angle of incidence (see Fig. 1).

The boundary conditions on the circular boundaries of the inclusions are Dirichlet clamping conditions:

$$W\Big|_{r=a} = 0, \quad \frac{\partial W}{\partial r}\Big|_{r=a} = 0. \tag{5}$$

We initially consider the limit as $a \to 0$, corresponding to an array of rigid pins constraining the plate. The radius of the inclusions is important for the implementation of the method that we use to solve the problem, as outlined in the papers by Movchan *et al.* [3-6] and Haslinger *et al.* [1-2].

We consider the scattering of plane waves, either of the Helmholtz type W_H or of the modified Helmholtz type W_M, by a grating of inclusions of radius a. The flexural displacement W is expanded for $y > a$ and $y < -a$ in terms of plane waves of type W_H and W_M. Above the grating the expansion has a down-going incident wave term and up-going reflected waves with amplitude R_p and $\hat{R}_p$, for the respective wave types. Below the grating the amplitudes of the down-going transmitted waves are denoted by T_p and $\hat{T}_p$. In order to connect these two types of expansions we introduce multipole expressions for W in the region $-a \le y \le a$. The multipole expansion for W_H involves cylindrical waves $J_n(\beta r)e^{in\theta}$ and $H_n^{(1)}(\beta r)e^{in\theta}$ with respective amplitudes A_n and E_n:

$$W_H(\mathbf{x}) = \sum_{n=-\infty}^{\infty} \{A_n J_n(\beta r) + E_n H_n^{(1)}(\beta r)\}e^{in\theta}. \tag{6}$$

The multipole expansion for W_M involves modified Bessel function terms $I_n(\beta r)e^{in\theta}$ and $K_n(\beta r)e^{in\theta}$ with respective amplitudes B_n and F_n:

$$W_M(\mathbf{x}) = \sum_{n=-\infty}^{\infty} \{B_n I_n(\beta r) + F_n K_n(\beta r)\}e^{in\theta}, \tag{7}$$

The amplitudes A_n, B_n, E_n, F_n are the multipole coefficients to be determined, and they are related by the boundary conditions (5).

The Rayleigh identity expresses the part of the expansion for W which is regular near the origin (i.e. the terms involving the Bessel functions J_n and I_n) as sums over the part which is irregular near the origin (i.e. the terms involving the Bessel functions $H_n^{(1)}$ and K_n), together with a term representing the expansion of the incident wave in multipoles. The resulting equations are uncoupled for solutions of the Helmholtz and the modified Helmholtz type. The necessary coupling between the two types of waves is provided by the boundary conditions.

The combination of the Rayleigh identity and the boundary conditions gives a system of linear equations, which is truncated (e.g. so that Bessel functions of orders $-L$ to L are retained) and solved to evaluate a set of multipole coefficients. These in turn are used to evaluate plane wave amplitude coefficients for the fields above and below the grating, using reconstruction equations (see Movchan *et al* [3-4]). The amplitudes of the reflected and transmitted waves are determined for the set of incident fields corresponding to a range of grating orders p, both propagating and evanescent. These are assembled into scattering matrices for reflection and transmission that completely characterise the grating's scattering action, and enable us to identify transmission resonances for a finite number of gratings.

2.2. Transmission resonances and Elasto-Dynamically Inhibited Transmission

The formulae for transmission and reflection matrices have already been obtained for a single grating, together with the formulae for the transmission and reflection matrices for a pair of identical gratings (see Movchan *et al.* [5])

$$\begin{aligned}\mathcal{T}_2 &= \mathcal{T}_1[\mathbf{I} - (\mathcal{R}_1)^2]^{-1}\mathcal{T}_1, \\ \mathcal{R}_2 &= \mathcal{R}_1 + \mathcal{T}_1\mathcal{R}_1[\mathbf{I} - (\mathcal{R}_1)^2]^{-1}\mathcal{T}_1.\end{aligned} \tag{8}$$

The matrices $\mathcal{T}_i$, $\mathcal{R}_i$ comprise the matrices $\mathbf{T}_i$ and $\mathbf{R}_i$ respectively, where $i = 1, 2$, which are formed using the coefficients for the plane wave representations, together with a diagonal propagation matrix $\mathcal{P}$:

$$\mathcal{R}_i = \mathcal{P}\mathbf{R}_i\mathcal{P}, \ \mathcal{T}_i = \mathcal{P}\mathbf{T}_i\mathcal{P}. \tag{9}$$

The propagation matrix incorporates the effect of both vertical separation of the gratings η and relative lateral shift ξ:

$$\mathcal{P} = \begin{pmatrix} P & 0 \\ 0 & P \end{pmatrix}, \qquad \text{where} \qquad P = [\delta_{tp} e^{i(\tilde{\chi}_p \eta/2 - \alpha_p \xi/2)}], \tag{10}$$

with $\tilde{\chi}_p = \chi_p$ if p corresponds to a Helmholtz type plane wave and $\tilde{\chi}_p = \hat{\chi}_p$ if p corresponds to a plane wave of modified Helmholtz type.

The idea of transmission resonances for gratings of rigid pins was addressed by Movchan *et al.* [4] and Haslinger *et al.* [1-2]. The increase of the radius of inclusions within the gratings affects the scattered fields, with a change in the frequency of the resonance mode as well as its quality factor Q. For a resonant peak of transmittance T occurring at $\beta = \beta_*$ with $T = T_{\max}$ there, if $T = T_{\max}/2$ for $\beta = \beta_+$ and β_-, then $Q = \beta_*/|\beta_+ - \beta_-|$.

We demonstrate the EDIT phenomenon with a triplet of gratings which typically supports two transmission resonances, an even and an odd mode. Haslinger *et al.* [2] described how the frequency of the odd mode is invariant with the relative lateral shift of the central grating, and this factor ensures that the EDIT shift ξ is simpler to tune by moving the symmetric mode to coincide with the anti-symmetric one. Illustrative examples for gratings of rigid pins are discussed by Haslinger *et al.* [1-2]. The latter paper also includes the case whereby the outer

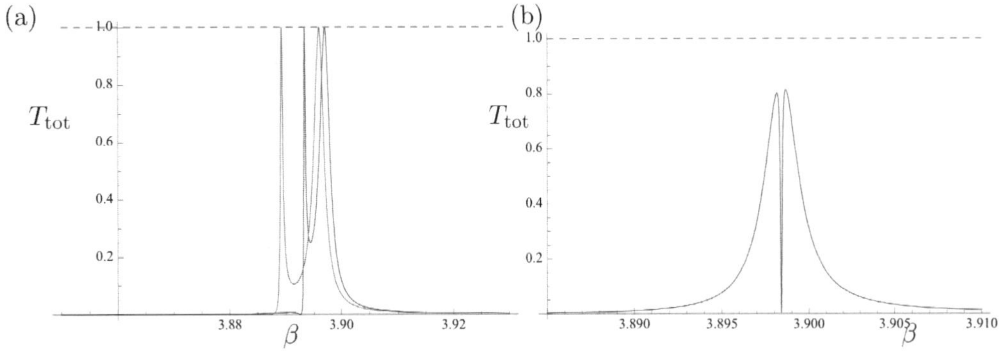

Figure 2. Total transmittance T_{tot} as a function of β for a triplet with a shifted central grating of finite radius inclusions with $a = 0.0035d$, and an outer pair of gratings of inclusions of radius $a = 0.01d$, $\theta_i = 26°$. Data used: (a) $\xi = 0.4d$ (red curve), $\xi = 0.42d$ (blue curve). (b) $\xi = 0.45556d$.

pair of rigid pins is replaced by inclusions of finite radius, and the central grating of rigid pins is shifted to facilitate the EDIT effect.

In Fig. 2(a), we consider a triplet with an outer pair consisting of inclusions of radius $a = 0.01d$, and a central grating with $a = 0.0035d$. Two shifts of the central grating are illustrated, $\xi = 0.4d$ shown by the red curve, and $\xi = 0.42d$ in blue. It is clear that there is no invariant mode, but the odd mode has a higher value of β and is being perturbed more slowly than the even mode, which is brought closer to its partner by an increase in ξ. By tuning this value of ξ carefully, we show the characteristic dip to zero transmittance of the EDIT phenomenon for the alignment of the two modes at $\xi = 0.45556d$ in Fig. 2(b).

3. Waveguides and quasi-periodic grating Green's function

We design a waveguide consisting of a triplet of rigid pin gratings. There are similarities with the scattering problem that we have previously analysed. In that case we identified extremely narrow frequency bands that supported transmission resonances for an incident plane wave characterised by its angle of incidence. The combination of the specific angle of incidence θ_i and its corresponding spectral parameter β means that each transmission peak is defined by a specific value of the Bloch parameter $\alpha_0 = \beta \sin\theta_i$. Therefore we pay particular attention to these Bloch values for the waveguide problem, with the expectation that they will support localisation within the grating structure.

3.1. Quasi-periodic grating Green's function and dispersion equation

We may consider a single grating of rigid pins as a line of point forces with constant separation d. Therefore we use a quasi-periodic Green's function of the form

$$G(\mathbf{r};\beta,\alpha_0) = \sum_{n=-\infty}^{\infty} g(\mathbf{r} - nd\mathbf{e^{(1)}},\beta)\, e^{i\alpha_0 nd} \tag{11}$$

to describe a single grating. Evans and Porter [7] give the fundamental Green's function for a time harmonic point source:

$$g\,(\mathbf{r};\beta) = -\frac{i}{8\beta^2}\left(H_0^{(1)}(\beta r) - H_0^{(1)}(i\beta r)\right), \tag{12}$$

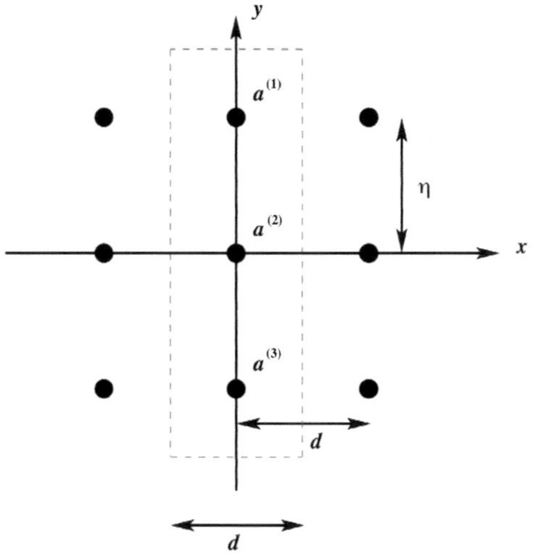

Figure 3. Elementary cell for the waveguide consisting of a triplet of unshifted rigid pin gratings.

which solves the equation

$$(\Delta^2 - \beta^4)g(\mathbf{r};\beta) + \delta(\mathbf{r}) = 0. \tag{13}$$

The corresponding quasi-periodic Green's function is finite at the origin, and can be written as follows

$$G(x,y;\beta,\alpha_0) = \frac{1}{4d\beta^2}\Big(i\sum_{n=-\infty}^{\infty}\chi_n^{-1}\exp(i[\alpha_n x + \chi_n|y|]) - \sum_{n=-\infty}^{\infty}\tau_n^{-1}\exp(i\alpha_n x - \tau_n|y|)\Big), \tag{14}$$

where

$$\alpha_n = \alpha_0 + \frac{2\pi n}{d},$$

$$\chi_n = \begin{cases} \sqrt{\beta^2-\alpha_n^2}, & \alpha_n^2 \le \beta^2 \\ i\sqrt{\alpha_n^2-\beta^2}, & \alpha_n^2 > \beta^2, \end{cases}$$

$$\tau_n = \sqrt{\beta^2+\alpha_n^2}.$$

For more details of their evaluation, see Movchan *et al.* [4-5], Haslinger *et al.* [1-2].

For convenience, we introduce grating Green's functions $G^{(j)}$ in such a way that

$$G^{(1)}(x,y;\beta,\alpha_0) = G(x,y-\eta;\beta,\alpha_0),\ G^{(2)}(x,y;\beta,\alpha_0) = G(x,y;\beta,\alpha_0),$$

$$G^{(3)}(x,y;\beta,\alpha_0) = G(x,y+\eta;\beta,\alpha_0).$$

For an elementary cell of the form shown in Fig. 3, we construct a matrix of Green's functions $\mathbf{M}$ which fully describes the triplet waveguide, and satisfies the dispersion equation

$$\mathbf{MA} = 0, \tag{15}$$

where $M_{ij} = G^{(j)}(0,(2-i)\eta;\beta,\alpha_0), i,j = 1,2,3$, $\mathbf{A}$ is a column vector of coefficients A_j to be determined. For the case of an unshifted triplet, we obtain a symmetric Toeplitz matrix, with

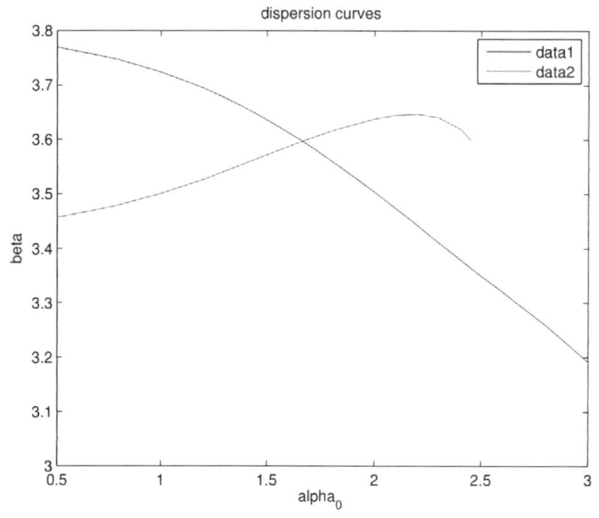

Figure 4. Dispersion diagram for a waveguide consisting of an unshifted triplet with the horizontal axis representing α_0 in the range $0 \leq \alpha_0 \leq \pi$, and β on the vertical axis in the range $2 \leq \beta \leq 4$. Produced using *Matlab*. The red curve represents the odd modes, and the blue curve, the even modes.

complex entries which describe the contribution of each grating layer to the displacement at the location of the pinned point under study:

$$\mathbf{M} = \begin{pmatrix} M_{11} & M_{12} & M_{13} \\ M_{12} & M_{11} & M_{12} \\ M_{13} & M_{12} & M_{11} \end{pmatrix}. \tag{16}$$

The matrix has three eigenvalues:

$$\lambda_1 = M_{11} - M_{13}, \quad \lambda_\pm = \frac{1}{2}(2M_{11} + M_{13} \pm \sqrt{8M_{12}^2 + M_{13}^2}). \tag{17}$$

The eigenvector corresponding to λ_1 is

$$v_o = \begin{bmatrix} +1 \\ 0 \\ -1 \end{bmatrix}, \tag{18}$$

and has odd symmetry. The other two vectors have even symmetry.

Dispersion curves for the modes are the trajectories along which the eigenvalues are zero (or more precisely, since we have complex entries, have a modulus close to zero). For the odd mode, we have the requirement

$$M_{11} - M_{13} = 0 \text{ or } |M_{11} - M_{13}| \simeq 0. \tag{19}$$

For the even modes, the dispersion curves correspond to

$$M_{12}^2 = \frac{1}{2}M_{11}(M_{11} + M_{13}). \tag{20}$$

We are primarily interested in the two lowest branches of the waveguide's dispersion curves, illustrated in Fig. 4.

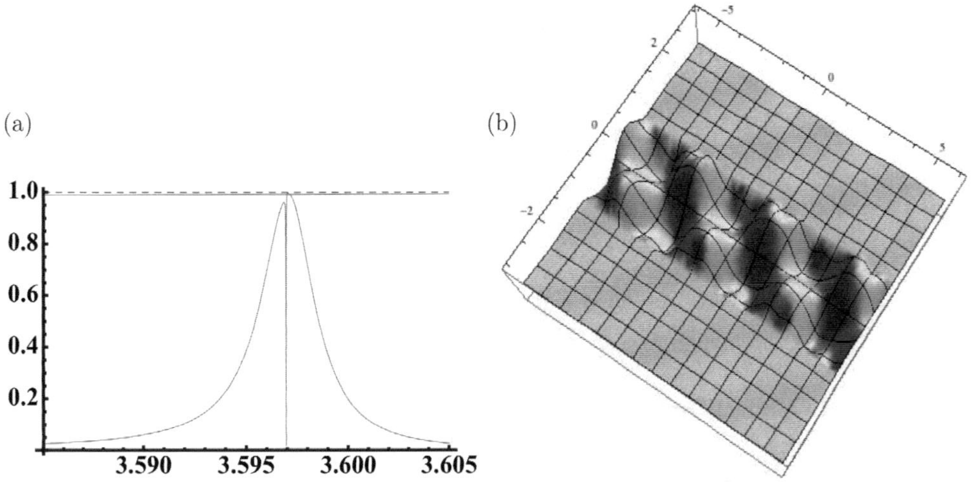

Figure 5. Normalised transmitted energy versus spectral parameter β for a triplet of rigid pins with angle of incidence $\theta_i = 0.481273$ and shift of the central grating $\xi = 0.012$ (b) Field plots for the example illustrated in (a).

We find that the odd and even modes cross. The even modes are represented by the curve with lower values of β after the intersection point, and they satisfy the condition

$$M_{12} = -\sqrt{M_{11}(M_{11} + M_{13})/2}. \tag{21}$$

The crossing of the two dispersion curves is synonymous with a double eigenvalue case, and is of great interest. In Fig. 5, we consider the corresponding transmission resonance example for a specific angle of incidence in the vicinity of the intersection point, and we observe an EDIT-like effect, when we introduce a small non-zero shift of the central grating $\xi = 0.012d$.

4. Tuning of the resonance transmission system

The two approaches outlined above present the grating stack as a periodic scatterer and as a waveguide. The results of analysis complement each other in a way that enables us to design a system possessing Elasto-Dynamically Inhibited Transmission (EDIT).

Based on a straightforward analysis of a pair of gratings of rigid pins, one evaluates the transmission coefficients and obtains resonance regimes as shown in Fig. 6. In part (a) we show the transmission resonances for a pair of rigid pin gratings with vertical separation $\eta = d$ (red curve), and $\eta = 2d$ (blue curve). In part (b) we illustrate the corresponding unshifted triplet with the same angle of incidence $\theta_i = 30°$. We note that the anti-symmetric mode with $\beta = 3.61747$ is common to both diagrams. The symmetric mode of the triplet is linked to the mode for $\eta = d$ featured in part (a) with $\beta = 3.58221$.

Furthermore, an addition of a third grating changes the symmetric resonance modes. The required tuning of the grating stack as a scattering system is rather complex, but it becomes a straightforward task with the knowledge of dispersion properties of trapped flexural waves, characterised by the dispersion curves in Fig. 4. The intersection point between these curves corresponds to the frequency in the neighbourhood of EDIT. Further refinement is achieved by the choice of the shift parameter ξ. An example for a rigid pin triplet is illustrated in Fig. 7.

Acknowledgements. S.G.H. gratefully acknowledges the financial support of the Duncan Norman Charitable Trust through the Duncan Norman Research Scholarship. N.V.M. and

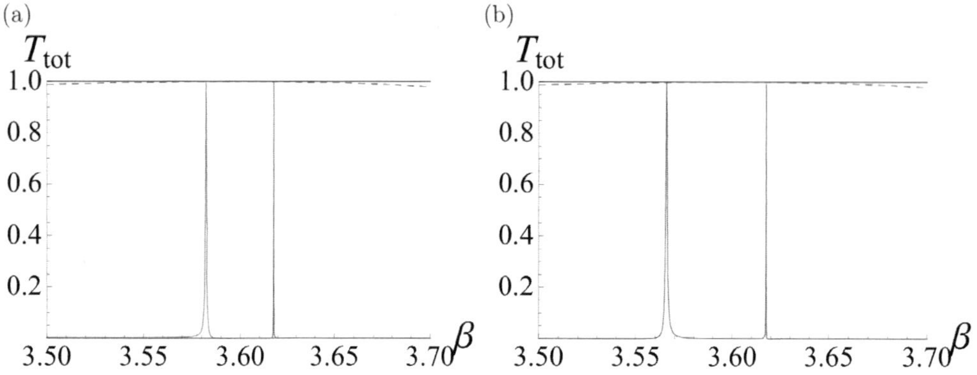

Figure 6. Normalised transmitted energy versus spectral parameter β for $\theta_i = 30°$ for (a) a pair of rigid pin gratings with $\eta = d$ (red curve) and $\eta = 2d$ (blue curve). (b) a triplet of rigid pin gratings.

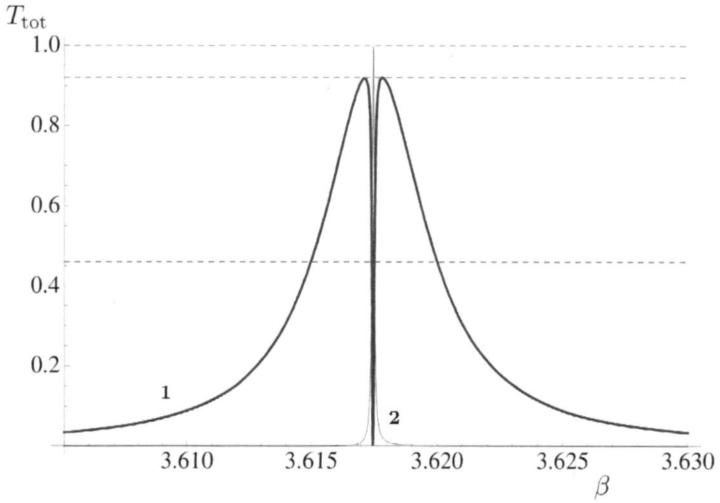

Figure 7. EDIT effect for a triplet of rigid pin gratings with the central grating shifted by $\xi = 0.25200d$ for the angle of incidence $\theta_i = 30°$. Total transmittance T_{tot} (curve **1**) as a function of β for the triplet. Curve **2** represents the total transmittance for the outer pair of gratings.

R.C.M. acknowledge the financial support of the European Community's Seven Framework Programme under the contract number PIAPP-GA-284544-PARM-2. R.C.M. also acknowledges support from the Australian Research Council through its Discovery Grants Scheme.

References

[1] Haslinger S G, Movchan N V, Movchan A B and McPhedran R C 2012 Transmission, trapping and filtering of waves in periodically constrained elastic plates *Proc. R. Soc. A* **468** 76-93

[2] Haslinger S G, McPhedran R C, Movchan N V and Movchan A B 2013 Localisation near defects and filtering of flexural waves in structured plates *Int. J. Fract.* 0376-9429

[3] McPhedran R C, Movchan A B and Movchan N V 2009 Platonic crystals: Bloch bands, neutrality and defects *Mechanics of Materials* **41** 356-363

[4] Movchan A B, Movchan N V and McPhedran R C 2007 Bloch-Floquet bending waves in perforated thin plates *Proc. R. Soc. A* **463** 2505-2518

[5] Movchan N V, McPhedran R C, Movchan A B and Poulton C G 2009 Wave scattering by platonic grating stacks *Proc. R. Soc. A* **465** 3383 - 3400

[6] Poulton C G, McPhedran R C, Movchan N V and Movchan A B 2010 Convergence properties and flat bands in platonic crystal band structures using the multipole formulation *Waves in Random and Complex Media* **20: 4** 702 - 716

[7] Evans D V and Porter R 2007 Penetration of flexural waves through a periodically constrained thin elastic plate in *vacuo* and floating on water *J. Eng. Maths.* **58** 317-337

Mathematical problems of interaction of different dimensional physical fields

G Chkadua

King's College London, Department of Mathematics, Strand, London WC2R 2LS, UK

E-mail: `george.chkadua@kcl.ac.uk`

Abstract. In the paper we consider mathematical problem of interaction of different dimensional physical fields for complex composite structures. We investigate the mixed transmission problem arising in the model of fluid-solid acoustic interaction when a piezo-ceramic elastic body (Ω^+) is embedded in an unbounded fluid (Ω^-). The corresponding physical process is described by boundary-transmission problems for second order partial differential equations. In particular, in the bounded domain Ω^+ we have 4×4 dimensional matrix strongly elliptic second order partial differential equation, while in the unbounded complement domain Ω^- we have a scalar Helmholtz equation describing an acoustic wave propagation. The physical kinematic and dynamic relations mathematically are described by appropriate boundary and transmission conditions. With the help of the potential method and theory of pseudodifferential equations the uniqueness and existence theorems are proved in Sobolev-Slobodetski spaces.

1. Introduction

In applications we rather often encounter multi-component composed bodies, where in different components we have different dimensional physical fields. In such cases, we actually have interaction problems for physical fields with different dimensions in the adjacent domains. The corresponding mathematical models for composed bodies are described by boundary, boundary-contact and transmission problems for systems of partial differential equations (PDE). The matrix operators generated by these PDEs have different orders in different domains. Furthermore, the situation becomes complicated because of the fact that on the interfaces between the adjacent domains of the composed body one needs to find appropriate interface conditions relating the different dimensional physical fields.

Due to the rapidly increasing use of composite materials in modern industrial and technological processes, on the one hand, and in biology and medicine, on the other hand, the mathematical modeling related to complex composite structures and their mathematical analysis became very important form the theoretical and practical points of views in recent years.

We consider the model of fluid-solid acoustic interaction when a piezo-ceramic elastic body (Ω^+) is embedded in an unbounded fluid (Ω^-). In this case in the domain Ω^+ we have four-dimensional piezoelectric field (a displacement vector with three components and an electric potential) while in the unbounded domain Ω^- we have a scalar acoustic wave field.

One such example of interacting of different dimensional fields is a piezoelectric transducer. Which can be used either to transform an electric current to an acoustic pressure field or, the opposite, to produce an electric current from an acoustic field. These devices are generally useful for applications that require the generation of sound in air and liquids. Examples of such

applications include phased array microphones, ultrasound equipment, inkjet droplet actuators, drug discovery, sonar transducers, bioimaging, and acousto-biotherapeutics (see [6], [19–22]).
Only few papers are devoted to the strict mathematical analysis of the above named problems. In the following papers mainly are obtained numerical results (see [13], [15], [16]).
In the present paper for boundary-transmission problems generated by the above mentioned mathematical model, we prove the existence and uniqueness theorems in Sobolev-Slobodetski spaces.
Same type problems for the classical model of elasticity (elastic body is embedded in fluid) are considered and studied in the papers: [1–5], [7–9], [11], [12], [14], [17], [18]. In these papers in the bounded domain Ω^+ is three dimensional elastic field (displacement vector with three components) and scalar wave field in unbounded domain Ω^-.
In our case addition of electric potentials complicates investigation and needs special analyses. We investigate the above mentioned problems with the use of the potential method and the theory of pseudodifferential equations on manifolds.

2. Piezoelectric field

Let Ω^+ be a bounded 3-dimensional domain in $\mathbb{R}^3$ with a compact, C^∞-smooth boundary $S = \partial\Omega^+$ and $\Omega^- := \mathbb{R}^3 \setminus \overline{\Omega^+}$. Assume that the domain Ω^+ is filled with an anisotropic homogeneous piezoelectric material.
The basic equation of steady state oscillation of the piezoelectricity for anisotropic homogeneous media is written as follows:

$$A(\partial,\omega)U + F = 0 \qquad \text{in } \Omega^+,$$

where $U = (u,\varphi)^\top$, $u = (u_1, u_2, u_3)$ is a displacement vector, $\varphi = u_4$ is an electric potential and $F = (F_1, F_2, F_3, F_4)^\top$. The three-dimensional vector (F_1, F_2, F_3) is mass force density, while $-F_4$ is charge density. $A(\partial,\omega)$ is the matrix differential operator,

$$A(\partial,\omega) = [A_{jk}(\partial,\omega)]_{4\times 4},$$

$$\begin{aligned} A_{jk}(\partial,\omega) &= c_{ijkl}\partial_i\partial_l + \rho_1\omega^2\delta_{jk}, \qquad A_{j4}(\partial,\omega) = e_{lij}\partial_l\partial_i \\ A_{4k}(\partial,\omega) &= -e_{ikl}\partial_i\partial_l \qquad\qquad\quad A_{44}(\partial,\omega) = \varepsilon_{il}\partial_i\partial_l \qquad j,k = 1,2,3, \end{aligned} \tag{1}$$

where $\omega > 0$ is the oscillation (frequency) parameter, ρ_1 is the density of piezoelectric material, $c_{ijlk}, e_{ikl}, \varepsilon_{il}$ are elastic, piezoelectric and dielectric constants respectively, δ_{jk} is the Kronecker symbol and summation is over repeated indices. These constants satisfy the following symmetry conditions:

$$c_{ijkl} = c_{jikl} = c_{klij}, \;\; e_{ijk} = e_{ikj}, \;\; \varepsilon_{ij} = \varepsilon_{ji}, \qquad i,j,k,l = 1,2,3.$$

Moreover from physical considerations it is assumed that the quadratic forms $c_{ijkl}\xi_{ij}\xi_{kl}$ and $\varepsilon_{ij}\eta_i\eta_j$ are positive definite, which provides positiveness of the internal energy:

$$c_{ijkl}\xi_{ij}\xi_{kl} \ge c_0\,\xi_{ij}\xi_{ij} \qquad \forall \xi_{ij} = \xi_{ji} \in \mathbb{R}, \tag{2}$$

$$\varepsilon_{ij}\eta_i\eta_j \ge c_1\,|\eta|^2 \qquad \forall \eta = (\eta_1,\eta_2,\eta_3) \in \mathbb{R}^3, \tag{3}$$

where c_0 and c_1 are positive constants.
$A(\partial,\omega)$ is a strongly elliptic nonselfadjoint operator.

In the theory of piezoelasticity the components of the three-dimensional mechanical stress vector acting on a surface element with a normal $n = (n_1, n_2, n_3)$ have the form

$$\sigma_{ij} n_i = c_{ijlk} n_i \partial_l u_k + e_{lij} n_i \partial_l \varphi \qquad j = 1, 2, 3,$$

while the normal components of the electric displacement vector $D = (D_1, D_2, D_3)^\top$ read as

$$-D_i n_i = -e_{ikl} n_i \partial_l u_k + \varepsilon_{il} n_i \partial_l \varphi.$$

Let us introduce the following matrix differential operator

$$T(\partial, n) = [T_{jk}(\partial, n)]_{4\times 4},$$

$$\begin{aligned} T_{jk}(\partial, n) &= c_{ijlk} n_i \partial_l, & T_{j4}(\partial, n) &= e_{lij} n_i \partial_l, & \\ T_{4k}(\partial, n) &= -e_{ikl} n_i \partial_l, & T_{44}(\partial, n) &= \varepsilon_{il} n_i \partial_l, & j, k = 1, 2, 3. \end{aligned}$$

For a vector $U = (u, \varphi)^\top$ we have

$$T(\partial, n) U = (\sigma_{1j} n_j,\ \sigma_{2j} n_j,\ \sigma_{3j} n_j,\ -D_i n_i)^\top. \tag{4}$$

The components of the vector TU given by (4) have the physical sense: the first three components correspond to the mechanical stress vector in the theory of electroelasticity, while the forth one is the normal component of the electric displacement vector.
In Green's formulae, there appears also the following boundary operator associated with the adjoint differential operator $A^*(\partial, \omega) = A^\top(-\partial, \omega)$,

$$\widetilde{T}(\partial, n) = [\widetilde{T}_{jk}(\partial, n)]_{4\times 4},$$

where

$$\begin{aligned} \widetilde{T}_{jk}(\partial, n) &= T_{jk}(\partial, n), & \widetilde{T}_{j4}(\partial, n) &= -T_{j4}(\partial, n), & \\ \widetilde{T}_{4k}(\partial, n) &= -T_{4k}(\partial, n), & \widetilde{T}_{44}(\partial, n) &= T_{44}(\partial, n), & j, k = 1, 2, 3. \end{aligned}$$

3. Scalar field

We assume that the exterior domain Ω^- is filled by a homogeneous isotropic (fluid) medium with the constant density ρ_2. Further, let some physical process (the propagation of acoustic wave) in Ω^- be described by a complex-valued scalar function (scalar field) w being a solution of the homogeneous wave equation (Helmholtz equation)

$$\Delta \mathrm{w} + \rho_2 \omega^2 \mathrm{w} = 0 \qquad \text{in } \Omega^-, \tag{5}$$

where $\Delta = \sum\limits_{j=1}^{3} \frac{\partial^2}{\partial x_j^2}$ is the Laplace operator.

We say that a solution w to the Helmholtz equation (5) satisfies the classical Sommerfeld radiation condition if the following relation holds

$$\frac{\partial \mathrm{w}(x)}{\partial |x|} - i\sqrt{\rho_2}\,\omega \mathrm{w}(x) = O(|x|^{-2}), \qquad |x| \to \infty. \tag{6}$$

Note that if solutions of Helmholtz equation (5) in Ω^- satisfy Sommerfeld radiation condition then (see [23])

$$\mathrm{w}(x) = O(|x|^{-1}), \qquad |x| \to \infty.$$

Denote by $Som(\Omega^-)$ a class of solutions of the Helmholtz equation (5) satisfying the Sommerfeld radiation condition.
Let Ω be a domain in $\mathbb{R}^3$ with a compact simply connected boundary $\partial\Omega \in C^\infty$.
By $H^s(\Omega)$ ($H^s_{loc}(\Omega)$) and $H^s(\partial\Omega)$, $s \in \mathbb{R}$, we denote the L_2 based Sobolev-Slobodetski spaces on a domain Ω and on a closed manifold $\partial\Omega$.
If $\mathcal{M}$ is a smooth proper submanifold of a manifold $\partial\Omega$, then we denote by $\widetilde{H}^s(\mathcal{M})$ the subspace of $H^s(\partial\Omega)$,

$$\widetilde{H}^s(\mathcal{M}) := \{g : g \in H^s(\partial\Omega), \operatorname{supp} g \subset \overline{\mathcal{M}}\},$$

while $H^s(\mathcal{M})$ denotes the space of restrictions on $\mathcal{M}$ of functions from $H^s(\partial\Omega)$,

$$H^s(\mathcal{M}) := \{r_{\mathcal{M}} g : g \in H^s(\partial\Omega)\},$$

where $r_{\mathcal{M}}$ is the restriction operator onto $\mathcal{M}$.

4. Formulation of mixed type interaction problem for steady state oscillation equations

Let us consider interaction problem of fluid and piezoelectric body. We assume that $S = \partial\Omega^+ = \partial\Omega^- \in C^\infty$.
Let the boundary $S = \partial\Omega^+ = \partial\Omega^-$ be divided into two disjoint parts S_D and S_N, i.e. $S = \overline{S_D} \cup \overline{S_N}$, $S_D \cap S_N = \varnothing$ and $l_m := \partial S_D = \partial S_N \in C^\infty$.
Mixed type problem (M_ω): Find a vector-function $U = (u, \varphi)^\top \in [H^1(\Omega^+)]^4$ and scalar function $\mathrm{w} \in H^1_{loc}(\Omega^-) \bigcap Som(\Omega^-)$ satisfying the following condition

$$\begin{aligned}
A(\partial, \omega)U &= 0 && \text{in } \Omega^+, && (7)\\
\Delta \mathrm{w} + \rho_2\omega^2 \mathrm{w} &= 0 && \text{in } \Omega^-, && (8)\\
\{u \cdot n\}^+ &= b_1\{\partial_n \mathrm{w}\}^- + f_0 && \text{on } S, && (9)\\
\{[T(\partial, n)U]_j\}^+ &= b_2\{\mathrm{w}\}^- n_j + f_j && \text{on } S, \quad j = 1, 2, 3, && (10)\\
\{\varphi\}^+ &= f^{(D)} && \text{on } S_D, && (11)\\
\{[T(\partial, n)U]_4\}^+ &= f^{(N)} && \text{on } S_N, && (12)
\end{aligned}$$

where $f_0 \in H^{-1/2}(S)$, $f_j \in H^{-1/2}(S)$, $j = 1, 2, 3$, $f^{(D)} \in H^{1/2}(S_D)$, $f^{(N)} \in H^{-1/2}(S_N)$. $b_1 b_2 \neq 0$, $\mathrm{Im}[\overline{b}_1 b_2] = 0$ and the symbols $\{\cdot\}^\pm$ denote the traces on S from $\Omega^\pm$.
Note that the transmission conditions (9)-(10) are kinematic and dynamic conditions. For interaction problem of fluid and piezoelectric body $\mathrm{w}(x) = P^{sc}(x)$ is pressure of scattered acoustic wave,

$$b_1 = [\rho_2\omega^2]^{-1}, \quad b_2 = -1, \quad f_0(x) = f_0^{inc}(x) = [\rho_2\omega^2]^{-1}\{\partial_n P^{inc}(x)\}^-,$$

$$(f_1, f_2, f_3) = b_2 n(x)\{P^{inc}(x)\}^-,$$

where P^{inc} is plane incident wave,

$$P^{inc}(x) = e^{id\cdot x}, \quad d = \omega\sqrt{\rho_2}\eta, \quad |\eta| = 1,$$

here $d = (d_1, d_2, d_3)$ denotes the direction of propagation of the plain incident wave.

5. Uniqueness of solution of the problem (M_ω)

We denote by $J_M(\Omega^+)$ the set of values of the frequency parameter $\omega > 0$ for which the following boundary value problem

$$A(\partial,\omega)U = 0 \quad \text{in } \Omega^+, \tag{13}$$
$$\{u\cdot n\}^+ = 0 \quad \text{on } S, \tag{14}$$
$$\{[T(\partial,n)U]_j\}^+ = 0 \quad \text{on } S, \quad j=1,2,3, \tag{15}$$
$$\{\varphi\}^+ = 0 \quad \text{on } S_D, \tag{16}$$
$$\{[T(\partial,n)U]_4\}^+ = 0 \quad \text{on } S_N, \tag{17}$$

has a nontrivial solution $U = (u,\varphi)^\top \in [H^1(\Omega^+)]^4$, where homogeneous mixed boundary conditions are given (see. [10]).
Solutions of the problem (13)-(17) are called Jones modes, while the corresponding values of ω are called Jones eigenfrequencies. The spaces of Jones modes corresponding to ω we denote by $X_{M,\omega}(\Omega^+)$. The spaces of Jones modes for the differential operator $A^*(\partial,\omega)$ denote by $X^*_{M,\omega}(\Omega^+)$.
Note that $J_M(\Omega^+)$ are at most enumerable, and for each ω corresponding space of Jones modes $X_{M,\omega}(\Omega^+)$ are of finite dimension.

THEOREM 5.1 *Let a pair (U,w) be a solution of the homogeneous problem (M_ω). Then $U \in X_{M,\omega}(\Omega^+)$ and $\mathrm{w}=0$ in Ω^-.*

COROLLARY 5.2 *Let $\omega \notin J_M(\Omega^+)$. Then the homogeneous problem (M_ω) has only the trivial solution.*

6. Existence results for the problem (M_ω)

THEOREM 6.1 *If $\omega \notin J_M(\Omega^+)$, then the problem (M_ω) is uniquely solvable.*

THEOREM 6.2 *If $\omega \in J_M(\Omega^+)$, then mixed type problem (M_ω) is solvable if and only if the orthogonality condition*

$$\sum_{j=1}^{3}\langle f_j,\{U_j\}^+\rangle_S - \langle\{[\overline{\widetilde{T}U}]_4\}^+,\overline{f^{(D)}}\rangle_{S_D} + \langle f^{(N)},\{U_4\}^+\rangle_{S_N} = 0, \qquad U\in X^*_{M,\omega}(\Omega^+), \tag{18}$$

holds, and the solution is defined modulo Jones modes $X_{M,\omega}(\Omega^+)$. Where the symbols $\langle\,,\,\rangle_S$, $\langle\,,\,\rangle_{S_D}$, $\langle\,,\,\rangle_{S_N}$ denote the duality between the spaces $H^{-1/2}(S)$ and $H^{1/2}(S)$, $\widetilde{H}^{-1/2}(S_D)$ and $H^{1/2}(S_D)$, $H^{-1/2}(S_N)$ and $\widetilde{H}^{1/2}(S_N)$ respectively.

REMARK 6.3 *Let $(f_1,f_2,f_3) = n\varphi$, where φ is scalar function and n is unit normal vector to S. Then necessary and sufficient condition (18), to be solvable the problem (M_ω), can be written*

$$-\langle\{[\overline{\widetilde{T}U}]_4\}^+,\overline{f^{(D)}}\rangle_{S_D} + \langle f^{(N)},\{U_4\}^+\rangle_{S_N} = 0.$$

References

[1] Bielak J and MacCamy R C 1991 Symmetric fnite element and boundary integral coupling methods for fluid-solid interaction *Quart. Appl. Math.* **49** 107-119
[2] Bielak J, MacCamy R C and Zeng X 1991 Stable coupling method for interface scattering problems (Research Report R-91-199 Dept. of Civil Engineering, Carnegie Mellon University)
[3] Boström A 1980 Scattering of stationary acoustic waves by an elastic obstacle immersed in a fluid *J. Acoust. Soc. Amer.* **67** 390-398

[4] Boström A 1984 Scattering of acoustic waves by a layered elastic obstacle in a fluid - an improved null field approach J. *Acoust. Soc. Amer.* **76** 588-593

[5] Goswami P P, Rudolphy T J, Rizzo F J and Shippy D J 1990 A boundary element method for acoustic-elastic interaction with applications in ultrasonic NDE *J. Nonde-struct. Eval.* **9** 101-112

[6] Gururaja T R 1992 Piezoelectric transducers for medical ultrasonic imaging *Proceedings of the Eighth IEEE International Symposium on applications of Ferroelectrics* 259-265

[7] Hsiao G C 1994 On the boundary-field equation methods for fluid-structure interactions *Proceedings of the 10.TMP, Teubner Texte zur Mathematik* Bd. **134** (Stuttgart-Leipzig) 79-88

[8] Hsiao G C, Kleinman R E and Schuetz L S1989 On variational formulation of boundary value problems for fluid-solid interactions. In: MacCarthy M.F. and Hayes M.A. (eds): Elastic Wave Propagation *IUTAM Symposium on Elastic Wave Propagation* (North-Holland-Amsterdam) 321-326

[9] Hsiao G C, Kleinman R E and Roach G F 1997 Weak solution of fluid-solid interaction problems (Technische Hochschule Darmstadt, Fachbereich Mathematik) Preprint-Nr.1917 May

[10] Jones D S 1983 Low-frequency scattering by a body in lubricated contact *Quart. J. Mech. Appl. Math.* **36** 111-137

[11] Jentsch L and Natroshvili D 1990 Non-local approach in mathematical problems of fluid-structure interaction *Math. Methods Appl. Sci.* **22** 13-42

[12] Junger M C and Fiet D 1986 Sound, Structures and Their Interaction (MIT Press, Cambridge, MA)

[13] Kagawa Y and Yamabuchi T 1979 Finite Element Simulation of a Composite Piezoelectric Ultrasonic Transducer *IEEE Transactions on Sonics and Ultrasonics* **26** (2) 81-87

[14] Luke C J and Martin P A 1995 Fluid-solid interaction: acoustic scattering by a smooth elastic obstacle *SIAM J. Appl. Math.* **55** (4) 904-922

[15] Lerch R 1988 Finite element analysis of piezoelectric transducers *IEEE 1988 Ultrasonics Symposium Proceedings* **2** 643-654

[16] Lerch R 1990 Simulation of piezoelectric devices by two- and three-dimensional finite elements *IEEE Transactions on Ultrasonics, Ferroelectrics, and Frequency Control* **37** (3) 233-247

[17] Natroshvili D and Sadunishvili G 1996 Interaction of Elastic and Scalar Fields *Mathematical Methods in the Applied Sciences* **19** 1445-1469

[18] Natroshvili D, Sadunishvili G, Sigua I and Tediashvili Z 2005 Fluid-Solid Interaction: Acoustic Scattering by an Elastic Obstacle with Lipschitz Boundary *Memoirs on Differential Equations and Mathematical Physics* **35** 91-127

[19] Neugschwandtner G S, Schwdiauer R, Bauer-Gogonea S, Bauer S, Paajanen M and Lekkala J 2001 Piezo- and pyroelectricity of a polymer-foam space-charge electret *Journal of Applied Physics* **89** (8) 4503 - 4511

[20] Nguyen-Dinh A, Ratsimandresy L, Mauchamp P, Dufait R, Flesch A and Lethiecq M 1996 High frequency piezo-composite transducer array designed for ultrasound scanning applications *1996 IEEE Ultrasonics Symposium Proceedings* **2** 943 - 947

[21] Smith W A 1990 The application of 1-3 piezocomposites in acoustic transducers *Proceedings 1990 IEEE 7th International Symposium on Applications of Ferroelectrics* 145 - 152

[22] Ting R Y 1992 A review on the development of piezoelectric composites for underwater acoustic transducer applications *IEEE Transactions on Instrumentation and Measurement* **41** (1) 64-67

[23] Vekua I 1943 On metaharmonic functions *Proceedings of Mathematical Institute of the Georgian Academy of Sciences* **12** 105-174

Journal of Physics: Conference Series **451** (2013) 012026 doi:10.1088/1742-6596/451/1/012026

Modeling of parametrically excited vibrating screen

L I Slepyan[1] **and V I Slepyan**[2]

[1] School of Mechanical Engineering, Tel Aviv University, 69978 Tel Aviv, Israel
[2] Loginov and Partners Mining Company, 2v Korolyova Avenue 03134 Kiev, Ukraine

E-mail: leonid@eng.tau.ac.il (Leonid Slepyan)

Abstract. The dynamic model of a RR-based rectangular vibrating screen is considered as an initially stretched system of two equal masses connected by a linearly elastic spring. Due to the geometric nonlinearity longitudinal oscillations of the masses and lateral oscillations of the spring are coupled. Under certain conditions, when the masses are subjected by a self-equilibrated periodic longitudinal action, the parametric resonance arises which amplitude is bounded by the nonlinearity. The dynamic problem is reduced to a system of two ordinary nonlinear equations. An exact analytical solution is found existing under some conditions. In a general case, the dynamics of this system is considered numerically. The dissipation in this process is estimated. A comparative analysis of the dynamics of conventional and PR-based screeners is presented. Based on the analysis of the model a vibrating screen machine was designed, built and set up in LPMC. The machine operated as predicted.

1. Introduction

The idea to create a PR-based oscillating screen came to us in 2007 while discussing drawbacks of existing types of the machines. In 2009, Ukrainian patent was issued on the excitation method of an oscillation screen and the corresponding structure of the latter [1]. A mathematical model of the PR-based machine was developed. By means of numerical simulations parameters of its structure and setting were determined, which ensure effective sustained PR-oscillation mode. Note that while the parametric resonance was mainly considered as an undesirable phenomenon, some attempts were made to use it to obtain a greater response to a moderate excitation [2], also see [3-5]. In this paper, the following topics are briefly considered: The conventional vibrating screen versus the PR-based screen. The PR-based screen mathematical model. Free and forced oscillations of the system. An exact solution for a forced oscillation regime. Some results of numerical simulations and, lastly, Estimation of the dissipation based on analysis of the collision of a particle with some types of the sieve.

2. Conventional model with the lateral excitation of the sieve

We consider a plane problem for a flexible initially stretched sieve, $-l < x < l$, excited by lateral harmonic oscillations of its edges. In the linear approximation, the sieve dynamics is described by the wave equation for a string

$$T\frac{\partial^2 v(x,t)}{\partial x^2} - \varrho\frac{\partial^2 v(x,t)}{\partial t^2} - \alpha\frac{\partial v(x,t)}{\partial t} = 0\,, \tag{1}$$

where $v(x,t)$ is the lateral displacement, T is the tensile force per unit length of the sieve, ϱ is the mass density per unit area and α is the respective viscosity number. For forced vibrations with frequency ω the oscillation amplitude is found to be distributed as

$$U(\xi) = U_0\sqrt{\frac{\cosh^2(k_1\xi) - \sin^2(k_2\xi)}{\cosh^2(k_1) - \sin^2(k_2)}}, \quad k_{1,2} = \frac{1}{\sqrt{2}}\sqrt{\sqrt{\lambda_1^2 + \lambda_2^2} \mp \lambda_1}\,, \tag{2}$$

where $\lambda_1 = \varrho l^2\omega^2/T, \lambda_2 = \alpha l^2\omega/T, \xi = x/l$.

Separation of wet granular materials, like sand or small-size gravel, meets a greater resistance doe to the size-dependent surface effects. In the considered model, this reflects in increase of the viscosity number, that is, in the increase of parameter λ_2, and in decrease of the oscillation amplitude. As a result, the conventional oscillation screens become inefficient. In contrast, the longitudinal oscillations of the sieve meet a very low inelastic resistance, and the PR excitation results in much greater amplitudes.The resonant regime corresponds to $\lambda_1 = \pi^2/4$. The results as the output-to-input ratio, $\Lambda_U = U(0)/U_0$ for the conventional screen and the lateral-to-longitudinal amplitude ratio for the PR-based screen, Λ_P, for $\lambda_2 = 1$ and $\lambda_2 = 2.5$ are as follows: $\Lambda_U = 3.137, \Lambda_P = 21.2, \mathbf{\Lambda_P/\Lambda_U = 6.76}$ $(\lambda_2 = 1)$; $\Lambda_U = 1.245, \Lambda_P = 13.3, \mathbf{\Lambda_P/\Lambda_U = 10.7}$ $(\lambda_2 = 2.5)$. This shows how the PR-based screen is efficient.

3. The PR-based screen model

The model considered here consists of two equal masses, M, connected by a spring of the mass density ϱ per unit length, the length $2l$ and the stiffness in tension $k/2$. Each of the masses is also connected with the rigid foundation by a side spring, which stiffness $\varkappa \ll k$, Fig. 1. The springs are initially stretched; the initial tensile force is denoted by T_0. The symmetric motion is considered, in which the masses can oscillate moving only along the initial spring line (horizontally), while the central spring (the sieve) can oscillate in normal direction. The oscillations are excited by vibrators acting synchronically on the left and the right masses in opposite horizontal directions. The amplitude and frequency of the action are denoted by P and ω, respectively. The lateral oscillation amplitude is assumed to be much less than the spring length. Also it is assumed that the inequalities $c = \sqrt{kl/\varrho} \gg \omega l, M \gg \varrho l$ are valid. This allows us to assume that the tensile force in spring, T, is independent of the coordinate x, whereas the spring inertia can be taken into account only in the lateral motion. So, if there is no dissipation the main natural longitudinal and lateral frequencies for small amplitudes are $\Omega_L = \sqrt{k/M}, \Omega_T = \pi/(2l)\sqrt{T_0/\varrho}$.

Under oscillations of the sieve the granular material separated mainly hangs at a distance from the sieve. We assume that the material-sieve interaction results in a force equal to the material weight and in dissipation, which is presented as linear viscosity.

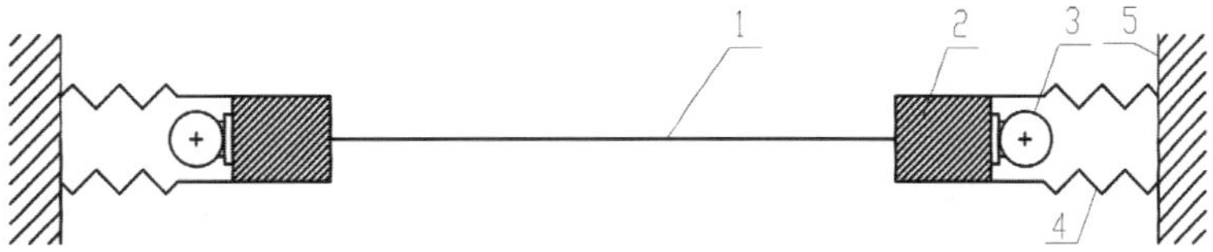

Figure 1. The PR-based oscillating screen model: the spring (the sieve) - 1, the edge mass - 2, the vibrator - 3, the side spring - 4 and the foundation - 5.

If ω is close enough to $2\Omega_T$ and P is large enough, the parametric resonance arises under which the amplitude of the lateral oscillations is bounded by the geometric nonlinearity. We based on nonlinear dynamic equations with respect to the longitudinal and lateral oscillations

$$M\frac{\mathrm{d}^2u(t)}{\mathrm{d}t^2} + \beta\frac{\mathrm{d}u(t)}{\mathrm{d}t} + T_1(t) = P\cos\omega t\,,$$
$$\varrho\frac{\partial^2 v(x,t)}{\partial t^2} + \alpha\frac{\partial v(x,t)}{\partial t} - T(t)\frac{\partial^2 v(x,t)}{\partial x^2} = q(x,t)\,, \tag{3}$$

which are coupled by the tensile force, $T_1(t)$, depending on both the displacement of the masses, $\pm u(t)\,(x=\pm l)$ and the displacement of the spring, $v(x,t)$

$$T(t) = T_0 + T_1(t)\,, \quad T_1(t) = k\left[u(t) + \frac{1}{2}\left(\int_0^l \sqrt{1+(v'(x,t))^2}\,\mathrm{d}x - l\right)\right]\,, \tag{4}$$

where $v'(x,t) = \partial v(x,t)/\partial x$. In addition, in Eq. (3), α and β are the viscosity numbers and the averaged value of q is the treated material weight per unit length. Expression (4) is valid, however, if it defines a nonnegative tensile force; otherwise, $T(t)=0$.

The left side of the second equation in (3) with the boundary conditions, $v(\pm l,t)=0$, admits the variables separation. For the main term of the series, $v(x,t) = w(t)\cos(\pi x/2l)$ we obtain

$$\varrho\frac{\mathrm{d}^2w(t)}{\mathrm{d}t^2} + \alpha\frac{\mathrm{d}w(t)}{\mathrm{d}t} + \frac{\pi^2}{4l^2}T(t)w(t) = q_0(t) = \frac{1}{l}\int_{-l}^{l} q(x,t)\cos\frac{\pi x}{2l}\,\mathrm{d}x\,,$$
$$M\frac{\mathrm{d}^2u(t)}{\mathrm{d}t^2} + \beta\frac{\mathrm{d}u(t)}{\mathrm{d}t} + T_1(t) = P\cos\omega t\,, \quad T_1(t) = k\left(u(t) + \frac{\pi^2}{16l}w^2(t)\right)\,. \tag{5}$$

4. Free oscillations

The total energy of the oscillations as the sum of the kinetic and potential energies is

$$\mathcal{E} = M[\,\mathrm{d}u(t)/\,\mathrm{d}t]^2 + \frac{1}{2}\varrho l[\,\mathrm{d}w(t)/\,\mathrm{d}t]^2 + T_0\frac{\pi^2}{8l}w^2(t) + \frac{1}{k}T_1^2\,. \tag{6}$$

It can be considered consisting of longitudinal and lateral parts as follows (note that this separation is in a sense conditional) :

$$\mathcal{E} = \mathcal{E}_u + \mathcal{E}_w\,,$$
$$\mathcal{E}_u = M[\,\mathrm{d}u(t)/\,\mathrm{d}t]^2 + \frac{1}{k}T_1^2(t) \quad \text{(longitudinal part)}\,,$$
$$\mathcal{E}_w = \frac{1}{2}\varrho l[\,\mathrm{d}w(t)/\,\mathrm{d}t]^2 + \frac{\pi^2}{8l}T_0 w^2(t) \quad \text{(transversal part)}\,. \tag{7}$$

From this and the dynamic equations with $P=q=\alpha=\beta=0$ we find the energy exchange rate between the longitudinal and transversal parts

$$\frac{\mathrm{d}\mathcal{E}_u}{\mathrm{d}t} = -\frac{\mathrm{d}\mathcal{E}_w}{\mathrm{d}t} = \frac{\pi^2}{4l}T_1 w(t)\frac{\mathrm{d}w(t)}{\mathrm{d}t}\,. \tag{8}$$

Another system of two-mode nonlinear oscillations with the PR-related ratio of frequencies, a spring pendulum, was considered long ago in [6] (also see [7]) and then in [8] assuming the angle amplitude small enough. In some respects, this system is similar to the oscillating screen, and we briefly consider it. The pendulum model consists of the point mass, M, connected to a fixed point by a massless spring of stiffness k. Let the initial spring length be l_0, $l(t) = l_0 + u(t)$

and $\alpha(t)$ be the angle ($\alpha = 0$ is statics). In the exact formulation, with no restrictions on values of the angle and the spring elongation, the energy is

$$\mathcal{E} = \Pi + \mathcal{K}\,, \quad \Pi = \frac{1}{2}ku(t)^2 + Mgl(t)[1-\cos\alpha(t)]\,, \quad \mathcal{K} = \frac{1}{2}M[\dot{u}^2(t) + l^2(t)\dot{\alpha}^2(t)]\,, \tag{9}$$

where g is the gravity acceleration. The dynamic equations follow from (6) as

$$\begin{aligned} M\ddot{u}(t) + ku(t) + Mg[1-\cos\alpha(t)] - Ml(t)\dot{\alpha}^2(t) = 0\,, \\ l(t)\ddot{\alpha} + g\sin\alpha(t) + 2\dot{l}(t)\dot{\alpha}(t) = 0\,. \end{aligned} \tag{10}$$

We define the energy of longitudinal and tangential oscillations to be

$$\mathcal{E}_u = \frac{1}{2}ku(t)^2 + \frac{1}{2}M\dot{u}^2(t)\,, \quad \mathcal{E}_\alpha = Mgl(t)[1-\cos\alpha(t)] + \frac{1}{2}Ml^2(t)\dot{\alpha}^2(t)\,, \tag{11}$$

respectively. Using the dynamic equations (10) we find the rate of the energy exchange between these parts of the total energy

$$\dot{\mathcal{E}}_u = -\dot{\mathcal{E}}_\alpha = \left[l(t)\dot{\alpha}^2(t) - g(1-\cos\alpha(t))\right] M\dot{u}(t)\,. \tag{12}$$

5. Forced oscillations

Consider equations (5) with $\alpha = q = 0$. There exists a periodic solution, in which the tensile force preserves its initial value, $T_1(t) \equiv 0$

$$\begin{gathered} u(t) = -\frac{P}{\omega(M^2\omega^2+\beta^2)}\left(M\omega\cos\omega t - \beta\sin\omega t + \sqrt{M^2\omega^2+\beta^2}\right), \\ w(t) = A\cos\frac{\omega t}{2} + B\sin\frac{\omega t}{2}\,, \quad T_1(t) = 0\,, \\ A = \sqrt{\left(\sqrt{M^2\omega^2+\beta^2} + M\omega\right)\psi}\,, \quad B = \sqrt{\left(\sqrt{M^2\omega^2+\beta^2} - M\omega\right)\psi}\,, \\ \psi = \frac{16lP}{\pi^2\omega(M^2\omega^2+\beta^2)}\,, \ (-u)_{max} = \frac{2P}{\omega\sqrt{M^2\omega^2+\beta^2}}\,, \ w_{max} = \sqrt{A^2+B^2} = \frac{4}{\pi}\sqrt{-lu_{max}}\,. \end{gathered} \tag{13}$$

Numerical calculations based on the dynamic equations (5) evidence that, in some range of the parameters, this solution is stable. It corresponds to the established regime for transient solutions with $T_1(0) \neq 0$.

For some other ranges of the parameters, including those with $\alpha \neq 0, q \neq 0$, other stable-oscillation regimes are revealed. An example is presented in Fig. 2

6. Dissipation

To estimate the energy consumption in screening we consider the collision of a particle with the sieve. The latter is represented in two models: an infinite elastic beam (the sieve consists of series of initially stretched parallel beams connected by weak links) and an elastic plate model. Using integral transforms, from the beam equation

$$Dw^{IV}(x,t) - Tw''(x,t) + \varrho\ddot{w}(x,t) = -m\ddot{w}(0,t)\delta(x)H(t)\,, \tag{14}$$

where D is the bending stiffness and m is the particle mass, we have found the acceleration

$$\ddot{w}(\tau) = -\frac{2v_0a(s_0+1)}{a+2\sqrt{s_0+1}}e^{s_0\tau} + \frac{v_0a}{\pi}\int_{-\infty}^{-1}\frac{s\sqrt{-s-1}}{s^2-a^2(s+1)}e^{s\tau}\,ds\,, \quad -1 < s_0 < 0, a > 0\,, \tag{15}$$

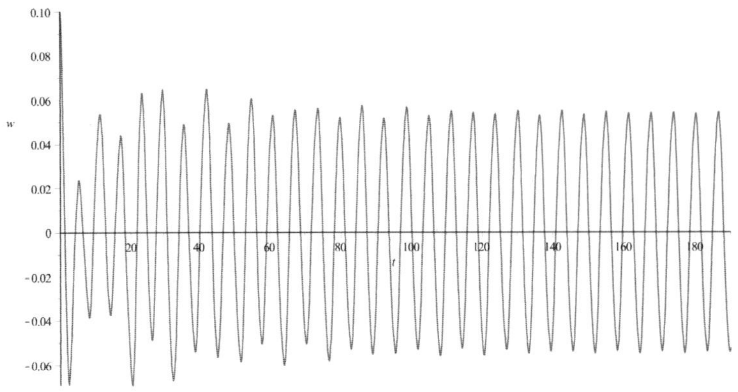

Figure 2. Forced PR-oscillations. The transversal oscillation at the center, $w(t)$; $M = 100, k = 250, T_0 = l = \varrho = 1, \alpha = 0.3, \beta = 0.05; w(0) = 0.1, u(0) = \dot{u}(0) = \dot{w}(0) = 0$.

where v_0 is the collision speed, τ is a nondimensional time. It can be seen that for $0 < t < \infty$ both terms are negative. This evidences that the collision is perfectly inelastic with zero restoration coefficient.

As the 2D model consider an infinite elastic plate stretched in x-direction and possessing the bending stiffness only corresponding to flexure in the (y, z)-plane. The dynamic equation is

$$\left(D\frac{\partial^4}{\partial y^4} - T\frac{\partial^2}{\partial x^2} + \varrho\frac{\partial^2}{\partial t^2} \right) w(x,y,t) = -m\frac{\partial^2 w(0,0,t)}{\partial t^2}\delta(x)\delta(y)H(t)\,, \tag{16}$$

From this equation the acceleration and the speed are found in the form

$$\ddot{w}(\tau) = v_0 b^{2/3} W(\tau)\,, \quad \dot{w}(\tau) = v_0\left(1 + \int_0^\tau W(x)\,\mathrm{d}x\right) \quad (b > 0)\,,$$
$$W(\tau) = -\frac{4}{3}\exp\left(-\frac{1}{2}\tau\right)\cos\left(\frac{\pi}{3} - \frac{\sqrt{3}}{2}\tau\right) + \frac{1}{\pi}\int \frac{x^{3/2}}{x^3+1}\exp(-x\tau)\,\mathrm{d}x\,. \tag{17}$$

It can be seen that, in this model, the speed and the nondimensional acceleration have unique expressions, which are independent of the plate and particle parameters. In this 2D case, the restoration coefficient is equal to 0.303 (see Fig. 3), that is, only 9% of the kinetic energy of the collision remains. Taking into account other possible factors of inelasticity, it can be concluded that the energy of perfectly inelastic collision represents the lower bound of the dissipation.

7. Conclusions

Based on the above considerations the dynamic equations were developed to reflect additional factors important for the PR-based screen setting. In particular, this concerns the sieve-material inelastic interaction, where the collisions were assumed to take place only under upward directed speed of the screen. Also note that in the vicinities of the end masses, where the sieve curvature is considerably large, the bending stiffness must be taken into account. This is important for the estimation of the sieve strength under the high-amplitude oscillations. The corresponding edge-effect solution was constructed to satisfy the clamped boundary condition end and to correspond to the mass-string solution 'at infinity'. With respect to the prospective research note that the exact analytical solution presented here is valid under certain conditions. In a more general case, analytical results can be obtained using harmonic analysis, that we are going to do.

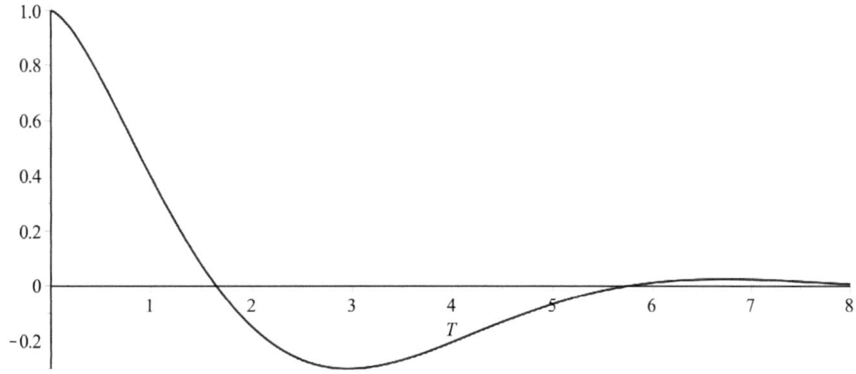

Figure 3. Collision of a particle with a plate: The nondimensional speed, $\dot{w}(\tau)/v_0$, based on (17).

Acknowledgments
We are thankful for the support provided by FP7-People-2011-IAPP European Union Grant No. 284544.

References

[1] Slepyan V I, Loginov I G and Slepyan L I 2009 The method of resonance excitation of a vibrating sieve and the vibrating screen for its implementation *Ukrainian patent on invention* No. 87369

[2] Rhoads J F and Shaw S W 2005 Tunable microelectromechanical filters that exploit parametric resonance *J. Bibrat. Acoust.* **127** 423-430

[3] Krylov S, Harari I and Cohen Y 2005 Stabilization of electrostatically actuated microstructures using parametric excitation *J. Micromech. Microeng* **15** 1188-1204

[4] Krylov S, Gerson Y, Nachmias T and Keren U 2010 Excitation of large-amplitude parametric resonance by the mechanical stiffness modulation of a microstructure *J. Micromech. Microeng* **20** 015041

[5] Plat H and Bucher I 2013 Optimizing parametric oscillators with tunable boundary conditions *J. Sound Vibrat.* **332** 487493

[6] Vitt A A and Gorelik G S 1933 *Zh. Tekh. Fiz.* **3** 294

[7] Gaponov-Grekhov A V and Rabinovich M I 1992 *Nonlinearities in action* (Berlin:Springer)

[8] Lai H M 1984 On the recurrence phenomenon of a resonant spring pendulum *Am. J. Phys.* **52(3)** 219-223

D2FAM 2013 IOP Publishing
Journal of Physics: Conference Series **451** (2013) 012027 doi:10.1088/1742-6596/451/1/012027

Diagnostic tool for structural health monitoring: effect of material nonlinearity and vibro-impact process

V R Hiwarkar[1], V I Babitsky[2] and V V Silberschmidt[2]

[1] Mechanical, Materials and Manufacturing Engineering, University of Nottingham, Nottingham, NG7 2RD, UK
[2] Wolfson School of Mechanical and Manufacturing Engineering, Loughborough University, Leicestershire,LE11 3TU, UK

E-mail: vikrant.hiwarkar@nottingham.ac.uk

Abstract. Numerous techniques are available for monitoring structural health. Most of these techniques are expensive and time-consuming. In this paper, vibration-based techniques are explored together with their use as diagnostic tools for structural health monitoring. Finite-element simulations are used to study the effect of material nonlinearity on dynamics of a cracked bar. Additionally, several experiments are performed to study the effect of vibro-impact behavior of crack on its dynamics. It was observed that a change in the natural frequency of the cracked bar due to crack-tip plasticity and vibro-impact behavior linked to interaction of crack faces, obtained from experiments, led to generation of higher harmonics; this can be used as a diagnostic tool for structural health monitoring.

1. Introduction

The presence of a crack in a structure can drastically change its general dynamic characteristics. The dynamic characteristics are affected by two possible effects occurring near the cracked zone. One of them is the crack-tip plasticity and another is a vibro-impact behavior of the crack, i.e. is the crack-face interaction. Few researchers have tried to address the issue of the effect of crack as discontinuity on dynamics of structures [1] [2]. In [3] it was assumed that the crack changes from a state of fully open to fully closed instantaneously, which gave a rise to bilinear stiffness nonlinearity that was approximated using a higher-order polynomial. Higher-order frequency response functions were defined using Volterra series for systems with polynomial type nonlinearity, and a damage-assessment procedure was suggested, based on a quantitative comparison of these functions. The finite-element scheme was developed for computing the Eigen frequencies for a cracked beam for different degrees of closure [4]. In this study, the finite-element method, the component-mode synthesis method and the linear-elastic fracture mechanics theory are integrated for modeling of the cracked structures. The method used by the authors was benchmarked against earlier results in the literature. A continuous cracked beam vibration theory developed in [5] was used for prediction of changes in transverse vibration of a simply supported beam with a breathing crack. The equation of motion and the boundary conditions of the cracked beam considered as a one-dimensional continuum were used. The Eigen frequency changes due to a breathing edge-crack were shown to depend on the bi-linear character of the system. The associated linear problems were solved over their respective domains of definition, and the solutions were matched through the initial conditions. The changes in vibration frequencies for a breathing crack were smaller than the ones caused by open cracks. Modal analysis was used to study the nonlinear dynamics of the cantilever beam with a transverse edge crack in [6]. The authors modelled nonlinearity produced by opening and closing of crack as a piece-wise linear system. To define effective natural frequencies for this system they used the idea of bilinear frequency along with a perturbation method to obtain the nonlinear normal modes of vibration and the associated period of motion. It was found that the bilinear frequency formula was a good approximation for the effective natural frequency.

In [7], a study was performed on a single-degree-of-freedom bilinear spring-mass system. Modal expansions were employed to compute the beam's response at any excitation with proper incorporation of continuity of displacement and velocity when the crack was in a state of transition. The Fourier transform of the steady-state time response showed the presence of nonlinearity, which was correlated with the location and depth of the crack. The beam was modeled as an oscillator with a bilinear restoring force and applied bi-spectral analysis to the response to study vibration of a beam with a breathing crack [8]. The bi-spectral analysis showed high sensitivity to the non-linear behavior of the system compared to other techniques. The dynamic behavior of a cantilever beam with a breathing crack was investigated under harmonic excitation theoretically and experimentally in [9]. In this work, a simple model with a single degree of freedom was developed with varying stiffness to simulate the dynamic behavior. The data for simulated and experimental responses were analyzed by applying empirical mode-decomposition and the Hilbert transform. The results obtained indicated that the instantaneous frequency oscillates between the values corresponding to the open and closed states of the crack. The variation of the instantaneous frequency increased with an increasing crack depth following a polynomial law, which could be used to estimate the crack size.

The problem of nonlinear dynamics of cracked cantilever beam under harmonic loading using the finite-element technique was investigated in [10]. The study revealed that the frequency did not change with the oscillation amplitude, but the steady-state response obtained was very rich of sub and super-harmonic components. Moreover, within the super-harmonic resonance ranges the phase portraits were characterized by significant wiggles due to impact between crack faces. In [11] numerical evaluation of a forced flexural vibration of a cantilever beam with a transverse surface crack extending uniformly along its width was performed to relate nonlinear resonances to the crack presence, location, and depth. The qualitative characteristics, namely, phase-portrait distortions, sub- and super-harmonic components in the Fourier spectrum, and curved shape of the modal line were considered. It was found that the acceleration record of the beam tip was sufficient to detect the existence of the crack and to identify its depth and site.The finite-element simulation was used to study the longitudinal harmonic loading for assessing damage in a cracked cantilever bar [12]. The nonlinear materials properties were introduced in terms of plastic behaviour. The effect of crack size and position along the beam length was studied. An observed change in the natural frequency was used to characterize the damage state of the component. In [13], [14] longitudinal vibration of the bar with a breathing crack was investigated. The results of simulations included higher-harmonics generation, a frequency shift and their dependency on crack parameters, as well as generation of low-frequency components, which can serve as a diagnostic tool for structural health monitoring.

2. Finite element model: Effect of crack tip plasticity on dynamic behavior of cracked structure

The 2D finite-element model of the un-cracked and cracked cantilever bars was implemented in commercial software MSC Marc. Figure 1 shows the methodology used for performing simulations to study the effect of crack-tip plasticity on the natural frequencies of the cracked bar. The cracked cantilever bar with length L was clamped at the left end and free at the right end. The crack was located at the distance d from the fixed end with d/L being a dimensionless crack position. Different crack depths – 2 mm, 4 mm, 6 mm, 8 mm and 10 mm – were used in analysis. The finite-element model of the cracked cantilever bar (Fig.2a) was meshed using four-node, isoparametric, arbitrary quadrilateral elements, in total,7500 elements were used. These elements use bilinear interpolation functions and the strain is constant throughout these elements. The stiffness of these elements is formed using four point Gaussian integration. Table 1 show the main parameters used in finite-element simulations. The first three natural frequencies were obtained for the cracked cantilever bar for two cases: with account for the effect of crack-tip plasticity and without it. Figure 2c presents results for the first case for which elastic-plastic material model was used while Fig. 2b for another one for which elastic material model was used.

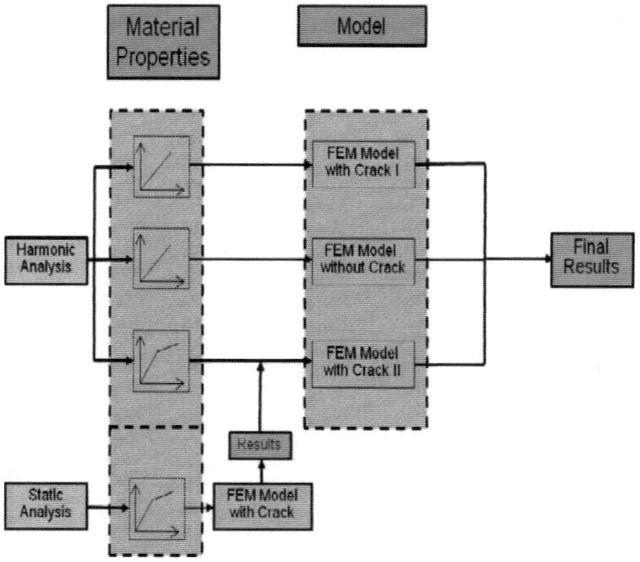

Figure 1. Methodology of simulations

Material	Aluminium (Grade 6082T6)
Length of bar (L) [mm]	300
Width of bar [mm]	10
Height of bar [mm]	25
Depth of crack [mm]	2, 4, 6, 8 and 10
Distance of crack from fixed end (d) [mm]	100, 150 and 250
Density[kg/mm^3]	2.70 x 10^{-6}
Modulus of Elasticity [N/mm^2]	7 x 10^4
Yield Strength [N/mm^2]	2.514 x 10^2
Ultimate tensile strength [N/mm^2]	3.434 x 10^5

Table 1. Simulation parameters

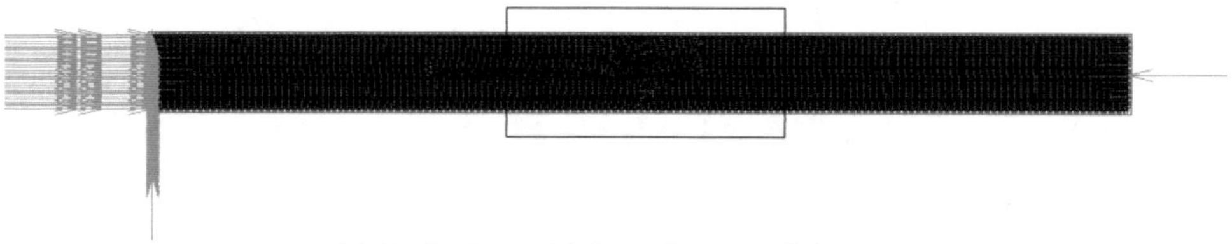

(a) Entire bar with boundary conditions

(b) Meshing near crack tip (shown by rectangle in (a)) (c) Stress concentration near crack tip

Figure 2. FEA model of a cracked cantilever bar

3. Simulation results: Effect of crack location on natural frequencies

Figure 3 shows a comparison between the first, second and third natural frequencies with and without plasticity at the crack tip for different d/L ratios and crack depths. Apparently, there was no change in the first natural frequency (Fig. 3a) whereas there were some marginal changes in the second and third natural frequencies as the d/L ratio increased (Figs. 3b and 3c)

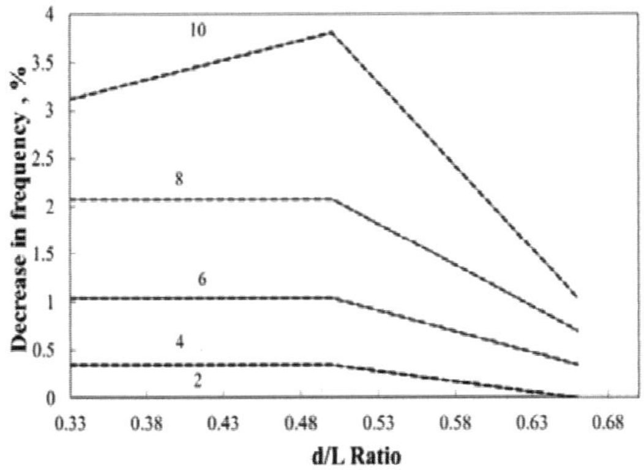

(a)

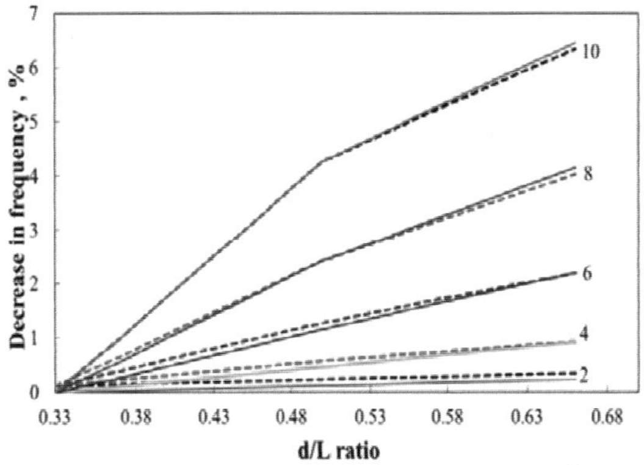

(b)

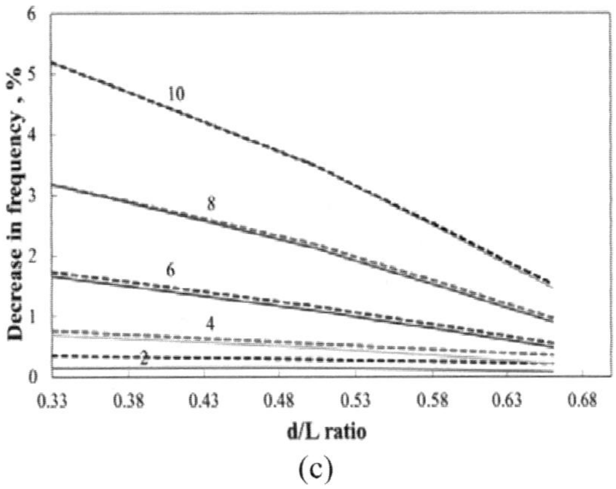

(c)

Figure 3. Effect of d/L ratio on frequencies for different crack depths (in mm): (a) First natural frequency; (b) second natural frequency; (c) third natural frequency (dotted line -with plasticity, continuous line - without plasticity)

4. Experimental Study: Effect of vibro-impact of behavior of dynamics of cracked bar

An aluminium bar with dimensions 300 mm x 25 mm x 10 mm with a fatigue crack at its centre, shown in Fig.4, was used in experiments. This aluminium bar was fixed at the left end and the other end of bar was excited in longitudinal direction by harmonic load using a shaker, which was driven by an oscillator (wave form generator). Piezoelectric strain gauges were glued at four different locations on the cracked bar to measure its dynamic response. Two of them were glued near the crack, one near the free end and the last near the fixed end of the bar. The frequency sweep was performed to obtain the resonant frequency of the bar. Then, the bar was excited by the harmonic load of 2.8 N using a shaker at this resonant frequency, and the responses were obtained at the four strain gauges in the longitudinal direction. These responses were processed using a picoscope taking into account the Nyquist criteria to avoid aliasing errors and the Hanning window to minimize a leakage error in the signal acquired.

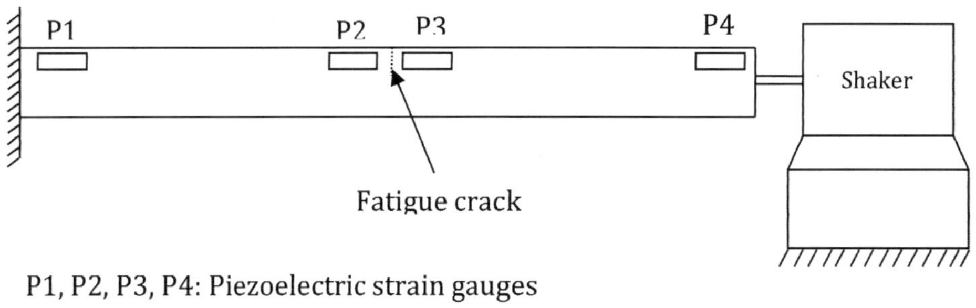

Figure 4. Schematic of experimental setup

5. Experimental results

From the frequency sweep it was found that the resonance occurred at the frequency of 2821 Hz. At this resonant frequency, the cracked bar was excited at the vibration amplitude of 2.8 N. It was found that the responses (Figs. 5b and 5c) obtained near the crack showed a higher harmonic of the forcing frequency at 8464 Hz. Similarly, the responses obtained far away from the crack showed only the resonant frequency of excitation. This gives an indication that the crack-induced nonlinearity has the localized effect on dynamics of the cracked bar leading to generation of additional perturbations in it.

6. Conclusions

This paper was focused on the effect of crack tip plasticity and the vibro-impact behavior of crack on dynamics of the cracked cantilever. The 2D finite-element simulations indicated that the presence of plasticity at the crack tip had no effect on the first natural frequency. On the contrary, as the d/L ratio increased, the second and third natural frequency of beams with different crack depths were affected. The results obtained from experiments revealed that the vibro-impact behavior of crack led to generation of higher harmonics in the response measured near the crack. It indicates that crack-induced nonlinearity had the localized effect on the dynamic behavior. The results obtained from simulations and experiments can be used as a good diagnostic tool for structural health monitoring.

7 Acknowledgement

The research leading to these results has received funding from the European Union's Seventh Framework Programme FP7 under grant agreement No. PIAPP-GA-284544-PARM-2, project PARM-2 (Vibro-Impact Machines Based on Parametric Resonance: Concepts, mathematical modelling, experimental verification and implementation).

References

[1] Dimarogonas A D 1996 Vibration of cracked structures: a state of the art review. *Eng. Fract. Mech.* **55** 831-857
[2] Doebling S W, Farrar C R and Prime M B 1998 A summary review of vibration-based damage identification methods *Shock and Vibration Digest.* **30** 91-105

[3] Ruotolo R, Surace C, Crespu P and Storer D 1996 Harmonic analysis of the vibrations of a cantilevered beam with a closing crack *Computers & Structures* **61** pp. 1057-1074

[4] Kisa M and Brandon J 2000 The effects of closure of cracks on the dynamics of a cracked cantilever beam. *J. Sound Vib.* **238** 1-18

[5] Chondros T G, Dimarogonas A D and Yao J 2001 Vibration of a beam with a breathing crack. *J. Sound Vib.* **239** 57-67

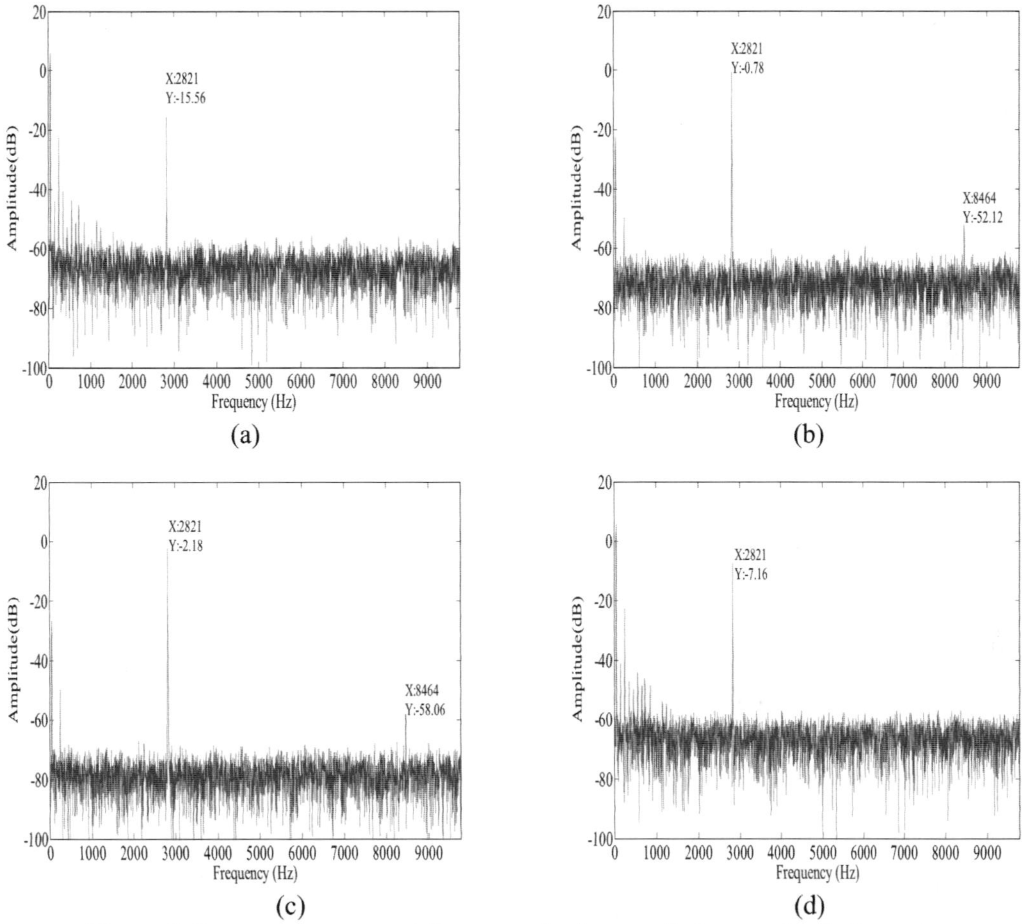

Figure 5. Response at resonant frequency for low excitation amplitude for various sensors: (a) P1; (b) P2; (c) P3; (d) P4

[6] Chati M, Rand R and Mukherjee S 1997 Modal analysis of a cracked beam *J. Sound Vib.* **207** 249-270

[7] Sundermeyer J N and Weaver R L 1995 On crack identification and characterization in a beam by non-linear vibration analysis *J. Sound Vib.* **183** 857-871

[8] Riviola A and White P R 1998 Bispectral analysis of the bilinear oscillator with application to the detection of fatigue cracks *J. Sound Vib.* **216** 889-910

[9] Loutridis S, Douka E and Hadjileontiadis L J 2005 Forced vibration behaviour and crack detection of cracked beams using instantaneous frequency *NDT&E International* **38** 411-419

[10] Andreaus U, Casini P and Vestroni F 2007 Nonlinear dynamics of a cracked cantilever beam under harmonic excitation *J. Nonlinear Mech.* **42** 566-575

[11] Andreaus U and Baragatti P 2011 Cracked beam identification by numerically analysing the nonlinear behaviour of the harmonically forced response *J. Sound Vib.* **330** 721-742

[12] Hiwarkar V R, Babitsky V I and Silberschmidt V V 2009 Damage assessment of a cracked bar: effect of material nonlinearity on vibro-impact response *Key Engng Mater.* **413-414** 237-244
[13] Hiwarkar V R, Babitsky V I and Silberschmidt V V 2012 On the modelling of dynamic structures with discontinuities *Nonlinear Dyn.* **67** 2651-2669
[14] Hiwarkar V R, Babitsky V I and Silberschmidt V V 2011 Crack as modulator, detector and amplifier in structural health *J. Sound Vib.* **331** 3587-3598

Dynamic analysis of vibrating screener system

Ł Bąk, S Noga, A Skrzat and F Stachowicz[1]

Rzeszow University of Technology, Powstańców Warszawy 8, 35-959 Rzeszów, PL

E-mail: stafel@prz.edu.pl

Abstract. Transversal vibrations of a screen for a fine granular material are studied using both analytical and numerical methods. In the analytical approach the motion of the screen is described by partial differential equations. The general solution of the screen free vibrations is derived from variables separation method. In the numerical computations the finite element method is applied. The screen compound geometry and the mass of the granular material are considered. Screen natural frequencies and natural mode shapes are determined. Numerical results are compared with the analytical solution. The screen geometry simplifications are proposed and validated by benchmark tests. The presented research allows determining the impact of the fine–granular material mass on the vibrating screener. The influence of the granular material on the screen natural frequencies is also investigated. It is important to note that the results of this research provide practical hints for vibrating screens designers.

1. Introduction

The large number of cut-outs and their distribution allows effective fine granular material sieving. A theoretical solution for a body of such geometry does not exist. Both static and mode shapes solutions exist only for a plate without openings. Many questions arise considering this model of sophisticated geometry. Is it possible to replace the screen geometry by a solid plate without the loss of accuracy in prediction of natural frequencies and natural mode shapes? What is the sufficient number and necessary concentration of openings? How to modify material data - Young's modulus and density - in order to obtain comparable results for the screen plate and the solid plate?

The answers for the above questions allow developing simplified vibrating screen model which has the same properties as considered screen. Numerical results for the solid plate can be compared with the ones derived from analytical methods. This way, the reliability of proposed simplified model can be validated.

2. Free vibrations of rectangular plate - theoretical solution

A vibrating system for fine-granular material is considered (figure 1). The vibrating screen is a rectangular plate 0.85 x 1.5 m one millimeter thick with the set of rectangular openings (figure 2). The solution of free vibrations of the rectangular plate simply supported (pinned) on all edges is commonly known. Here, the analytical solution of natural frequencies and mode shapes for the rectangular plate in which two opposite edges are pinned while the other two are free is presented. The considered plate in the Cartesian coordinate system is shown in figure 3.

[1] To whom any correspondence should be addressed

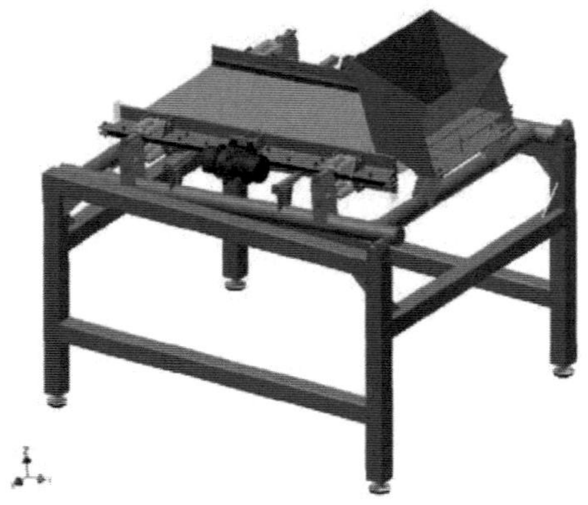

Figure 1. The general view of the vibrating screen system.

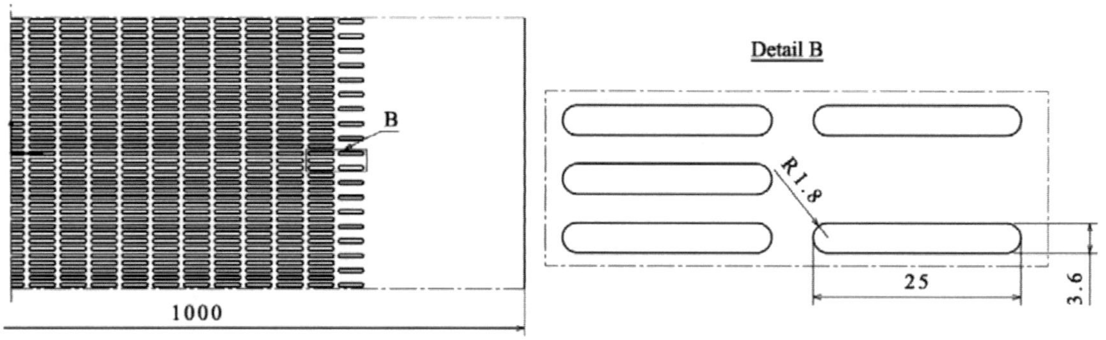

Figure 2. The enlargement of the vibrating screen geometry.

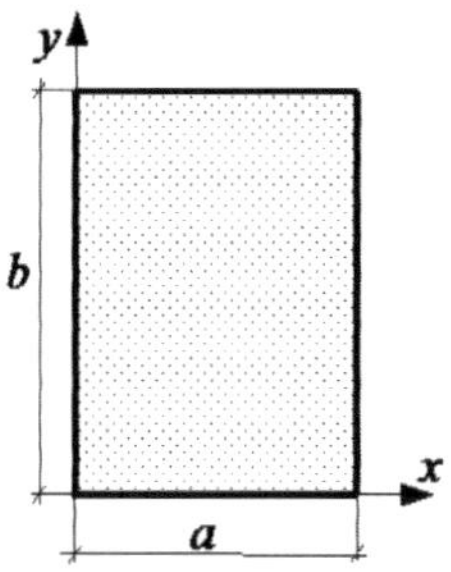

Figure 3. Rectangular plate simply supported on two edges.

The differential equation for plate natural vibrations is:

$$\frac{\partial^2 w}{\partial t^2}+\frac{D}{\rho h}\nabla^4 w=0 \tag{1}$$

where ρ is the density, h is the plate thickness, D is the plate bending stiffness

$$D=\frac{Eh^3}{12\left(1-\nu^2\right)} \tag{2}$$

where: E - Young's modulus, ν - Poisson ratio

The boundary conditions for the plate are:

$$w(0,y,t)=w(a,y,t)=0;\quad \left.\frac{\partial^2 w(x,y,t)}{\partial x^2}\right|_{x=0}=\left.\frac{\partial^2 w(x,y,t)}{\partial x^2}\right|_{x=a}=0 \tag{3}$$

and

$$\left.\left(\frac{\partial^2 w}{\partial y^2}+\nu\frac{\partial^2 w}{\partial x^2}\right)\right|_{\substack{y=0\\y=b}}=0;\quad \left.\frac{\partial}{\partial y}\left(\frac{\partial^2 w}{\partial y^2}+(2-\nu)\frac{\partial^2 w}{\partial x^2}\right)\right|_{\substack{y=0\\y=b}}=0 \tag{4}$$

In the variables separation method the solution $w(x,y,t)$ is assumed to be a product of spatial and time functions:

$$w(x,y,t)=W(x,y)T(t) \tag{5}$$

After substituting (5) into (1), (3) and (4) one obtains:

$$\begin{aligned}&\ddot{T}(t)+\omega^2 T(t)=0\\&\nabla^4 W(x,y)-p^4 W(x,y)=0\end{aligned} \tag{6}$$

where: ω is the natural frequency. After introducing parameter p:

$$p^4=\frac{\rho h\omega^2}{D} \tag{7}$$

By application of Levy's method for considered boundary conditions, the natural frequencies can be obtained from the following equation:

$$\begin{aligned}&\sin(\delta_1 b)\sinh(\delta_2 b)\left[\delta_2^2\left(p^2-\alpha_m^2(1-\nu)\right)^4-\delta_1^2\left(p^2+\alpha_m^2(1-\nu)\right)^4\right]\\&-2(\cosh(\delta_2 b)\cos(\delta_1 b)-1)\delta_1\delta_2\left[p^4-\alpha_m^4(1-\nu)^2\right]^2=0\end{aligned} \tag{8}$$

where:

$$\alpha_m=\frac{m\pi}{a},\quad m=1,2,\ldots,\quad \delta_1^2=p^2-\alpha_m^2,\quad \delta_2^2=p^2+\alpha_m^2. \tag{9}$$

Equation (8) represents highly non-linear and extremely ill-conditioned problem. Very small variations of p parameter can cause rapid jumps on the left-hand-side of (8) e.g. if $f(p)$ is the left-hand-side of (8) for the considered plate f(5.21) = 3.24e9 while f(5.22) = -1.69e9. Moreover, if $p<\alpha_m$, then the function (8) returns a complex number.

After setting m the n values the p parameter can be found i.e. $p=p_{mn}$ (m,n = 1, 2, ...). When p is known the natural frequencies can be obtained from (7)

$$\omega_{mn}=p_{mn}^2\sqrt{\frac{D}{\rho h}} \tag{10}$$

The mode shapes for the considered plate are [1]:

$$\begin{aligned} W_{mn}(x,y) = & \Big[\Big\{-(\cosh(\delta_2 b)-\cos(\delta_1 b))\big[p_{mn}^4-\alpha_m^4(1-\nu)^2\big]\Big\}\Big\{\delta_1\big[p_{mn}^2+\alpha_m^2(1-\nu)\big]\sinh(\delta_2 y) \\ & +\delta_2\big[p_{mn}^2-\alpha_m^2(1-\nu)\big]\sin(\delta_1 y)\Big\}+\Big\{\delta_1\big[p_{mn}^2+\alpha_m^2(1-\nu)\big]^2\sinh(\delta_2 b) \\ & -\delta_2\big[p_{mn}^2-\alpha_m^2(1-\nu)\big]^2\sin(\delta_1 b)\Big\}\Big\{\big[p_{mn}^2-\alpha_m^2(1-\nu)\big]\cosh(\delta_2 y) \\ & +\big[p_{mn}^2+\alpha_m^2(1-\nu)\big]\cos(\delta_1 y)\Big\}\Big]\sin\left(\frac{m\pi}{a}x\right). \end{aligned} \tag{11}$$

For the considered plate the magnitude of natural frequencies found analytically are presented in table 1.

Table 1. Natural frequencies of rectangular plate.

Parameter	Frequency (s^{-1})
m=1, n=1	3.3516
m=1, n=2	4.1974
m=1, n=3	6.7790
m=1, n=4	11.1764
m=2, n=1	13.4999

3. Free vibrations of rectangular plate - numerical solution

The solution for free vibrations of the rectangular plate described in the previous chapter is also found numerically using the finite element method [2]. Two commercial FEM programs are used independently: ANSYS and ABAQUS. In both programs the rectangular plate is modeled as a mesh of shell elements. The mesh consisting of higher-order rectangular elements is generated in both programs. An example of the solution obtained in ANSYS is presented in figure 4.

The magnitudes of natural frequencies of the rectangular plate computed by ANSYS and ABAQUS programs are summarised in table 2. Results presented in table 2 show excellent accuracy and convergence to the theoretical results for the first two-three natural frequencies. For higher natural frequencies the precision is still acceptable but the approximation errors are slightly larger. Fortunately, these frequencies are not important from the point of view of presented research.

Table 2. Natural frequencies (s^{-1}) - numerical solution.

ANSYS	ABAQUS
3.2719	3.3514
4.0972	4.1959
6.6171	6.7751
10.9110	11.1710
13.1930	13.4990

4. Free vibrations of rectangular plate with cut-outs

Following numerical simulations for the solid rectangular plate, similar computations are made for the rectangular screen. This problem cannot be solved analytically. In the numerical computations very dense finite elements meshes are used. In the best-fitted model each gap visible in figure 2 is covered by three layers of shell elements. The size of the problem and computation time is much larger now than those in previous example. The mesh generation is a very laborious task because of sophisticated geometry which includes thousands of openings. An exemplar solution obtained by ANSYS program is presented in figure 5 (a mesh of holes is clearly visible).

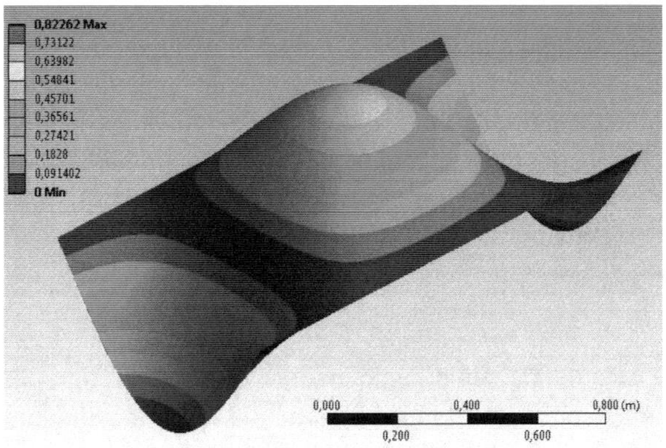

Figure 4. Free vibrations of rectangular plate - mode 3.

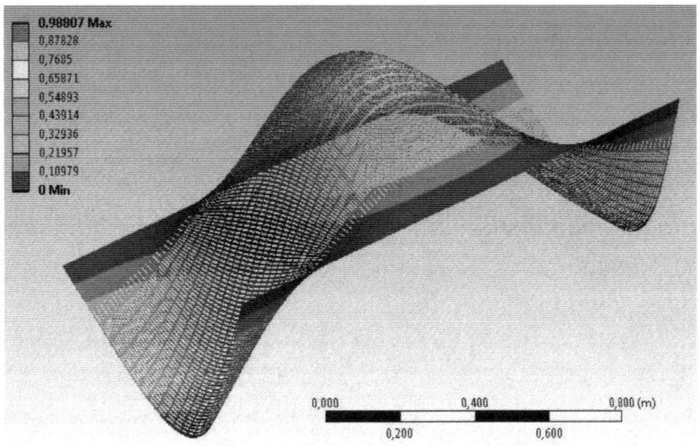

Figure 5. Free vibrations of rectangular plate - numerical solution, mode 3.

The natural frequencies of the rectangular screen plate computed by ANSYS and ABAQUS programs are presented in table 3. Comparison of the natural frequencies of solid and screen rectangular plates shows that for the first three frequencies obtained results are comparable and they are similar to the analytical solution of the solid plate. Thus, the question arises if the solid model of rectangular plate can be used instead of screen one? It is very important from the point of view of the total cost of numerical computations [3]. The answer for the above question is positive if material data is properly modified. Two material constants influence natural frequencies radically i.e. Young's modulus (representing stiffness) and the density (representing mass). If the ratio of E/ρ is preserved, the solution obtained for the same geometry and the same boundary conditions is the same. In this case the geometries of solid and screen plates are different.

Table 3. Natural frequencies (s^{-1}) - numerical solution.

ANSYS	ABAQUS
3.0745	3.1440
4.0239	4.0568
6.2825	6.2399

9.4290	9.2835
12.2850	12.5600

In the numerical tests considering replacing the model of the screen plate with the solid one the mass of the plate should remain constant. Thus, the density of the solid plate is corrected. The magnitude of the Young's modulus is also modified. This modification is based on the accuracy of the first natural frequency which is the main frequency of interest. During the series of computations the optimal Young's modulus was found. The magnitude E = 1.06e11 Pa is selected in the analysis of solid plate model used instead of the screen plate model. Of course, if other natural frequencies were also important, the choice of Young's modulus may have taken them into account and the least squares based-methods could have been applied.

5. Forced vibrations of rectangular plate

Another problem solved within presented research is the prediction of rectangular plate forced vibrations. In this case the plate is loaded by the mass of stones. The layer of stones covering the plate is shown in figure 6.

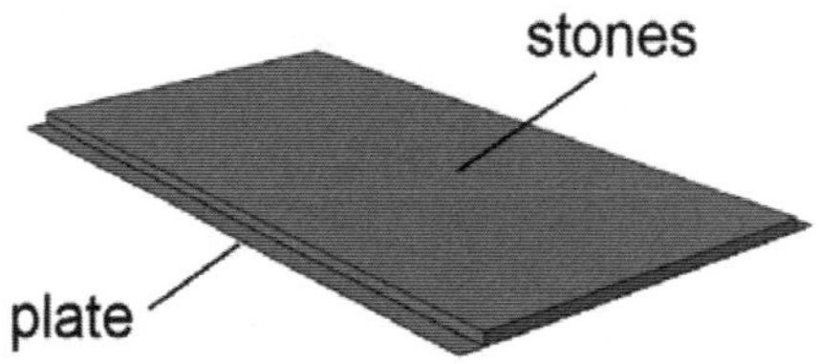

Figure 6. Forced vibrations model of rectangular plate.

The contact between the plate (modeled as a shell) and stones (modeled as a solid body) is assumed to be of type "bonded" (ANSYS program). It means that transverse displacements of both entities are common. The presented analysis is the first attempt to solve the screen forced vibration problem [4]. The assumed numerical model is as simple as possible. Instead of the screen model the solid plate model is used. Material data is modified as described in the previous chapter. Unlike in the reality, the thickness of stones layer is constant here. Of course, this model does not allow individual stones to pass through the screen.

The main goal of this test is the investigation of the influence of stones mass on the natural frequencies of the screen. The results of analysis are summarised in table 4. As expected, the natural frequencies have decreased because of the larger vibrating mass. Although presented model is relatively simple, the results of numerical simulations are promising. They should be verified experimentally. Appropriate research is planned in the future.

Table 4. Natural frequencies (s^{-1}) - screen and stones.

Freq. number	ANSYS
1	1.3264
2	1.7369
3	2.6806
4	3.9253
5	5.3594

6. Conclusions

Numerical investigations presented in this paper are a part of a larger project considering development of effective vibrating screen system. Frequencies of vibration inductors applied in this system should be far away from the resonance i.e. far from the natural and forced frequencies of the vibrating screen. The set of free and forced natural frequencies can be found by the numerical analysis. The most important are the first few frequencies.

In this research the finite element method is used in order to provide free/forced frequencies and appropriate mode shapes. Results of numerical analyses are compared, if possible, with the theoretical solution. The simplified model of the solid plate based on the modified material data is proposed. In the future research the experimental investigations will be undertaken to help validate and improve proposed numerical models.

Acknowledgement

This work was supported by the European Research Agency - FP7-PEOPLE-2011-IAPP - Marie Curie Industry-Academia Partnerships and Pathways, grant agreement No. 284544.

References

[1] Rao S 2007 *Vibration of Continuous Systems* (Hoboken, NJ: John Wiley & Sons)
[2] Friswell M and Mottershead J 1995 *Finite element model updating in structural dynamics* (Dordrecht: Kluwer Academic Publishers)
[3] Leissa AW 1993 *Vibration of plates* (NASA SP 160: US Government Printing Office)
[4] Bogacz R and Noga S 2011 *Proceedings of the XVIIIth PTSK Conference pp* 14-17

Analysis of the effect of shot peening on mechanical properties of steel sheets used as screener sieve materials

M Śledź, Ł Bąk, F Stachowicz and W Zielecki

Rzeszow University of Technology, Powstańców Warszawy 8, 35-959 Rzeszów, PL

E-mail: stafel@prz.edu.pl

Abstract. Numerous studies have been performed in an attempt to improve fatigue strength of materials by creating compressive residual stresses in the surface layers as a result of the shot peening process. It was shown that the process is highly effective when dealing with particular parts of the body (detail) needed special attention. In this case, those are parts of screen sieve plate situated near the fixed edge and undergo the largest deformation caused by impact bending and thus need a special treatment. In this paper, the results of experimental tests are presented to analyse the effect of microshot peening on mechanical properties, surface layer characteristics and fatigue strength of steel sheet specimens. The data obtained from our experiments can be used for numerical simulations to design details with improved fatigue resistance.

1. Introduction

Screening is one of the basic technology in mechanical processing of granular materials. Vibrating screen is widely used for grading and screening materials in the following fields: minerals, quarry, building materials, etc. The parts of screen sieve plate which are located near the fixed edge undergo the largest deformation, caused by impact bending during the separating screener exploitation. Taking also into account energy dissipation associated with the process, it is extremely important to have a suitable material for screen sieve plate. One of the methods to enhance the fatigue strength of material used for screen sieve is shot peening.

Shot peening is a process where the surface of a detail of its component is bombarded with small spherical particles called shot. It is widely used to improve the fatigue strength of material through the creation of a compressive residual stress in their surface layers [2-6]. Depending on the character of the forces acting on the surface during the process, one distinguishes static and dynamic peening [7]. The shots impart small indentations (or dimples) on the striking surface leading to a deformation of the surface. In an attempt to return yield surface to its initial shape, residual compressive stresses appear in the surrounding elastic material forming the cold work hardened surface layer. Compressive residual stresses in the surface layer increase the fatigue strength of the sieve material, and therefore may be beneficial in reducing the fatigue damage. Due to the shot peening, the surface layer of material is caused to yield plastically under the impact of shot. Residual compressive stresses result from the inability of the plastically deformed material to reaccommodate itself on the elastic subsurface. The following mechanical properties of the design material maybe improved as a result of the shot peening process: resistance to fatigue, stress corrosion cracking, fretting, galling, erosion and closing of pores [4]. The tall orders of the burnishing process are materials with hardness above 45 HRC [2]. The experimental investigations show, that the shot peening has the good influence on fatigue strength of the constructional steels [3], spring steels [4, 6], titanium alloys [7]. The shot peening results depend on various parameters such as shot size, velocity, shape, material, coverage, hardness, intensity, etc. The shot peening results have been widely studied theoretically and experimentally [2-9].

In this work, one of the burnishing methods (pneumatic shot peening was used. The machine for pneumatic shot peening has been developed in the Rzeszow University of Technology with the aim to increase the fatigue strength of low and medium hardness in medium and small lot production. This process is characterized by small overall dimensions, low costs, quiet operation and a perfect quality of constituted surface layer. It has been shown [10] that application of the pneumatic shot peening process to ground surfaces may cause even elimination of subsequent polishing operation. The research has shown its great effectiveness when it comes to the fatigue of constructional steels about 10-40%, titanium alloys about 10-20 %, steels with layers about 5-25 % [11], spring steels about 30-40 % [12]. Parameters of the burnishing done by the pneumatic shot peening play however a decisive role to provide desirable properties. Moreover, without proper tuning, it might cause the effects opposite to those one wants to achieve. This has been the main reason to conduct the research which main goal is to determine optimum parameters of pneumatic shot peening providing the best properties for the each piece of machine [12].

In the presented experimental investigations, the particular part of the screen is shot peened. We use the 3 mm bearing balls accelerated to a suitable speed, by the blast of the compressed air. The pneumatic shot peening process has been realized using one or a few nozzles consisting of lower and upper nozzle (figure 1). As usual, the lower nozzle is installed directly on working chamber bottom and made in form de La Valla nozzle, supported compressed air by pressure of 0.15-0.6 MPa. The nozzle task is produce stream air about large linear velocity which captures balls from chamber bottom and directed through upper nozzle on workpiece. Balls bang on the reinforcement surface with the speed limit of 3-8 m/s, and after the returning, they go down to the bottom of the chamber and then they are reused for the process. During the burnishing, the reinforced piece of the machine moves in order to provide the full coverage of the element with the signs of the treatment [11]. Kinematics movements should make possible achieving 100% coverage of the workpiece. This process itself is affected by ejector nozzle geometry and: ball diameter, air pressure, working nozzle to workpiece, and peening time [10].

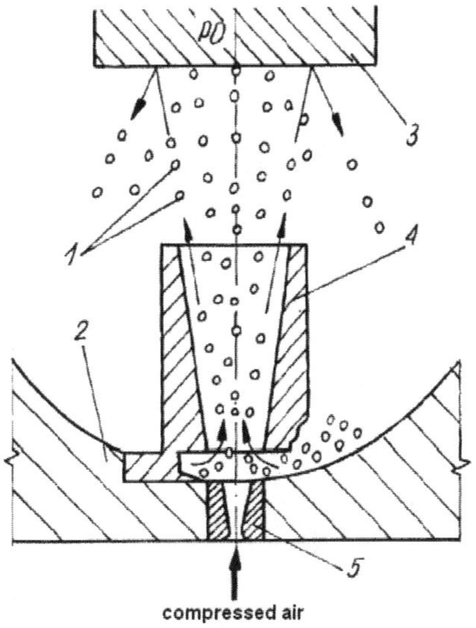

Figure 1. Scheme of ejector nozzle and pneumatic shot peening: 1 - balls, 2 - working chamber bottom, 3 - workpiece, 4 - upper nozzle, 5 - lower nozzle.

2. Material and experimental procedure

In the modified version of screener, 1 mm thick perforated steel sheet metal is used as its sieve (figure 2). Four different kinds of the steel sheets were used in experimental investigation: 1.0530; 1.4301; 1.8159 and 30HGSA. Both specimen surfaces have been subjected to pneumatic shot peening. The uniaxial tension test and rotating bending fatigue tests were performed with two types of specimens: peened and unpeened to examine the effect of the process.

When the mechanical testing is concerned, tensile specimens of 240 mm gauge length and 12.5 mm width were prepared from strips cut perpendicular to the rolling direction of the sheet. During experiments, we recorded simultaneously the tensile load, the current length and the current width of the specimens.

Specimens dimensions used for fatigue testing were 30 x 100 mm. The rotating bending fatigue test was conducted on the special test stand which was built on the basis of a Wöhler type testing machine ETS at a frequency of 300 cycles/s. During the investigation the specimen was placed in the hydraulic handle (figure 3) which has a constant strength of interlocking and can lead to the resonance vibration. Specimen undergoes vibration until the cracking and/or gaining final number of cycles. During the research there were 2 x 10^6 accepted cycles.

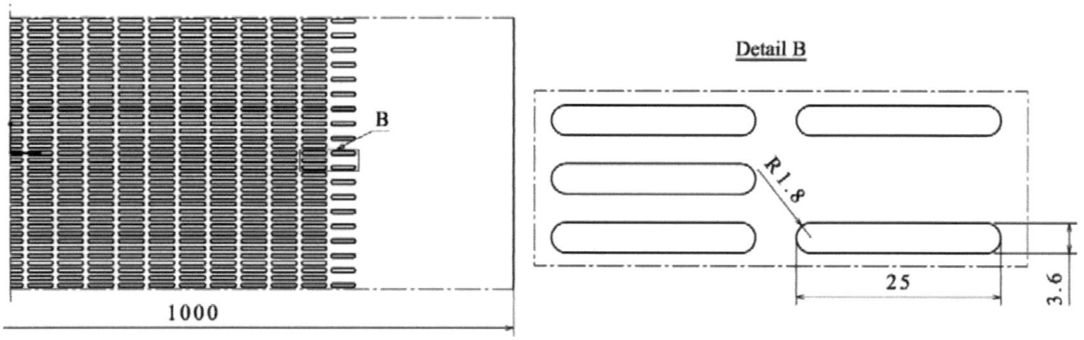

Figure 2. Scheme of the vibrating screen geometry.

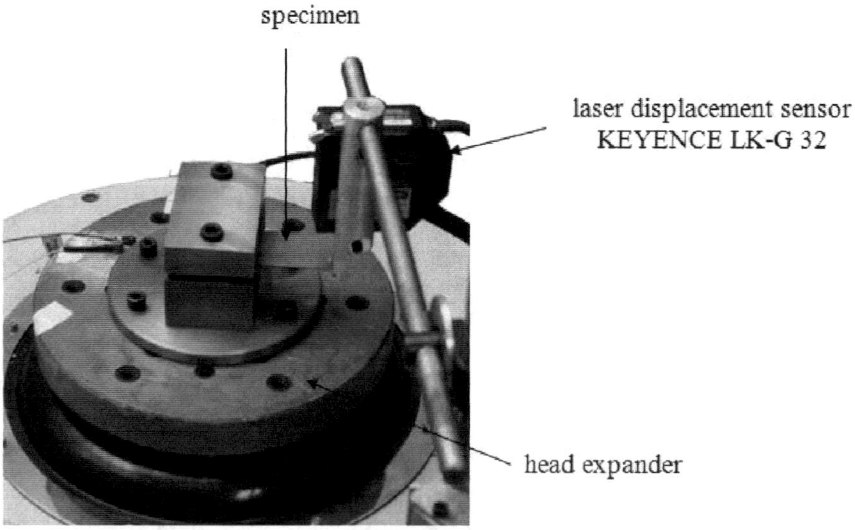

Figure 3. Vibratory stand used for fatigue test.

3. Results

As a result of pneumatic shot peening, the surface of the sheet specimen has been covered with small dimples (figure 4). The results of uniaxial tension test (table 1 and 2) demonstrate visible change in the mechanical parameters after the shot peening. Among others, the following conclusions can be drawn:

- The value strain hardening exponent obtained for peened specimens is lower than that of unpeened sheet specimens for all tested materials.
- Shot peening process has no evident effect on the value of yield stress and ultimate strength.
- Large difference, however, can be observed for tensile characteristic, especially in the case of materials with physical yield point - 1.8159 steel and 30HGSA steel (figure 5).
- Shot peening of sheet surface resulted in decreasing in the value of total elongation.
- The process has the lowest effect on the value of plastic anisotropy factor.

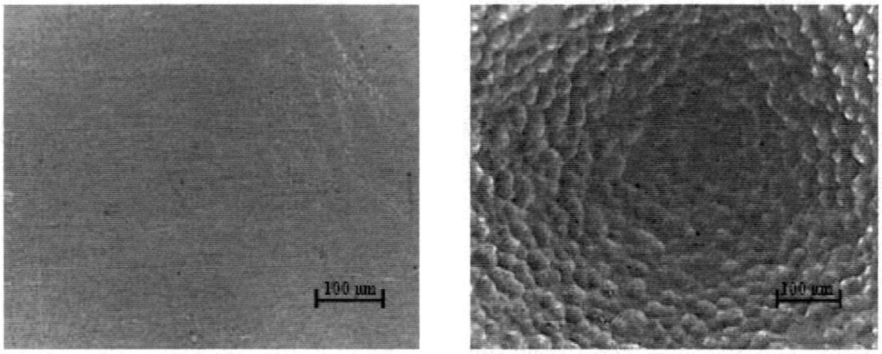

Figure 4. Surface of 1.4301 steel sheet: before pneumatic shot peening (left) and after pneumatic shot peening (right).

Table 1. Mechanical properties of the sheet metal tested before shot peening

Sheet material	Yield stress	Ultimate strength	Total elongation	Strain hardening parameters		Anisotropy factor
	$R_{p0,2/eH*}$ MPa	R_m MPa	A_{50} %	C MPa	n	r
1.0530	316	516	30	864	0.193	1.356
1.4301	315	708	51	1547	0.441	0.779
1.8159	212*	409	24	1105	0.234	1.083
30HGSA	377*	580	27	1082	0.242	1.606

Table 2. Mechanical properties of the sheet metal tested after shot peening.

Sheet material	Yield stress	Ultimate strength	Total elongation	Strain hardening parameters		Anisotropy factor
	$R_{p0,2}$ MPa	R_m MPa	A_{50} %	C MPa	n	r
1.0530	332	495	28	784	0.147	1.334
1.4301	357	733	44	1522	0.387	0.757
1.8159	208	590	23	896	0.141	1.104
30HGSA	390	580	24	887	0.156	1.750

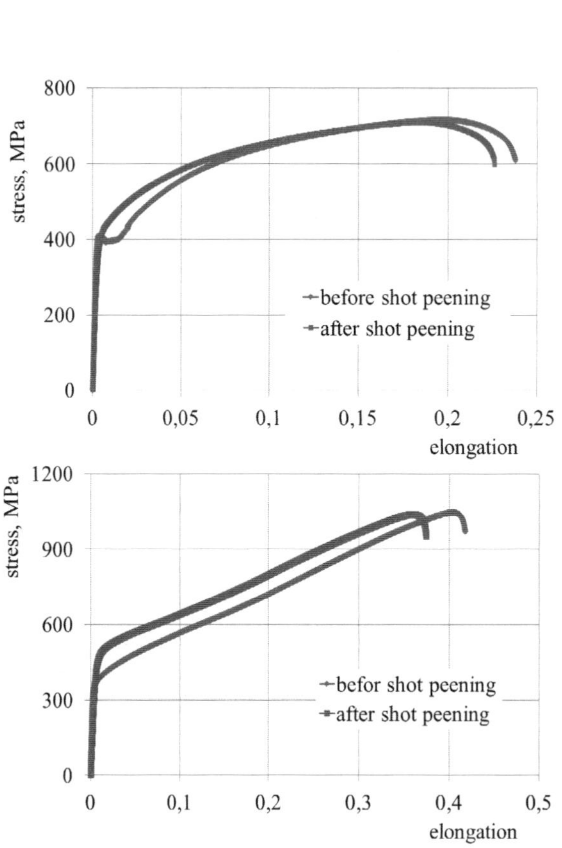

Figure 5. Effect of shot peening on uniaxial tensile testing characteristics in the case of materials with (upper - 30HGSA) and without (bottom - 1.4301) physical yield point.

Based on results of the uniaxial tension test one can conclude that the 1.4301 steel seemed to be a good material for screener sieve. It is also noteworthy that the stainless austenitic chromium-nickel steel 1.4301 shows good corrosion resistance (particularly in natural environment) and weldability. For this steel sheet, fatigue test has been also performed. Results of the bending fatigue in the form S-N curves (figure 6) clearly demonstrate the shot peening benefits. Thus/for example, fatigue strength of peened specimens increased above 100 MPa in comparison to that of unpeened specimens. The reason for such an improvement can be attributed to the formation of compressive residual stresses in the surface layer of the material as the result of the shot peening [4]. As compressive stresses are introduced into the surface, fatigue cracks not easily initiate in neither propagate through an area under compression. As a result, the improvement in bending fatigue strength has been achieved. Due to increase their fatigue strength, the shot peening process is specifically recommended to burnishing of metallic parts undergoing dynamic loading. Comparison of the results presented in figure 5 and 6 confirms that the effect of the shot peening is more pronounced in the case of fatigue strength testing than in the uniaxial tensile.

4. Conclusion

The influence of a shot peening treatment on the mechanical properties of different steel sheets have been investigated. The results of uniaxial tension test demonstrate that shot peening resulted in significant change in the mechanical characteristics and improve fatigue strength. Thus, the use of shot peening of sheet surface made it possible to increase fatigue life of screener sieve.

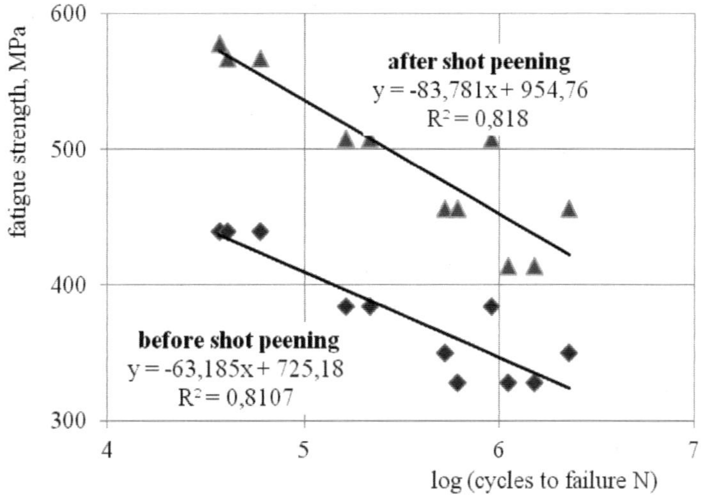

Figure 6. Effect of shot peening on fatigue testing of the 1.4301 steel sheet.

Acknowledgement
This work was supported by the European Research Agency - FP7-PEOPLE-2011-IAPP - Marie Curie Industry-Academia Partnerships and Pathways, grant agreement No. 284544.

References

[1] Dyr T, Wodziński P 2002 Model particle velocity on a vibrating surface *Physicochemical Problems of Mineral Processing* **36** 147-157
[2] Przybylski W 1987 *Technologia obróbki nagniataniem* (WNT Warszawa)
[3] Torres M A S, Voorwald H J C 2002 An evaluation of shot peening, residual stress and stress relaxation on the fatigue life of AISI 4340 steel *Intern. J. of Fatigue* **24** 877-886
[4] Tekeli S 2002 Enhancement of fatigue strength of SAE 9245 steel by shot peening *Mater. Letters* **57** 604-608
[5] Gao Y K, Lu F, Yin Y F, Yao M 2003 Effects of shot peening on fatigue properties of OCr13Ni8Mo2Al steel *Mater. Sci. and Technol.* **14** 372-374
[6] Aggarwal M L, Agrawal V P, Khan R A 2006 A stress approach model for prediction of fatique life bz shot peeling of EN45A spring steel *Int. J. Fatigue* **28** 1845-1853
[7] Zaleski K 2009 The effect of shot peening on the fatigue life of parts made of titanium alloy Ti-6Al-4V *Maintenance and Reliability* **4** 65-71
[8] Nakonieczny A 2002 *Dynamiczna powierzchniowa obróbka plastyczna. Kulowanie. Shot peening* (Instytut Mechaniki Precyzyjnej Warszawa)
[9] Bagherifard S, Ghelichi R, Gaugliano M 2012 Numerical and experimental analysis of surface roughness generetad by shot peening *Appl. Surface Sci.* **258** 6831-6840
[10] Łunarski J, Zielecki W 1990 Possibilities of improving fatigue properties of machine elements by pneumatic shot peening *Proc. ICSP 4 Tokio* pp 263-272
[11] Zielecki W 1990 Improvement fatigue properties after pneumatic ball peening *ZN PRz, Mechanika* **22** 59-67
[12] Zielecki W 1987 Influence of pneumatic shot peening on 50HF steel fatigue strength and surface condition *ZN PRz, Mechanika* **14** 35-40

Modelling the dynamic behaviour of hard-to-cut alloys under conditions of vibro-impact cutting

R Muhammad[1], M Demiral, A Roy and V V Silberschmidt
Wolfson School of Mechanical and Manufacturing Engineering, Loughborough University, Loughborough, Leicestershire, UK

E-mail: R.Muhammad@lboro.ac.uk; R.Muhammad@outlook.com

Abstract. Industrial applications of titanium alloys have grown significantly over the years primarily due to their high strength, light weight as well as good fatigue and corrosion-resistance properties. Since these alloys are hard-to-machine, there is an obvious demand to develop simulation tools in order to analyse the material's behaviour during machining and, thus, optimise the entire cutting process. Deformation processes in the machining of Ti alloys are typically characterized by high strains, strain rates and temperatures; the material behaviour under such conditions was derived from split Hopkinson pressure-bar (SHPB) tests.
In the current study, a computational three-dimensional finite element model of the SHPB test is developed to assess the material's behaviour at high-strain, strain-rate and temperature regimes. The material behaviour derived for high strain rates was used in three-dimensional finite-element modelling of conventional turning and ultrasonically assisted turning of Ti-15333 alloy. A parametric analysis of the studied alloy was carried out using the developed model. Comparisons of simulations with experimental results demonstrate their predictive capability.

1. Introduction

Beta-titanium alloys have an exceptional combination of low density, high strength and good corrosion resistance properties. These properties make beta-titanium alloys an ideal material for aerospace applications. However, these alloys are hard-to-cut, producing high cutting forces and excessive temperature in various machining processes[1-3]. Since high levels of strains and strain rates accompanied by high temperatures characterize the deformation process in machining, the material behaviour at these extreme conditions is vital for accurate computational modelling of machining processes [4] . There are several experimental methods available to predict the material behaviour at high strain rates and temperatures; the most commonly used is split Hopkinson pressure-bar (SHPB) test. This method is applicable at strain rates between 10^2 and 10^4 s^{-1}. However, in machining processes, the strain rate can exceed these levels and there is a need to predict material behaviour at such high strain magnitudes to be used in order to model accurately the machining processes.[1]

Some basic assumptions are employed in the assessment of stress strain relationship in SHPB tests including the use of a one-dimensional theory of elastic wave propagation, homogeneity of stress and strain fields in the axial direction and neglection of a frictional effect. A thorough assessment of significance of these assumptions the relevance and accuracy of SHPB is available in [5]. The non-

[1] To whom any correspondence should be addressed.

uniformity of axial stress and strain filed was investigated numerically and experimentally by various researchers [6-8]. Thus, there is a need to critically reassess the validity of the assumptions in the determination of the mechanical response of recently introduced advanced alloys such as Ti-15333.

In this study, a thermo-mechanically coupled three-dimensional finite-element (FE) simulation of the SHPB test is carried out to better understand the deformation mechanisms in a specimen during the impact. Various levels of strain rates and temperatures are used to investigate the response of the beta titanium alloys. Additionally, a thermomechanically coupled three-dimensional FE model of oblique turning is developed for conventional turning (CT) and ultrasonically assisted turning (UAT) [9-10]. The models allow investigation of the effect of ultrasonic vibration on machining cutting forces and temperature at the cutting region in turning of Ti-15333.

2. Split Hopkinson pressure-bar technique

A schematic presentation of the SHPB system is shown in Figure 1. The system used for our tests consist of two 1.2 m long maraging steel rods with diameters of 21.77 mm called *incident* and *transmitter* bar, a striker bar made of a same material and diameter with a length of 200 mm and a specimen with a length of 6 mm and a diameter of 8 mm. The striker bar is shot against the free end of the incident stress bar, generating a stress pulse that propagates in the latter bar towards the specimen that is sandwiched between the incident and transmitter bars [11]. The stress pulse is divided into two parts after reaching the specimen interface. A part of stress is reflected back to the incident bar as an elastic wave while the rest of pulse transmits through the specimen to the transmitter bar, resulting in plastic deformation in the specimen. Strain gauges were used in the setup to measure the strains on the incident and transmitted bar. The data obtained from the strain gauges were then used to calculate the levels of strain, strain rate and stress in the specimen. The 1-D wave analysis of SHPB data for the strain rate, strain and stress can be expressed as follows [12]:

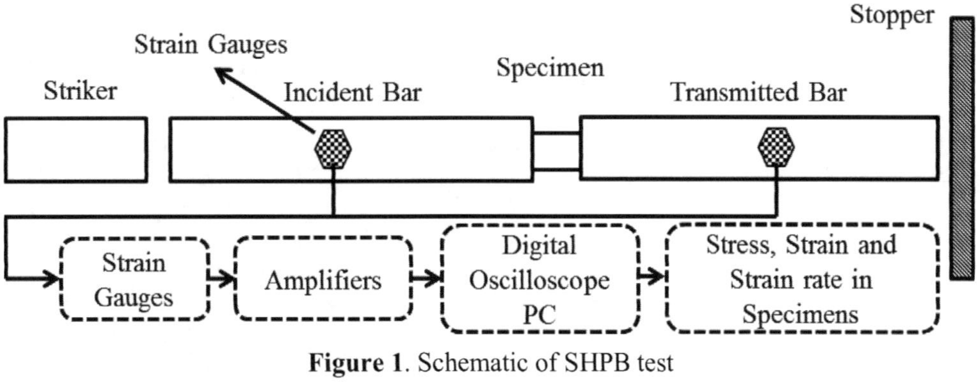

Figure 1. Schematic of SHPB test

$$\varepsilon(t) = -\frac{2C}{L_s}\int_0^t \varepsilon_R(t), \quad (1)$$

$$\dot{\varepsilon}(t) = -\frac{2C}{L_s}\varepsilon_R(t), \quad (2)$$

$$\sigma(t) = \frac{A_b E_b}{A_s}\varepsilon_T(t), \quad (3)$$

where C is the elastic wave velocity, L_s is the initial length of the specimen, ε_R and ε_T are the reflected and transmitted strain pulses, respectively, A_b and A_s are the cross sectional areas of the bars and the specimen, respectively, E_b is the Young's modulus of the bar material, ε (t) is the strain, $\dot{\varepsilon}$ (t) is the strain rate and σ (t) is the time dependant stress level.

The SHPB tests were carried out for Ti-15333 at strain rates ranging from 600 s^{-1} to 3300 s^{-1} at room temperature. An ingot of Ti-15333 was wire cut into specimens with dimension requested for the tests. The ratio of the length to diameter was 0.75 (within the established range of 0.5-1.0) in the current tests. The material properties of the specimen and the setup used in both experiments and simulations with relative dimension are listed in Table 1.

Table 1. Dimensions and material properties of SHPB system

	Material	Length, mm	Diameter, mm	Young's modulus, GPa	Passion's ratio	Density, kg/m^3
Striker	Maraging Steel	200	21.77	198	0.329	8470
Bars	Maraging Steel	1200	21.77	198	0.329	8470
Specimen	Ti-15333	6	8	87	0.3	4900

3. Oblique turning test

A Harrison 300 lathe machine was suitably modified to mount a transducer on the tool post, which imposed vibrations on the tool in the primary cutting direction. The system is capable of carry out conventional machining and UAT on the same setup. A Kistler dynamometer was used to record cutting forces at various cutting conditions. A FLIR ThermaCAM™ SC3000 camera was employed to carry out thermal analysis in real time during the cutting operation. The new Stirling-cooled Quantum Well Infrared Photon (QWIP) photo-detector enables the system to capture images with a spectral response between 8 and 9 µm, temperature range of -20°C to 2000°C, 320 × 240 pixel resolution, low noise detection and high image stability and uniformity. In our experiments, a temperature resolution of 30°C to 650°C was used. The ThermaCAM® QuickView™ software was used to analyse the data of the FLIR ThermaCAM™ SC3000 system. The experimental setup used for turning experiments is shown in Figure 2.

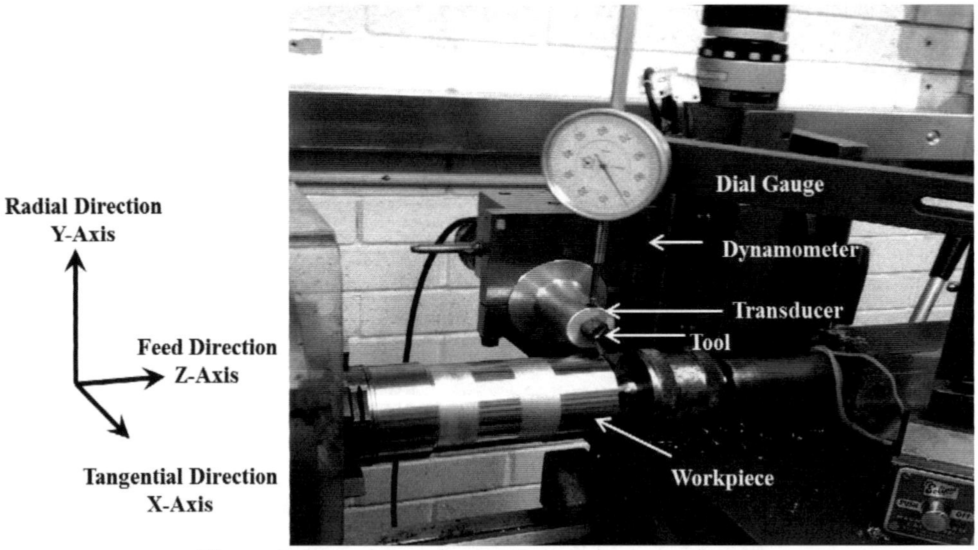

Figure 2. Experimental setup for turning experiments

A workpiece material was an ingot of Ti-15333, and a commercial tungsten carbide insert, PVD coated with TiAlN (DNMG-150608 supplied by Seco) was employed as a cutting tool. All experiments were carried out in dry-cutting conditions without the use of any lubricants or coolants. The experiments and simulations were carried out for a cutting speed of 10 m/min, depth of cut of 300 µm and feed rate of 0.1 mm/rev. A series of tests for conventional turning (CT) and UAT were performed, and the tangential and radial components of cutting forces (see Figure 2) were measured. Each experiment was repeated at least five times to get good repeatability. The measured cutting forces reduced significantly during ultrasonically assisted turning when compared to those obtained with CT.

In addition to force reduction, other substantial benefits were observed, in particular, elimination of the chatter noise during operations and better surface finish of the machined workpiece [2, 9, 13].

4. Finite-element modelling

A part of deformable workpiece with the following dimensions: a length of 5 mm, width of 1.3 mm and height of 0.5 mm + a_p (depth of cut) was modelled in oblique turning of Ti-15333 (Figure 3). Initially, 8-node hexahedral elements with a minimum length of 0.05 mm were used to model the workpiece. Later on, 5-node tetrahedron elements were used due to lack of support of global re-meshing for 8-node elements in simulations. A rigid cutting tool with a nose radius of 0.8 mm, rake angle of 14.5° and end relief angle of 0° was modelled reflecting the experiments conducted. Four-noded tetrahedral elements were used to discretise the cutting tool. A constant velocity (V) was imposed on the workpiece in the tangential direction. The tool was made to move in the feed direction with a feed rate of 0.1 mm/rev. In UAT, vibration was superimposed on the cutting tool in the tangential direction with an amplitude of 8 µm and frequency of 20 kHz.

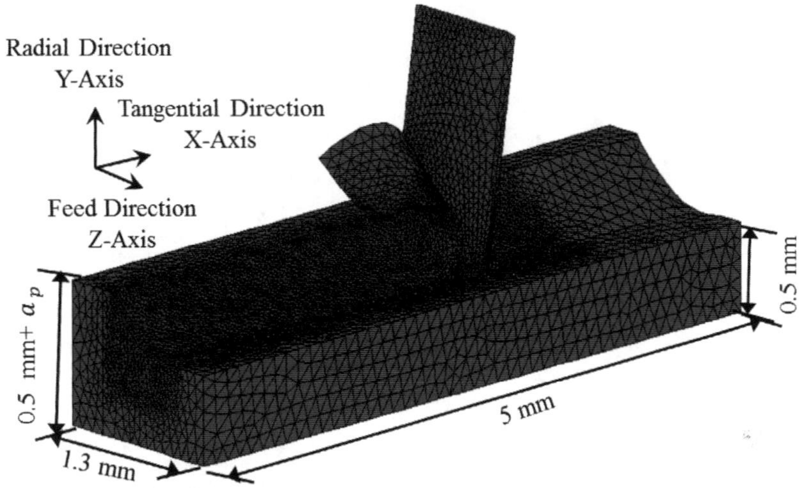

Figure 3. Oblique turning model

5. Material model

A piecewise linear material model was used to represent the material behaviour at high strain rates and elevated temperature in FE simulation of SHPB and oblique turning. The data comprised a combination of four different strain rates (0.1 s^{-1}, 1 s^{-1}, 3300 s^{-1}, 10^5 s^{-1}) and four different temperature values (20°C, 200°C, 600°C, 950°C) (Figure 4). Details of the material data are available elsewhere [14]. The data were obtained from a series of the SHPB tests carried out at the Tempere University of Technology, Finland. In order to account for higher strain rates beyond the experimentally characterized ones, the stress-strain data for a higher strain rate level (10^5 s^{-1}) need to be assumed based on the corresponding values for the highest strain rate in experiments 3300 s^{-1} (Figure 4b). The extrapolation was introduced assuming a 20% offset from the stress levels observed at 3300 s^{-1} and was verified using FE modelling of the SHPB setup.

A thermomechanically coupled finite-element model for SHPB test was developed using commercially available code MSC Marc/Mentat [15]. Eight-noded, isoparametric, hexahedral elements (Element type 7) were used to discretize the striker, incident bar, transmitted bar and specimen (see Figure 5). The incident and transmitter bars were discretized with 5760 elements each and the striker bar with 960 elements. The initial number of elements used to mesh the workpiece was 75600. Global remeshing and rezoning technique was used to avoid any distortion in element of the workpiece material. The right hand side of the transmission bar is constrained in all directions to model the end stopper used in the experiments.

The modified shear-friction model was used to simulate friction between the materials of tool, chip and workpiece in FE simulation of oblique turning process,

$$\sigma_{\mathrm{fr}} \leq -m_{\mathrm{fr}} \frac{\bar{\sigma}}{\sqrt{3}} \frac{2}{\pi} \operatorname{sgn}(V_{\mathrm{r}}) \operatorname{artan}\left(\frac{V_{\mathrm{r}}}{V_{\mathrm{cr}}}\right), \quad (4)$$

where σ_{fr} is a friction stress, $\bar{\sigma}$ is the equivalent stress, V_{r} is the relative sliding velocity, V_{cr} is the critical sliding velocity below which sticking is simulated, m_{fr} is the friction coefficient and sgn(x) is the signum function.

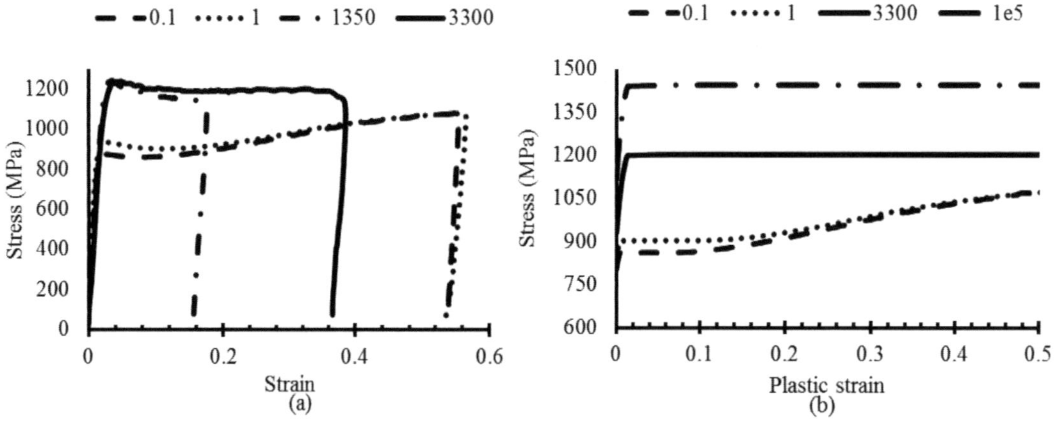

Figure 4. Experimental stress-strain curves for various strain rates (a) experiment; (b) modified flow stresses used in simulations

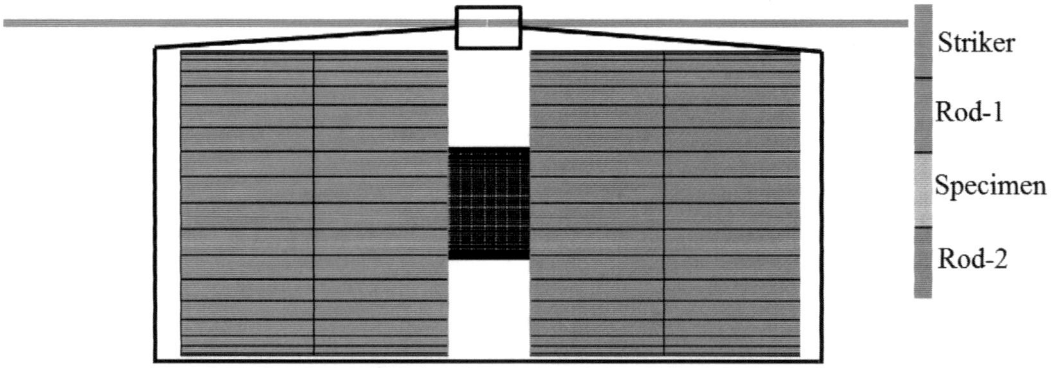

Figure 5. FE model of SHPB test

The extrapolated data for two high strain levels 10^5 and 10^{10} s^{-1} assumed in simulation was used to interpolate the observed experimental data at lower strain rate level (at 1300 s^{-1}), as shown in Figure 6, The observed stress levels obtained from simulations demonstrate that the interpolated flow stress level 10^{10} s^{-1} was lower to that observed in SHPB tests. On the other hand, a good agreement between the experimental data and calculated results was for the 20%-offset assumption for a strain rate of 10^5 s^{-1}. The results obtained in FE simulation of the SHPB test, verified our assumption that was used in the FE modelling of CT and UAT. In addition to that, the SHPB model was used for additional studies that will be discussed in the following section.

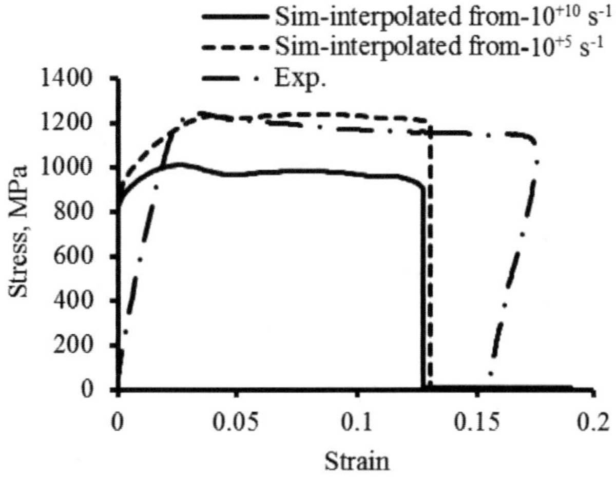

Figure 6. Interpolation of stress level at lower strain rate of 1300 s^{-1} from data at higher strain rates

6. Results and discussions

6.1. Numerical results of SHPB test

The values of the axial strain calculated for locations of the strain gauges at the middle span in tests on the incident and transmitted bar was calculated with the FE model (see Figure 7a). A compressive wave pulse propagating in the incident bar resulted from an impact of the striker bar with it. This response of the bar was achieved before compression of the specimen, while a dynamic plastic response of the workpiece was gained through the transmitted and reflected pulses measured from the transmitted and incident bars, respectively. The transmitted pulse was compressive whereas the reflected pulse was tensile. The flow stresses for Ti-15333 were analysed for a strain-rate 3300 s^{-1} and room temperature (see Figure 7b). The stress-strain relations obtained from simulations were in reasonably a good agreement with experimental stress magnitudes.

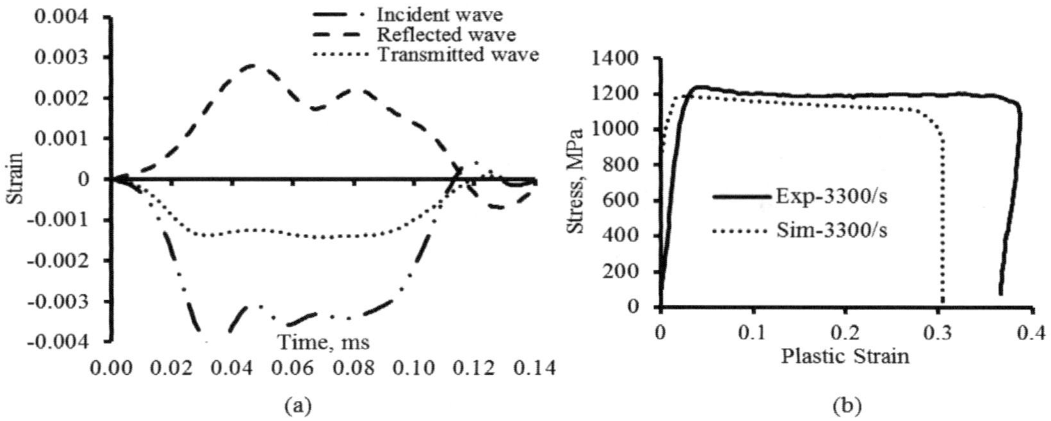

Figure 7. (a) Strain evaluation at midspan: for Incident, transmitted and reflected waves; (b) stresses observed on midpoint

An inhomogeneous deformation of the specimen was observed at the start of the compression process in the FE simulations; at 3% plastic strain, a homogeneous stress state was reached. After the first contact between the incident bar and the specimen the levels of equivalent and the axial stresses varied from 138 MPa to approx. 400 MPa as shown in Figure 8. The same was observed in the equivalent plastic strain rate (1350 s^{-1} to 3000 s^{-1}) at the same cross-sectional plane of the workpiece (see Figure 9a). This may be of some concern when deriving the relevant stress-strain relations from the experimental procedure since one of the fundamental assumptions of the SHPB theory was that of a homogeneous stress state in the body during the entire deformation evaluation.

A temperature increase in the specimen was observed due to low thermal conductivity and adiabatic heat condition of the studied cases. A rise of 117°C was caused by plastic deformation of the workpiece material; the thermal history of the midpoint of the workpiece material is shown in Figure 9c. The effect of adiabatic heating should be included in the materials stress-strain response in actual experiments. On the other hand, including this effect may cause non-negligible errors in the data and could ultimately affect the accuracy of numerical simulations. The same discussion can be extended to the reference strain-rate value, which was found to be non-constant during the initial stages of the test. Using a reference strain-rate value for experiments conducted without accounting for its variability during the compression may also lead to errors.

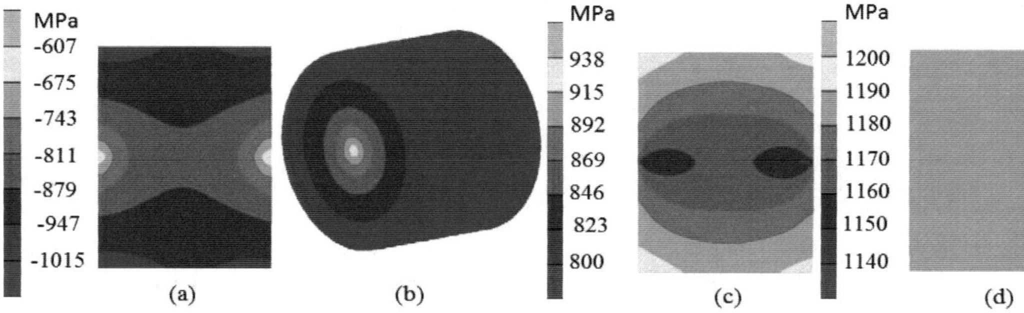

Figure 8. Inhomogeneous distribution of axial stress in central cross sectional plane (a) and specimen, t = 12 μs (b); inhomogeneous stress distribution, , t = 12 μs (c); homogeneous stress distribution in central cross sectional plane, t = 30 μs (d)

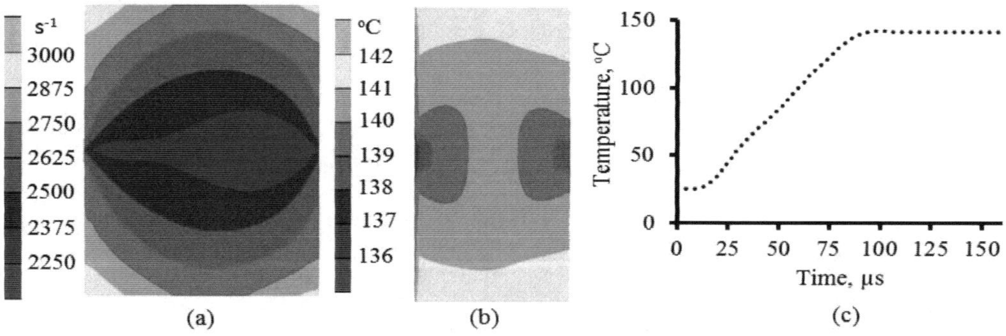

Figure 9. (a) Inhomogeneous distribution of strain rate, t = 12 μs; (b) temperature distribution in specimen, t = 125 μs; (c) temperature history for midspan of specimen

6.2. Numerical results of oblique turning process

The developed model of oblique turning provides all the three components of cutting forces in FE simulations. The tangential cutting force (*Fx*) of magnitude 110 N was observed as shown in Figure 10a. A cutting velocity of 10 m/min, feed rate of 0.1 mm/rev and depth of cut of 300 μm were used for these simulations. A radial component (*Fy*) of 48 N and an axial component of (*Fz*) 20 N were calculated.

Application of ultrasonic vibration to the cutting tool produced a significant reduction in average cutting force. In the process of vibration, the cutting tool separated from the workpiece in each cycle, resulting in considerable reduction of the average cutting force. During the penetration stage of the tool in UAT, the tool experiences a higher magnitude of cutting force in the tangential, radial and axial directions. However, during the retardation stage, the cutting forces in all three directions started to decline, ultimately vanished at complete tool disengagement (see Figure 10b). The average cutting force component *Fx* declined from 110 N to 35 N (68%). Also, a considerable reduction of 68% (from 48 N to 15 N) was observed for average *Fy* in UAT. In addition to this, ultrasonically assisted turning yielded a significant reduction of 55% (20 N to 9 N) in average *Fz*. The simulations of CT and UAT were carried out for various depths of cuts, and a considerable reduction in forces was consistently observed in UAT when compared to CT. The obtained results of simulations were in good agreement with experimental data (Figure 11).

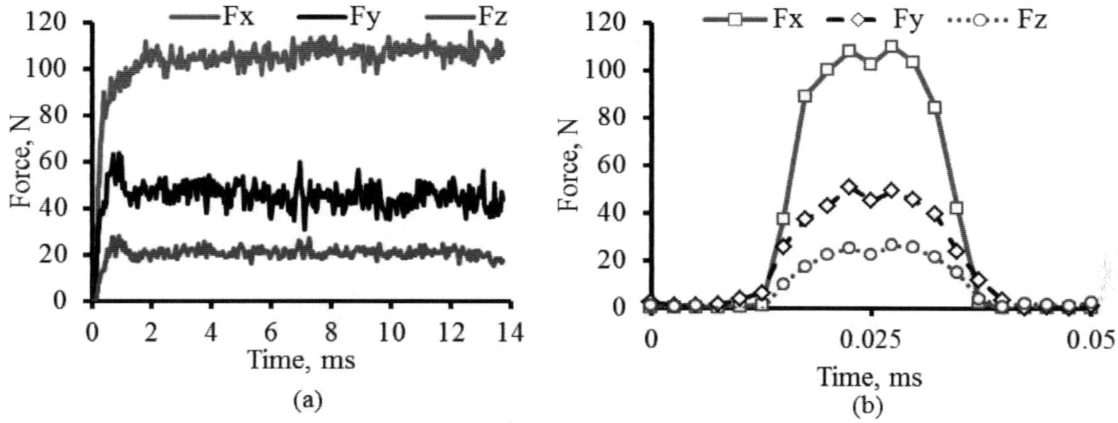

Figure 10. Cutting forces in CT (a); and UAT (b)

The temperature in the process zone was also investigated with the developed 3D FE model of oblique turning in both CT and UAT, as shown in Figure 12. In UAT, a relatively higher temperature was observed along the contact area of the tool-workpiece interface. Temperature in UAT was higher at the maximum penetration of the tool; the possible reason for that was application of extra energy to the cutting tool in the form of vibration. The application of vibration to the cutting tool increased its relative velocity causing high plastic deformation of the workpiece material, resulting in high temperature (approx. 105°C) when compared to CT. The process zone temperature in both CT and UAT was higher in magnitude when compared to experimental results; this could be due to a limitation of the infrared thermography in measurements. There was often an obstruction to the clear line-of-sight due to chips, which ultimately yielded a lower estimate of the process zone temperature. However, the temperature magnitudes in regions visible to the thermal camera were in a good agreement with the obtained simulation results. Hence, the FE models have the added advantage of elucidating the temperature characteristics in regions of the workpiece not visible to thermal imaging.

A distribution of von-Mises stress in the cutting region of the studied region in both CT and UAT was investigated with the developed oblique turning model (Figure 13). A uniform distribution of stresses in the primary and secondary deformation zones was observed in CT. while in UAT, a distribution of stresses was transient during each cycle of vibration, as expected. The stress magnitude in the process zone at maximum penetration in UAT was 1050 MPa which reduced significantly at the retardation stage when the tool separated from the workpiece as shown in Figure 13c. The stress level in CT was nearly the same as the peak stress in UAT. The average levels of the von-Mises stress in the cutting region for one cycle in UAT was significantly lower than that of CT.

Similarly, the strain rate calculated in CT and UAT in oblique turning was also studied. The strain rate in UAT was significantly higher almost four-times when compared to CT (see Figure 14). The high strain rate in UAT was due to the vibro–impact interaction of the cutting tool with the workpiece material resulting in strain hardening of machined surface.

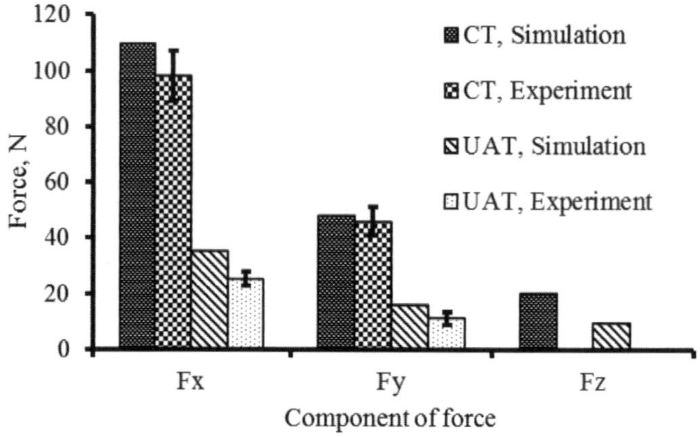

Figure 11. Comparison of cutting forces calculated in experiments and simulation

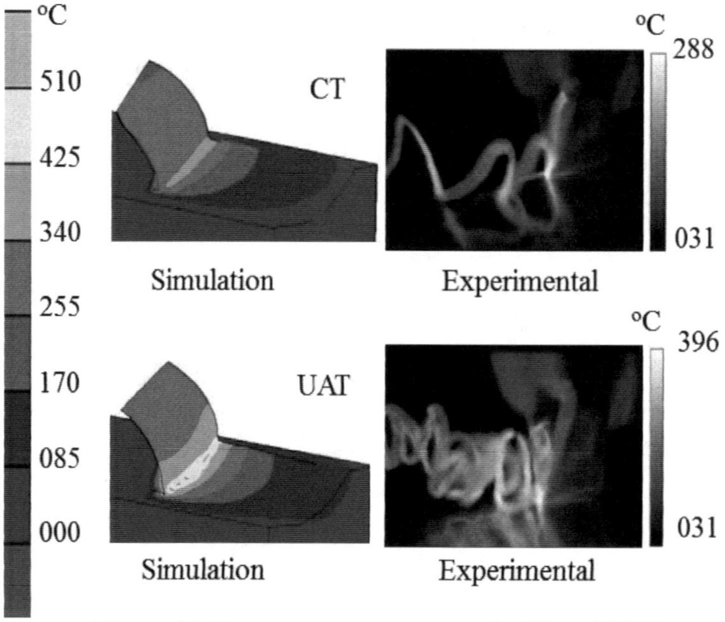

Figure 12. Process zone temperature in CT and UAT

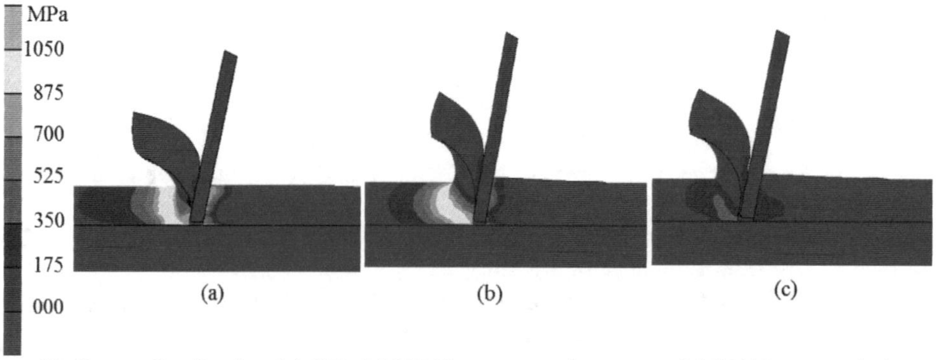

Figure 13. Stress distribution (a) CT; (b) UAT at penetration stage; (c) UAT at retardation stage

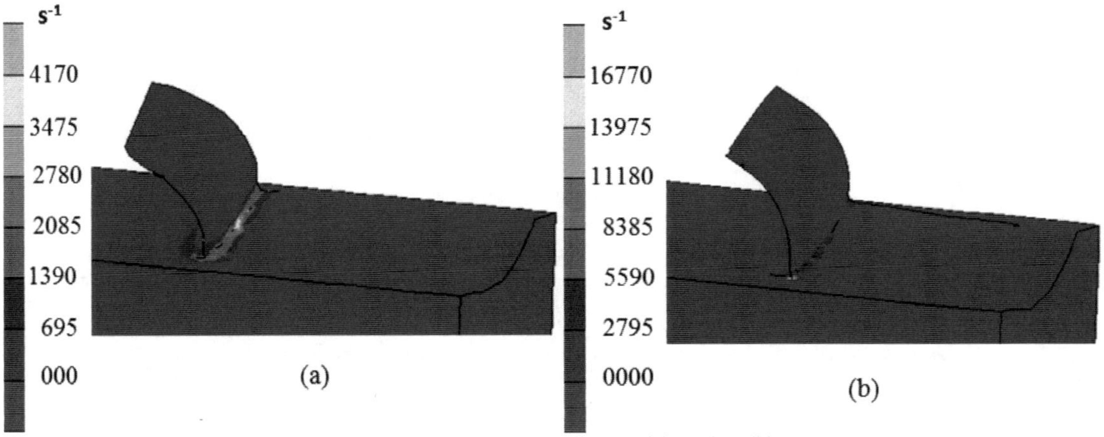

Figure 14. Strain rate in CT (a); UAT (b)

7. Conclusions

In this study, a 3D thermo-mechanically coupled finite-element model was developed for SHPB test. The model was used to validate the FE simulations of CT and UAT based on assumptions of the deformation behaviour of the studied beta titanium alloy (Ti-15333). In the FE simulation of the SHPB experiment, an inhomogeneous deformation behaviour was observed in the specimen at the initial stages of compression contrary to a standard assumption of homogeneity employed in stress-strain calculations in SHPB. Thus, the validity of such assumptions should be critically assessed for each combination of material and high-strain-rate loading conditions. The adiabatic heating regime and variable strain rate during the compression process should also be taken into consideration in the identification of the material model`s parameters, since this can result in non-adequate extrapolation of the future real-life material properties.

The validated SHPB tests data were used in the FE models of CT and UAT of Ti-15333. The models were capable to analyse all the three components of cutting forces in both CT and UAT. The reduction in the radial and the axial components of cutting forces was explained with the help of the developed models. The level of cutting forces in UAT was significantly lower when compared to that in CT; this reduction can be attributed to the tool separation from the workpiece material in each cycle of vibration. A relatively high temperature of the process zone was observed in UAT together with changing in the process zone due to the tool movement. However, the level of average stress was lower in UAT when compared to CT.

Acknowledgements

This work was supported by the European Union Seventh Framework Program (FP7/2007-2013) under grant agreement No. PITN-GA-2008-211536, project MaMiNa.

The authors would also like to acknowledge Tampere University of Science and Technology for providing the SHPB test data for simulations.

References

[1] Muhammad R, Maurotto A, Roy A and Silberschmidt V V 2011 Analysis of forces in vibro-impact and hot vibro-impact turning of advanced alloys *Appl. Mech. Mater.* **70** 315-320

[2] Muhammad R, Maurotto A, Roy A and Silberschmidt V V 2012 Ultrasonically assisted turning of Ti-6Al-2Sn-4Zr-6Mo *J. Phy. Conf. Ser.* **382** 012016

[3] Donachie M J 2004 *Titanium- A technical guide* (ASM International Ohio)

[4] Muhammad R, Ahmed N, Abid M and Silberschmidt V V 2010 3D modeling of drilling process of AISI 1010 steel *J. Mach. Form. Tech.* **2(3/4)** pp 201-216

[5] Zhao H and Gary G 1996 On the use of SHPB techniques to determine the dynamic behavior of materials in the range of small strains *Int. J. Solid. Strut.* **33** 3363
[6] Yang L M and Shim V P W 2005 An analysis of stress uniformity in split Hopkinson bar test specimens *Int. J. Imp. Eng.* **31** 129
[7] Davies E D H and Hunter S C 1963 The dynamic compression testing of solids by the method of the split Hopkinson pressure bar *J. Mech and Phy. Solid* **11** 155-179
[8] Bell J F 1966 An experimental diffraction grating study of the quasi-static hypothesis of the split hopkinson bar experiment *J. Mech and Phy. Solid* **14** 309-327
[9] Muhammad R, Ahmed N, Roy A and Silberschmidt V V 2012 Numerical modelling of vibration-assisted turning of Ti-15333 *Pro. CIRP* **1** 347-352
[10] Maurotto A, Muhammad R, Roy A and Silberschmidt V V 2013 Enhanced ultrasonically assisted turning of a β-titanium alloy *Ultrasonics* **53** 1242-1250
[11] Kolsky H 1949 An Investigation of the Mechanical Properties of Materials at Very High Rates of Loading *Proc. Phy. Soc. London* 676
[12] Li Z and Lambros J 1999 Determination of the dynamic response of brittle composites by the use of the split Hopkinson pressure bar *Comp. Sci. Tech.* **59** 1097
[13] Muhammad R, Ahmed N, Roy A and Silberschmidt V V 2012 Turning of advanced alloys with vibrating cutting tool *Sol. Stat. Phen.* **188** 277-284
[14] Demiral M, Roy A and Silberschmidt V V 2010 Effects of loading conditions on deformation process in indentation *Comput. Mater. Conti.* **475(1)** 1-18
[15] MSC *Marc User's Guide Version 2012 MSC Software Corporation LA*

Journal of Physics: Conference Series **451** (2013) 012031 doi:10.1088/1742-6596/451/1/012031

Numerical fatigue life assessment of cardiovascular stents: A two-scale plasticity-damage model

HAFA Santos[1], F Auricchio[2] and M Conti[2]

[1] Wolfson School of Mechanical and Manufacturing Engineering, Loughborough University, Loughborough, Leicestershire LE11 3TU, UK

[2] Dipartimento di Ingegneria Civile ed Architettura, Universit degli studi di Pavia, Pavia, Italy

E-mail: hugofreixialsantos@gmail.com

Abstract. Cardiovascular disease has become a major global health care problem in the last decades. To tackle this problem, the use of cardiovascular stents has been considered a promising and effective approach. Numerical simulations to evaluate the *in vivo* behavior of stents are becoming more and more important to assess potential failures. As the material failure of a stent device has been often associated with fatigue issues, numerical approaches for fatigue life assessment of stents have gained special interest in the engineering community. Numerical fatigue life predictions can be used to modify the design and prevent failure without making and testing numerous physical devices, thus preventing from undesired fatigue failures. We present a numerical fatigue life model for the analysis of cardiovascular balloon-expandable stainless steel stents that can hopefully provide useful information either to be used for product improvement or for clinicians to make life-saving decisions. This model incorporates a two-scale continuum damage mechanics model and the so-called Soderberg fatigue failure criterion. We provide numerical results for both Palmaz-Schatz and Cypher stent designs and demonstrate that a good agreement is found between the numerical and the available experimental results.

1. Introduction

Stents are small tube-like medical devices used to restore patency of blood vessels where the lumen area is reduced due to atherosclerosis, a degenerative disease of the vessel wall. A stent acts as a mechanical scaffold for the vessel and its implant is performed by a minimally-invasive procedure inserting a catheter through a small incision in the femoral artery. Stenting is nowadays a common clinical practice especially for the treatment of cardiovascular stenosis, which has a high economical and social impact in the Western countries.

As the heart beats, and hence the arteries pulse, at typically 70 times per minute (around 40 million times per year), stents are subjected to long-term cyclic loading conditions. The US regulatory authority over these devices (FDA) recommends that stents must withstand 10-15 years of pulsatile loading *in vivo*, the equivalent to 400-600 million loading cycles. Therefore, the long-term structural integrity of a stent, in particular its mechanical fatigue behavior, must be one of the major design considerations.

The fatigue behavior of a stent results from a complex interaction of two stress states. The first of these is the state to which the device is statically loaded and deformed as a result from the positioning of the stent within the vessel. This state is typically characterized by high plastic strain levels in some parts of the stent and, therefore, by high residual stresses. A second state is then imposed, which corresponds to the cyclic component of the loading as a consequence of the pulsatile blood-pressure variations in the vessel. The first state will be hereafter termed the mean stress state, whereas the second will be referred to as the alternating stress state.

For stents made of standard metallic materials, the fatigue damage induced by the alternating stress state can be divided into three main stages: crack initiation, stable crack growth and final failure [1]. Approaches to fatigue life prediction traditionally focus on one of the two first stages of the fatigue damage process. Total-life approaches define failure as the initiation of a crack and the number of loading cycles before crack initiation is assumed as the total fatigue life. These approaches are based either on $S - N$ or $\varepsilon - N$ experiments, both giving the number of loading cycles until a plain specimen fails. While $S - N$ curves are suited for materials in the high-cycle fatigue regime, where typically more than 10^5 cycles at low stresses are required to failure, and where deformation is primarily elastic, $\varepsilon - N$ curves are usually employed for materials in the low-cycle regime, where less than 10^5 cycles characterized by stresses that are high enough for plastic deformations to occur are required to failure.

Stents are typically designed to work in the high-cycle fatigue regime. In this regime, the equivalent von Mises stresses are above the fatigue limit (usually defined as the stress amplitude requiring 10^6 load cycles up to rupture), but considerably below the yield stress [1]. This leads very often to a number of loading cycles to failure greater than 10^5. In this regime, damage can be considered as a microscale process of progressive deterioration of the material with no influence on the mesoscopic behavior, where, up to crack initiation, the material deforms primarily elastically at the scale of the representative volume element (mesoscale), and coupling between plasticity and damage may be neglected everywhere but at the microscale [2].

The material degradation caused by the initiation, growth and coalescence of microcracks due to cyclic loading is well suited for characterization by the theory of continuum damage mechanics, *e.g.* [3]. Centered on microcrack development, the continuum damage mechanics theory provides a good understanding of the mechanisms of fatigue failure by means of damage variables. Different continuum damage mechanics based models have been proposed in the literature for the high-cycle fatigue analysis of metallic structures: (i) models based on two-scale approaches in which micro and meso scales are linked by means of a localization law [4, 5, 2, 6]; and (ii) models relying on a single (macroscopic) scale [7, 8].

Due to its inherent multi-scale manifestation, high-cycle fatigue damage can only be properly captured by resorting to models which are capable of dealing with plasticity and damage on a scale smaller than the macroscopic scale. Furthermore, as the experiments for a stent device are challenging, due to its small size, the fatigue life assessment of a cardiovascular stent is not a trivial task. Computational tools, such as the finite element method, have proved extensively their usefulness to analyze cardiovascular stents, see *e.g.* [9, 10, 11, 12]. None of these works have however addressed the fatigue life assessment of stents. Exceptions to this can be found in [13, 14, 15, 16].

We present a numerical model for the fatigue life assessment of cardiovascular balloon-expandable stainless steel stents [17]. This model incorporates the two-scale plasticity-damage model proposed by [4, 5], modified by means of the so-called Soderberg fatigue relation [18].

2. Overview of the proposed model

One of the first attempts to extend the framework of continuum damage mechanics to the fatigue field including its multi-scale aspect is the model proposed in [19], which was later on further extended and improved in [20, 4, 5, 2]. This model considers two phases. One represents a microscopic spherical inclusion with an elasto-plastic coupled with isotropic damage behavior, and the other represents a mesoscopic elastic, possibly elasto-plastic, matrix represented through the representative volume element in which the inclusion is embedded. The two phases are assumed to have the same elastic behavior. The matrix behavior is characterized by its yield stress, its ultimate stress and its asymptotic fatigue limit, the latter viewed as the asymptote of a $S-N$ curve. As for the inclusion, an elasto-plastic material behavior with strain hardening is assumed. This allows to produce results which are coherent with most of the experimental results obtained in the high-cycle fatigue regime of metals. A weakness of the inclusion is considered by assuming that its corresponding yield stress is identical to the asymptotic fatigue limit of the matrix material.

The two-scale plasticity-damage model was conceived to capture degradation by means of a locally coupled strain-driven analysis [20]. This basically consists in performing first a global elastic structural analysis, using for instance the finite element method to compute stress and strain variables defined at the mesoscale, followed by a time integration of the elasto-plastic-damage constitutive equations, required to compute the microscopic stress, strain, hardening and damage variables. The time integration is only performed at critical points, as a post-processing scheme using the history of mesoscopic stresses and strains as inputs. This corresponds to a separated multi-scale model, as micro and meso calculations are performed independently. Time integration is performed step by step until a stabilized cycle is reached, after which a jump in cycles procedure is considered to avoid a large number of steps. This post-processing procedure is repeated until the critical damage value is reached at the critical point, indicating microcrack initiation. The critical point is assumed as the point featuring the maximum damage equivalent mesoscopic stress.

Experiments have shown that, at fixed amplitude the fatigue life decreases as the mean stress increases [1]. As pointed out in [21], the crack closure parameter of the two-scale plasticity-damage model is not sufficient to realistically represent the material behavior at high mean stresses. Hence, and as cardiovascular stents work under high levels of mean stresses, the two-scale model must be modified in order to be applicable to the fatigue life prediction of stents. The strategy adopted in this work consisted in, rather than using a fixed asymptotic fatigue limit within the yield potential, as was considered in [4, 5], taking an asymptotic fatigue limit which varies with respect to the mean stress defined at the mesoscale. This was accomplished in the present work by resorting to the Soderberg relation [18]. A Goodman relation could alternatively have been employed. For further details on the model, the reader is referred to [17].

3. Application of the model to the fatigue life prediction of cardiovascular balloon-expandable stents

Two stent design models were analyzed. We first analyzed a *Palmaz-Schatz* stent, as this was the one for which we could find some fatigue experimental results, indeed required to validate the numerical results provided by the proposed model. We then analyzed a *Cypher* stent, one of the most adopted stent designs nowadays.

3.1. Finite element analysis

In the unexpanded configuration, the *Palmaz-Schatz* stent was assumed to be a tube with rectangular slots along its length. Its initial length, inner and outer diameters were taken as $16mm$, $1mm$ and $1.2mm$, respectively (strut thickness is $200\mu m$). The stent has 5 slots in the longitudinal direction and 12 slots in the circumferential direction, each slot measuring $2.88mm$ and $0.24mm$, respectively. The mesh was built from the CAD geometry constructed using Rhinoceros v. 4.0 Educational (McNeel & associates, Seattle, WA, USA). As for the *Cypher* stent, the length, inner diameter and outer diameter were set to $8.4mm$, $0.85mm$ and $1.15mm$, respectively, *i.e.*, the thickness of the struts were taken as $300\mu m$. This geometry resembles the *Cypher* stent geometry (Boston Scientific Co., Natick, MA, USA) with a nominal diameter and length of $3mm$ and $8mm$, respectively.

In both stent designs, the cardiovascular artery was modeled as an idealized vessel represented by a thin-walled pipe with an initial inner diameter of $2.5mm$, wall thickness $0.5mm$ and length $12mm$.

The *Palmaz-Schatz* stent was assumed to be made of a SS316LN stainless steel, similar to the SS316L steel used in many commercial balloon-expandable stents, such as the original J & J *PalmazSchatz*. As for the *Cypher* stent, it was assumed to be made of a AISI 316L stainless steel. We assumed an elasto-plastic perfect constitutive behavior for both stent designs, characterized by Young's modulus $E = 196GPa$ and Poisson's ratio $\nu = 0.3$. As for the yield stress, we adopted $\sigma_y = 205MPa$ for the *Palmaz-Schatz* stent [22] and $\sigma_y = 375MPa$ for the *Cypher* stent [23].

The material of the artery wall was modeled using a 5-parameter second-order Mooney-Rivlin hyperelastic model suitable for incompressible isotropic materials [10].

The loadings on the stents were considered in different steps to simulate the loading conditions they experience in service, namely, balloon-inflation, recoil and physiological loading within the artery. For both stent designs, the stent expansion was modeled as a displacement driven process, by enforcing the radial displacements of a rigid cylinder (previously introduced into the stent) to expand the stent to an inner diameter of 3 mm. The expansion of the stents was performed into a hyperelastic thin-walled pipe. The stent expansion step was accomplished by modeling contact between the expansion cylinder and the stent, as well as between the stent and the internal surface of the tube. After the expansion step, the stent/vessel systems were allowed to recoil by removing the deployment boundary conditions. This step was taken to simulate the balloon deflation and retraction of the catheter. Maximal and minimal uniform pressure loads of $120mmHg = 0.015MPa$ and $80mmHg = 0.010MPa$ were then sequentially applied to the inner surface of the pipes to conservatively represent physiological systolic and diastolic blood-pressure loads within the arteries. Appropriate kinematic boundary conditions were set.

The analyses of the stents under the mean stress state were performed using ABAQUS/Standard finite element code (Simulia, Dassault Systems, Providence, RI, USA).

In the *Palmaz-Schatz* test case, the stent and the thin-walled pipe were modeled using 10716 and 6080 C3D8R (8-node three-dimensional brick 'reduced-integration') elements, respectively. The rigid cylinder was modeled using 408 SFM3D4 (4-node quadrilateral surface) elements. As for the *Cypher* case, the stent and the thin-walled pipe were modeled using 31848 and 15808 C3D8R elements, respectively, whereas the rigid cylinder was modeled using 1909 SFM3D4 elements.

The finite element analyses performed over the *Cypher* stent showed that, although some parts of the stent were characterized by relatively high plastic strains, which is indeed a consequence of the expansion and subsequent recoil of the balloon catheter, leading in turn to high residual stresses in the material, the material response was found to be linear-elastic during the fatigue cycles. As for the *Palmaz-Schatz* stent case, the material response was found to be elasto-plastic, even during the fatigue cycles.

3.2. Fatigue life assessment

3.2.1. Material Parameters Identification There were 10 parameters to be identified: 4 on the mesoscale (asymptotic fatigue limit σ_f, ultimate stress σ_u, rupture stress σ_R and plastic modulus C), 5 damage parameters (damage strength S, damage exponent s, damage threshold ε^p_D, critical damage D_c and correction parameter m), and the microdefects closure parameter h.

The mesoscale parameters were identified on a monotonic tensile curve [2], whereas the identification of the parameters S, s, p_D, D_c, m and h was carried out using some fatigue tests, more specifically, an experimental $S - N$ curve and some low-cycle fatigue tests.

In order to take into account the size effect, the parameters σ_u, σ_R and ε^p_D were explicitly identified from the tensile stress-strain curve obtained in [24] for a stent of $100\mu m$ of thickness.

σ_f was identified from the experimental fully reversed $S - N$ tests conducted in [25] on a non-treated stainless steel 316L. D_c, h and C were obtained following [2]. s, S and m were calibrated using the experimental $S - N$ results given in [25]

The adopted parameters are given in Table 1.

Table 1. Mesoscale material parameters

$\sigma_f(MPa)$	$\sigma_u(MPa)$	$\sigma_R(MPa)$	$C(MPa)$	$S(MPa)$	s	ε_D^p	D_c	m	h
200	820	650	1520	0.8	0.5	0.28	1.0	1.2	0.2

3.2.2. Prediction of microcrack initiation The critical points of the stents were taken as the Gauss points in which the damage equivalent mesoscopic stress was maximum. As for the *Cypher* stent model, the material behavior was found to be linear elastic during the cyclic loading, *i.e.*, the critical point did not change from cycle to cycle, and thus it was sufficient to perform the post-processing scheme at this single point. As for the *Palmaz-Schatz* stent model, the material response was found to be elasto-plastic during the cyclic loading. For this reason, the post-processing scheme was applied in the latter case over all Gauss points of the stent model. The critical point was selected as the one for which the damage variable first reached the critical damage parameter.

Application of the post-processing scheme based on the locally coupled analysis to modeling the stents under the alternating stress states led to a number of cycles up to microcrack initiation of 64 millions for the *Palmaz-Schatz* stent model, and 53 millions for the *Cypher* stent model. This is equivalent to approximately 22 and 18 months of the heart pumping in an average human adult, respectively.

Although the obtained results correspond to very short fatigue lives, our prediction for the *Palmaz-Schatz* stent design seems to be in accordance with the experiments carried out in [26], in which real stainless steel Palmaz-Schatz^{TM} stents ($15mm$ long stents from Johnson and Johnson Corporation) were tested to fatigue in a saline solution at $37°C$ to simulate real *in vivo* physiological conditions. Therein, the criterion for fatigue failure was considered as the rupture of the first strut of the stent. The results given in [26] show that, although 2 out of 2 tested stainless steel stents have survived to 1 million cyclic loadings, 0 out of 2 stents have survived to 10 million cycles, only 1 out of 2 stents have survived to 40 million cycles. The experimental results also indicate that 0 out of 4 stents have survived to 100 million cycles.

Clearly, the stent designs under analysis would not satisfy the requirements imposed by the US Food and Drug Administration, which recommends that a cardiovascular stent must be able to withstand at least 400 million cardiac cycles (equivalent of approximately ten years) without exhibiting fatigue-associated failure. It is worth mentioning however that the proposed fatigue life assessment model was conceived to predict crack initiation, and not fatigue rupture. To predict the total life of the stents up to final failure, the analyses should have taken into account the crack propagation effects up to final failure. Despite this, we remind that, in the high-cycle fatigue regime, the crack initiation stages may cover a large percentage of the fatigue life and, thus, we would not expect the total fatigue lives of the stent models under analysis to be considerably higher than the ones we obtained.

4. Conclusions

A numerical model was presented for the numerical fatigue life assessment of cardiovascular balloon-expandable stainless steel stents. The model relies on a two-scale continuum damage mechanics theory and assumes the Soderberg relation as the fatigue failure criterion, which allows the model to take into account the high mean stress effects inherent to cardiovascular stents. The model was applied to the analysis of two common stent geometries, more specifically, *Palmaz-Schatz* and *Cypher* designs. For both stent geometries, the obtained results indicated a limited fatigue life. In particular, the results obtained for the *Palmaz-Schatz* stent model were shown to be in close agreement with the available experimental results. We hope that the present model can be used to modify the design and prevent from undesired fatigue failures without testing numerous physical devices.

References

[1] Suresh S 1998 *Fatigue of materials* (New York: Cambridge University Press)
[2] Lemaitre J and Desmorat R 2005 *Engineering damage mechanics: ductile, creep, fatigue and brittle failures* (New York: Springer Verlag)
[3] Lemaitre J 1992 *A course on damage mechanics* (New York: Springer-Verlag)
[4] Lemaitre J, Sermage J P and Desmorat R 1999 *Int J Fracture* **97** 67–81
[5] Desmorat R and Lemaitre J 2001 *Handbook of Materials Behavior Models* vol 1: Deformations of Materials (Academic Press) chap Continuous Damage: A two-scale model for quasi-brittle and fatigue damage, pp 525–535
[6] Desmorat R, Kanea A, Seyedi M and Sermageb J P 2007 *Eur J Mech A-Solid* **26** 909–935
[7] Oller S, Salomon O and Onate E 2005 *Comp Mater Sci* **32** 175–195
[8] Ottosen N S, Stenstrom R and Ristinmaa M 2008 *Int J Fatigue* **30** 996–1006
[9] Migliavacca F, Petrini L, Colombo M, Auricchio F and Pietrabissa R 2002 *J Biomech* **35** 803–811
[10] Lally C, Dolan F and Prendergast P J 2005 *J Biomech* **38** 1574–1581
[11] Beule M D, Mortier P, Carlier S, Verhegghe B, Impe R V and Verdonck P 2008 *J Biomech* **41** 383–389
[12] Mortier P, Holzapfel G, Beule M D, Loo D V, Taeymans Y, Segers P, Verdonck P and Verhegghe B 2010 *Ann Biomed Eng* **38** 88–99
[13] Perry M, Oktay S and Muskivitch J C 2002 *Minim Invasiv Ther* **11** 165–171
[14] Marrey R V, Burgermeister R, Grishaber R B and Ritchie R O 2006 *Biomaterials* **27** 1988–2000
[15] Pelton A R, Schroeder V, Mitchell M R, Gong X Y, Barney M and Robertson S W 2008 *J Mech Behav Biomed Mater* **1** 153–164
[16] Li J, Luo Q, Xie Z, Li Y and Zeng Y 2010 *Heart Vessels* **25** 333–337
[17] Santos H A F A, Auricchio F and Conti M 2012 *J Mech Behav Biomed Mater* **15** 78–92
[18] Soderberg C R 1939 *Transactions of the American Society of Mechanical Engineers* **52** 13–28
[19] Lemaitre J 1985 *J Eng Mater-T ASME* **107** 83–89
[20] Lemaitre J and Doghri I 1994 *Comput Method Appl M* **115** 197–232
[21] Barbier G, Desmorat R, Sermage J P, du Tertre A, Courtin S, Dehouve J and Tchou-Kien D 2008 *LCF6 - 6th International Conference on Low Cycle Fatigue* (Berlin, Germany)
[22] Auricchio F, Loreto M D and Sacco E 2001 *Comput Method Biomec* **4** 249–263
[23] Murphy B P, Savage P, McHugh P E and Quinn D F 2003 *Ann Biomed Eng* **31** 686–691
[24] Harewood F J and McHugh P E 2007 *Ann Biomed Eng* **35** 1539–1553
[25] Agarwal N, Kahn H, Avishai A, Michal G, Ernst F and Heuer A H 2007 *Acta Materialia* **55** 5572–5580
[26] Glenn R and Lee J 1997 *SMST-97: Proc. of the Second International Conference on Shape Memory and Superelastic Technologies* ed Pelton A, Hodgson D, Russell S and Duerig T (California, USA: Pacific Grove, CA: SMST Society)

Journal of Physics: Conference Series **451** (2013) 012032 doi:10.1088/1742-6596/451/1/012032

Dynamic simulation of stent deployment - effects of design, material and coating

A Schiavone, L G Zhao and A A Abdel-Wahab
Wolfson School of Mechanical and Manufacturing Engineering, Loughborough University, LE11 3TU, Loughborough, UK

E-mail: a.schiavone@lboro.ac.uk

Abstract. Dynamic finite-element simulations have been carried out to study the effects of cell design, material choice and drug eluting coating on the mechanical behaviour of stents during deployment. Four representative stent designs have been considered, i.e., Palmaz-Schatz, Cypher, Xience and Endeavor. The former two are made of stainless steel while the latter two made of Co-Cr alloy. Geometric model for each design was created using ProEngineer software, and then imported into Abaqus for simulation of the full process of stent deployment within a diseased artery. In all cases, the delivery system was based on the dynamic expansion of a polyurethane balloon under applied internal pressure. Results showed that the expansion is mainly governed by the design, in particular open-cell design (e.g. Endeavor) tends to have greater expansion than closed-cell design (e.g. Cypher). Dogboning effect was strong for slotted tube design (e.g. Palmaz-Schatz) but reduced significantly for sinusoidal design (e.g. Cypher). Under the same pressure, the maximum von Mises stress in the stent was higher for the open-cell designs and located mostly at the inner corners of each cell. For given deformation, stents made of Co-Cr alloys tend to experience higher stress level than those made of stainless steels, mainly due to the difference in material properties. For artery-plaque system, the maximum stress occurred on the stenosis and dogboning led to stress concentration at the ends of the plaque. The drug eluting coating affected the stent expansion by reducing the recoiling phenomenon considerably, but also raised the stress level on the stent due to property mismatch.

1. Introduction

Finite-element method has been widely employed to assess the biomechanical behaviour of stents by modelling their deployment inside diseased arteries. Chua *et al.* [1] simulated the expansion of a Palmaz-Schatz stent inside an artery with stenotic plaque. The results demonstrated the capability of slotted-tube stents in resisting the recoil of blood vessel and preventing restenosis after expansion. Zahedmanesh *et al.* [2] modelled the deployment of an ACS Multi-Link RX DUET stent in a stenotic artery based on the digitalization of 3D angiography images of a human coronary artery. They suggested that direct application of pressure on the inner surface of stent may be used as an optimal modelling strategy for complex arterial geometries. Pericevic *et al.* [3] investigated the influence of stenotic plaque compositions (i.e., hypecellular, hypocellular and calcified) on stent deployment, and showed that a calcified plaque exhibited less stress level but higher resistance to expansion compared to other types of plaque. Lally *et al.* [4] performed a comparative study of two stents (i.e. S7 and NIR stents) by simulating their deployment inside a stenotic artery. Their results showed that the S7 stent caused lower stress on the blood vessel wall compared to the NIR, which was consistent with clinical observations that the NIR stent has a higher restenosis rate than S7. Tambaca *et al.* [5] studied the mechanical behaviour of different fully expanded stent designs under uniform compression and bending forces. It was shown that the latest stent designs (i.e. Xience) exhibited good radial stiffness and bending flexibility.

In addition, Gu *et al.* [6] studied the influence of an ultra-thin silicone coating during deployment of drug eluting stent in an artery with long fusiform aneurysm. It was shown that the coated stent needed an appreciatively 30% higher pressure to expand the artery than its bare metal version. On the other hand, Hopkins *et al.* [7] investigated the 2D delamination effect of the stent coating during deployment, which occurred mainly in the region of hinge with high plastic strain. The initiation of coating debonding depended on the coating thickness, the coating material and the curvature of the hinge.

Currently, there is a lack of research in the modelling of the full deployment process of the latest stent designs, in particular the effects of stent materials and coatings on the expansion behaviour of stents. It is also not well understood how the deformation and stress of the artery-plaque system are influenced by different stent designs during deployment.

In this paper, finite-element analyses have been carried out to simulate stent deployment for four representative stent designs (Palmaz-Schatz, Cypher, Xience and Endeavor). Comparisons have been made to assess the effects of cell design, material and coating on stent expansion, stress distribution and geometrical change during the deployment process inside a diseased artery.

2. Finite element simulation

2.1. Geometrical models

Models of Palmaz-Schatz, Cypher, Xience and Endeavor stents were produced using ProEngineer software according to their real geometrical features. The length of all stents was chosen to be 10mm and the diameter at the crimpled state was fixed at 3mm. The strut thickness is 100μm, 140μm, 81μm and 91μm for Palmaz-Schatz, Cypher, Xience and Endeavor, respectively. To study the coating effect, a two-ring model was created for Xience stent, with a length of 4mm and a diameter of 3mm, coated with a thin layer of polymer (7.8μm). The balloon was modelled as a simple cylinder with an outer diameter that matched the inner diameter of the stent. The balloon was always 2mm longer than the stent (i.e., one millimetre longer at each end). The blood vessel was modelled as a cylinder with axisymmetric stenotic plaque inside. The artery has a length of 20mm and a wall thickness of 0.4mm, while the plaque has a length of 5mm and a maximum thickness of 0.58mm. The outer diameter was 5.4 mm and 5.0 mm for the artery and the plaque, respectively.

2.2. Materials

The materials used to manufacture the stents are 316L stainless steel for Palmaz-Schatz and Cypher and Co-Cr alloy for Xience (type L605) and Endeavor (type F562). These materials were modelled with a bilinear elastic-plastic stress-strain relationship with parameters given in Chua *et al.* [1] and Pochrząst *et al.* [10]. The stent coating was made of Phosphorylcholine (PC) polymer with bilinear model parameters obtained from Hopkins *et al.* [7]. The balloon was modelled using a Mooney-Rivlin hyperelastic strain energy potential [1] and the blood vessel was described by 5-parameter hyperelastic strain energy potential and considered uncompressible [4]. Material properties are summarized in Tables 1, 2 and 3.

2.4. Load and constraint

A uniform and linearly increasing pressure was applied to the inner surface of the balloon to simulate the inflation process, followed by deflation (pressure decreased to zero). The step time was chosen to be 25ms for inflation and 5ms for deflation. Implicit dynamic analyses have been used to simulate the whole process using Abaqus. In all simulations, the ends of the balloon were fully constrained to avoid rigid body motion.

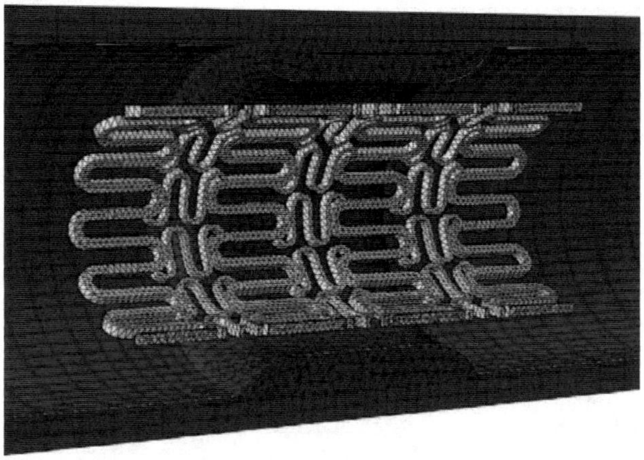

Figure 1. Finite element mesh for Cypher stent and stenotic artery.

Table 1. Parameters for stent and coating materials.

Material	ρ (kg/mm³)	E (MPa)	ν	S_y (MPa)	E_t (MPa)
SS316L	$7.8 \cdot 10^{-6}$	193000	0.27	207	692
Co-Cr L605	$9.7 \cdot 10^{-6}$	243000	0.3	476	680
Co-Cr F562	$8.8 \cdot 10^{-6}$	233000	0.35	414	744
PC-Polymer	$1.2 \cdot 10^{-6}$	240	0.5	16	7

Table 2. Mooney-Rivlin coefficients for polyurethane balloon.

Material	ρ (kg/mm³)	C_{10}	C_{01}	D_1
Polyurethane	$1.07 \cdot 10^{-6}$	1.03176	3.69266	0

Table 3. Coefficients of the hyperelastic model for the blood vessel.

Material	ρ (kg/mm³)	a_{10}	a_{01}	a_{20}	a_{11}	a_{30}
Artery Wall	$1.066 \cdot 10^{-6}$	18.90	2.75	590.42	857.18	0
Calcified Plaque	$1.45 \cdot 10^{-6}$	495.90	506.61	1193.53	3637.80	4737.25

2.3. Mesh

Stents were meshed with tetrahedral elements (C3D10), and the number of elements was 58428, 44222, 46221 and 63076 for Palmaz-Schatz, Cypher, Xience and Endeavor, respectively. The balloons were meshed using hexahedral elements with reduced integration and hybrid formulation (type C3D8RH), and the number of elements was about 10000 for all simulations. The artery wall was also meshed using hexahedral elements with reduced integration and hybrid formulation (C3D8RH), while the plaque was meshed with tetrahedral elements with hybrid formulation (C3D10H). The artery and plaque system has 40247 elements in total. Figure 1 shows the generated mesh for the whole system (Cypher stent). Interaction between the interfaces (balloon-stent-artery/plaque) was modelled as surface-to-surface hard contact. Mesh sensitivity study has confirmed the convergence of results for the current mesh.

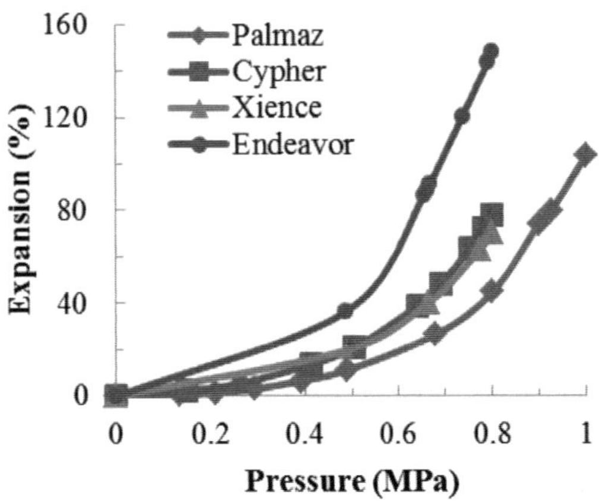

Figure 2. Radial expansion (in percentage) against inflation pressure for the four stents under free expansion.

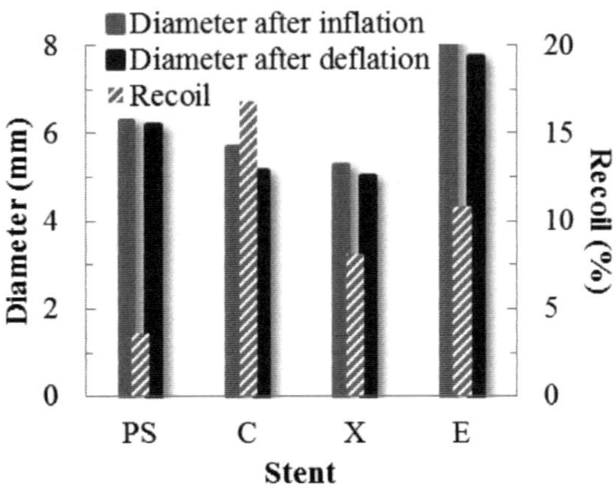

Figure 3. Diameter change and recoil effect for Palmaz-Schatz (PS), Cypher (C), Xience (X) and Endeavor (E) under free expansion.

2.5. Model validation

The model was first validated by reproducing the Palmaz-Schatz expansion and obtained results are in close agreement with those in Ju *et al.* [8] and Chua *et al.* [9]. This was purely for model validation purposes, to ensure that the mesh, boundary/loading conditions and solvers were correctly adopted and the materials and contacts were also properly simulated.

3. Results and discussion

3.1. Free expansion

Figure 2 shows the change of diameter against the pressure, applied inside the balloon, for all four stent designs during free expansion. The diameter change is mainly controlled by the radial stiffness of stent. Among the four designs, Endeavor is the most flexible stent, which expanded 143% of its original diameter at 0.8 MPa pressure, due to its open-cell design and peak-to-peak strut connection. Xience, with a spiral-cell design and peak-to-valley strut connection, and Cypher, with a closed-cell design and peak-to-peak strut connections, exhibited similar flexibility and expanded about 71% and 78% of their original diameter, respectively, under the same pressure. Palmaz-Schatz, with a slotted tube design, showed the highest stiffness with an expansion of only 45% at 0.8MPa. In fact, Palmaz-Schatz stent required a pressure of 1MPa in order to reach an expansion of 103%.

Recoiling effect after deflation is compared in Figure 3 for the four stents. Palmaz-Schatz stent showed the lowest recoiling effect (4%) due to its high radial stiffness, followed by Xience (8%) and Endeavor (11%). In terms of dogboning, Xience exhibited the strongest negative dogboning effect (-24%), followed by Endeavor (-34%), Palmaz-Schatz (-19%) and Cypher (-18%). Negative dogboning effect is usually appreciated, as it helps with the dilation of the stenotic plaque which is in direct contact with the middle section of the stent. In addition, Palmaz-Schatz had the highest foreshortening effect (6.4%), followed by Cypher (2.0%), Xience (1.0%) and Endeavor (0.4%). It is important to make foreshortening effect as low as possible to ensure a full coverage of the diseased part of the artery.

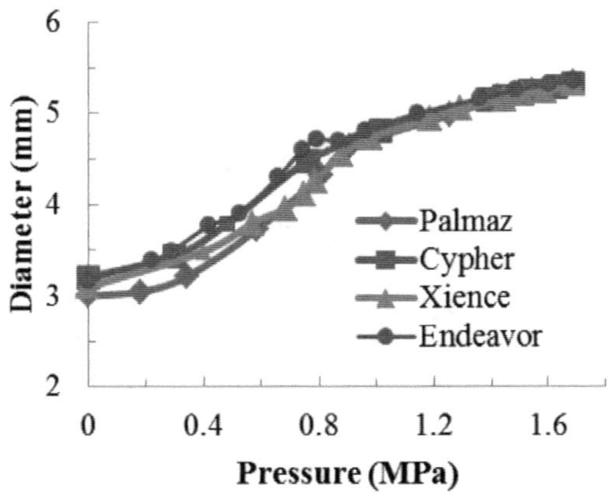

Figure 4. Change of diameter against inflation pressure for the four stents during expansion inside a stenotic artery.

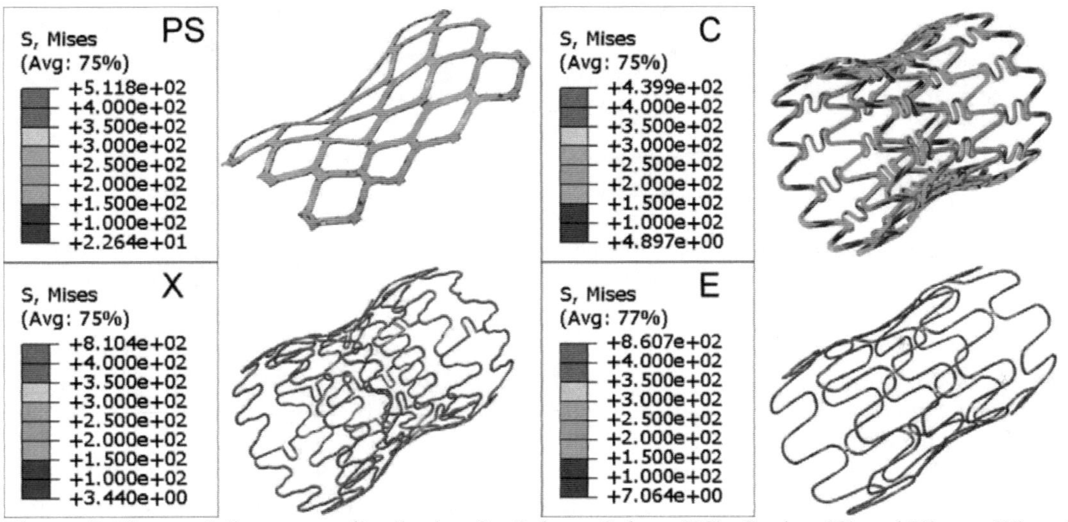

Figure 5. The von Mises stress distribution for Palmaz-Schatz (PS), Cypher (C) and Xience (X) and Endeavor (E) in a diseased artery at an inflation pressure of 1.6MPa.

All four stents showed severe stress concentrations at the corners of struts. At the same deformation level, stents made of Co-Cr alloys tend to have higher stress level compared to those made of stainless steel, in line with material properties. After deflation, residual von Mises stresses were obtained for all simulated stents. For Cypher, residual stress was 58.1% (211MPa) of the stress at the peak pressure, while for Palmaz-Schatz it was 52.9% (230MPa). Xience and Endeavor have a residual stress level similar to that at the maximum inflation pressure, i.e., 84.7% (466MPa) and 90.6% (690MPa) of the peak stress, respectively. The high stresses in some regions also indicate a potential fracture during the deployment process, which has been also frequently reported by clinical studies.

3.2. Expansion inside a stenotic artery

For stent expansion inside a diseased artery, the diameter change during the inflation and deflation for each stent is shown in Figure 4. All four stents had similar expansion behaviour with a maximum diameter of 5.35mm achieved at 1.7MPa pressure. Endeavor was the stent that expanded the artery slightly more than others while Cypher was the stent that expanded less.

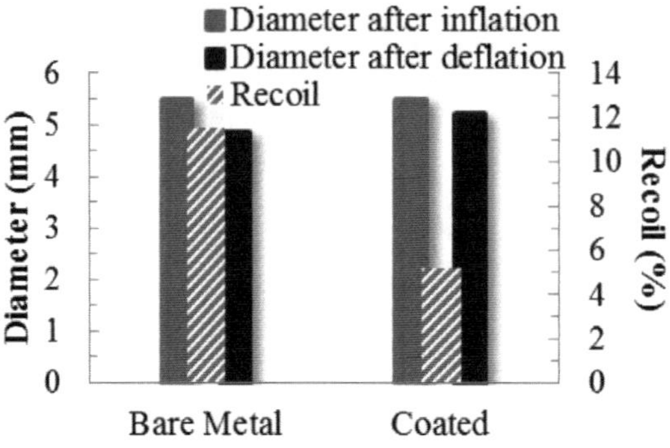

Figure 6. Diameter change and recoil effect for coated and bare metal Xience stent.

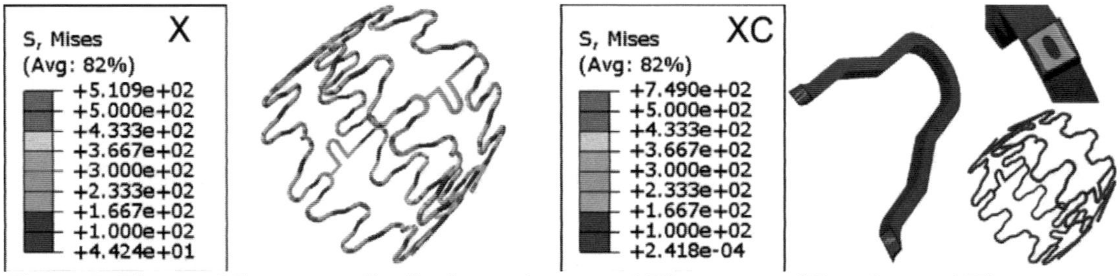

Figure 7. The von Mises stress distribution on bare metal Xience stent (X) and coated Xience stent (XC) at the end of inflation (i.e., peak pressure).

The von Mises stress distribution within the stents is shown in Figure 5 at the same pressure 1.6MPa. The location of the maximum stress is consistent with the case of free expansion. Endeavor and Xience showed higher magnitudes (861 MPa and 810MPa respectively) compared to the Cypher and the Palmaz-Schatz (440MPa and 512MPa, respectively), which is also consistent with free expansion simulations.

For dogboning effect, Xience exhibited the lowest level (1.8%), Endeavor showed also a very low level (3.7%), while Cypher and Palmaz-Schatz had higher levels (12% and 13%, respectively). This indicates the advantage of the latest stent designs which can expand the artery more uniformly and prevent damage to the healthy section of blood vessel. Furthermore, at the same diameter, Xience had the lowest foreshortening effect (1.2%), followed by Cypher and Endeavor (2.0% and 2.1%), whilst the highest value was ehibited by Palmaz-Schatz (6.2%). All these results are again consistent with free expansion simulations, showing that Xience tends to preserve its length better during deployment.

3.3. Effect of stent coating

To study the influence of stent coating, expansion of 2-ring Xience stent (coated and uncoated) was simulated under an inflation pressure of 1MPa. We used Xience as an example study and we assumed the effects of coating were similar for other types of stent.

Figure 6 shows the diameter change during expansion process. The maximum diameter reached 5.5mm in both simulations, but the coated stent recoiled only 7.3% while the bare metal one had a recoil effect of 12.7%. Results suggest that drug eluting coatings do not affect stent expansion (i.e. radial stiffness) but reduce recoil effect significantly. For both cases, the maximum von Mises stress was located at the peaks and the valleys of the sinusoidal struts, as shown in Figure 7, at the maximum pressure. The coated stent had a higher stress level (749MPa) compared to the bare metal stent (510MPa). After recoil, the residual stresses were found to be 691MPa and 448MPa for the coated and uncoated stents, respectively. The increase of stress level on the coated stent is due to property mismatch between the stent and the coating.

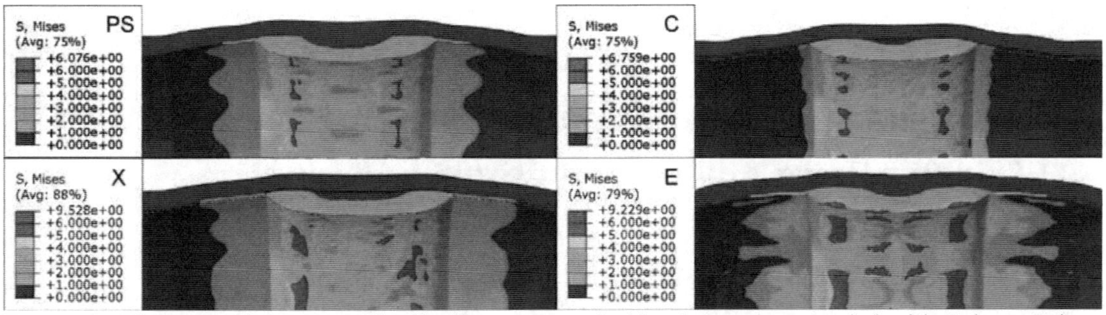

Figure 8. The von Mises stress distribution on stenotic blood vessel, expanded with Palmaz-Schatz (PS), Cypher (C), Xience (X) and Endeavor (E) stents, respectively, under a pressure of 1MPa.

3.4. Stresses on plaque/artery

The von Mises stress distribution on the plaque/artery had a similar pattern for all stents (Figure 8), with the maximum stress located towards the ends of the plaque where the stent and the plaque were directly in contact. At a pressure of 1MPa, Xience and Endeavor caused higher stress magnitude on the plaque (18.7MPa and 16.1MPa, respectively) than Cypher (6.8MPa) and Palmaz-Schatz (6.1MPa).

4. Conclusions

Effects of cell design, material choice and drug eluting coating on the expansion behaviour of stents have been studied using finite element simulations. Results showed that stent expansion is mainly controlled by its design. Stents with open-cell design (e.g. Endeavor) tend to have greater expansion than those with closed-cell design (e.g. Cypher). Among the four designs, Xience, with a spiral-cell design and peak-to-valley strut connection, seems to have the lowest dogboning and foreshortening effects. For given deformation, stents made of Co-Cr alloys tend to have higher stress level than those made of stainless steels due to the difference in material properties. The drug eluting coating affected the stent expansion by reducing the recoiling phenomenon considerably, but also raised the stress level on the stent due to material property mismatch. The von Mises stress distribution on the plaque/artery had similar pattern for all four stents, with stress concentrations located towards the ends of plaque where it experienced a strong contact with the stent due to the dogboning effect.

References

[1] Chua S N D, MacDonald B J and Hashimi M S J 2004 *J. Mater. Process. Technol.* **155-156** 1772
[2] Zahedmanesh H, Kelly D J and Lally C 2010 *J. Biomech.* **43** 2126
[3] Pericevic I, Lally C, Toner D and Kelly D J 2009 *Med. Eng. Phys.* **31** 428
[4] Lally C, Dolan F and Prendergast P J 2005 *J. Biomech.* **38** 1574
[5] Tambaca J, Canic S, Kosor M, Fish R D and Paniagua D 2011 *Tex. Heart I. J.* **38** 491
[6] Gu L, Santra S, Mericle R A and Kumar A V 2004 *J. Biomech.* **38** 1221
[7] Hopkins C G, McHugh P E and McGarry J P 2010 *Annu. Rev. Biomed. Eng.* **38** 2263
[8] Ju F, Xia Z and Sasaki K 2008 *J. Mech. Behav. Biomed. Mat.* **1** 86
[9] Chua S N D, McDonald B J and Hashimi M S J 2003 *J. Mat. Process. Technol.* **143-144** 591
[10] Pochrząst M, Walke W and Kaczmarek M 2009 *J. Achiev. Mater. Manuf. Eng.* **32** 340

Effect of random microstructure on crack propagation in cortical bone tissue under dynamic loading

X Gao[1], S Li, A Adel-Wahab and V Silberschmidt

Wolfson School of Mechanical and Manufacturing Engineering
Loughborough University, UK

E-mail: X.Gao@lboro.ac.uk

Abstract. A fracture process in a cortical bone tissue depends on various factors, such as bone loss, heterogeneous microstructure, variation of its material properties and accumulation of microcracks. Therefore, it is crucial to comprehend and describe the effect of microstructure and material properties of the components of cortical bone on crack propagation in a dynamic loading regime. At the microscale level, osteonal bone demonstrates a random distribution of osteons imbedded in an interstitial matrix and surrounded by a thin layer known as *cement line*. Such a distribution of osteons can lead to localization of deformation processes. The global mechanical behavior of bone and the crack-propagation process are affected by such localization under external loads. Hence, the random distribution of microstructural features plays a key role in the fracture process of cortical bone. The purpose of this study is two-fold: firstly, to develop two-dimensional microstructured numerical models of cortical bone tissue in order to examine the interaction between the propagating crack and bone microstructure using an extended finite-element method under both quasi-static and dynamic loading conditions; secondly, to investigate the effect of randomly distributed microstructural constituents on the crack propagation processes and crack paths. The obtained results of numerical simulations showed the influence of random microstructure on the global response of bone tissue at macroscale and on the crack-propagation process for quasi-static and dynamic loading conditions.

Introduction

Each year, some 730,000 road casualties happen in the UK [1], and 8.4 million sport injuries were reported in the EU in 2009 [2]. More often, such traffic accidents and sport injuries would lead to orthopaedic traumas, among which, the femur fracture is one of the critical ones because femurs are the strongest bones in the human body [3]. In order to develop strategies for treatment, understanding of bone fracture processes is required. The fracture process of cortical bone tissue is directly related to factors such as mineral-density reduction (bone loss), microstructure changes, variations of its material properties, and microcracks' accumulation [4]. Although bone loss has been considered as one of the key factors that lead to the fracture of cortical bone, detailed understanding of crack propagation processes for cortical bone with regards to its microstructural constituents is still lacking [5] and in-depth investigations of effects of microstructure, material properties and propagation of microcracks are still needed [6]. The mechanical properties of cortical bone as well as crack initiation and propagation are largely affected by histological variations, such as porosity, mineralization, orientation and diameters [4, 6-7]. In order to better assess the fracture risk of cortical bone, it is important to understand the bone's resistance to crack initiation and propagation with regards to its heterogeneous microstructure.

Due to these complex features of heterogeneous microstructure and their mechanical properties, there were only limited developments of finite-element (FE) models to investigate the effect of randomly distributed microstructural constituents on the deformation and fracture behavior of a cortical bone tissue. Abdel-Wahab *et al.* [4] used the extended finite-element method (X-FEM) to investigate the effect of microstructure constituents, especially the cement line, on the fracture process in bone. Hamed *et al.* [8] used a micromechanics method and composite-materials laminate theories to model a cortical bone's elastic response from macro-scale to micro-scale for obtaining its effective elastic constants. In another study, Budyn *et al.* [9] utilized a multiple-scale statistical X-FEM method to simulate multiple crack propagation in human cortical bone under tension loading condition. Li and his co-authors [5] evaluated the effect of microstructural transitions on macroscopic mechanical behaviors of cortical bone using three different microstructural realizations at quasi-static loading conditions. To the authors' knowledge, there is no available model to investigate the effect of random bone microstructure on crack growth in it under dynamic loading conditions. Therefore, the aim of this study is to develop such models to study the crack propagation processes for different microstructural realizations in dynamic regime and, consequently, improve our understanding of the role of microstructure in bone's failure. In order to achieve this, optical measurement was used to collect statistical data on random distribution of microstructural features in real cortical bone structures; then, a MATLAB program was designed to generate random distributions of the microstructural realizations based on quantification of real images. Finally, three FE models based on different realizations were created to investigate the effect of random microstructure on the fracture process and global responses under both quasi-static and dynamic loading conditions.

2. Material and methods

2.1. Random Microstructure

Micrographs were captured from radial-transverse cross-section of anterior cortex of bovine femur. They were used for collecting the data on statistical distribution of random microstructural features (Fig. 1). At microscale level, four microstructural features were analyzed: Haversian canals located at the center of each osteon, which are surrounded by cement line and interstitial matrix the remnants of

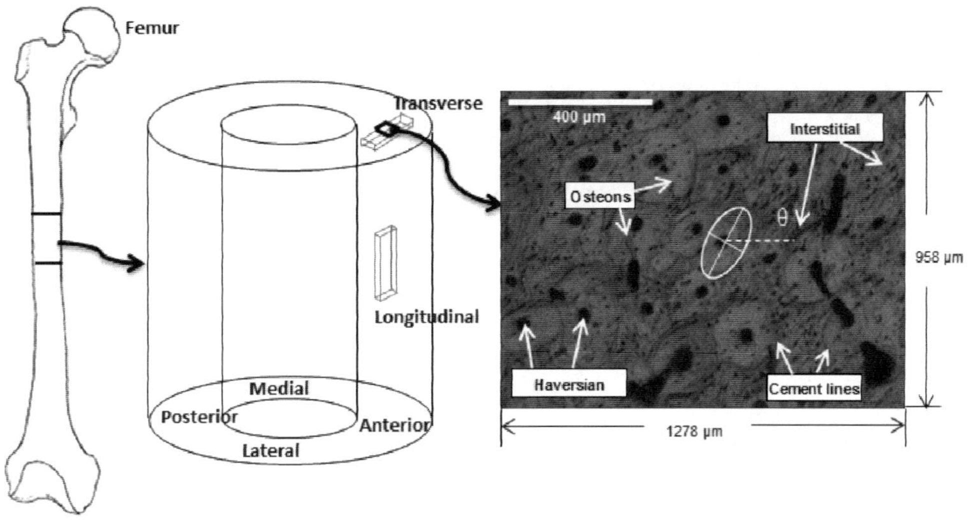

Figure 1. Microstructure of cortical bone at radial-transverse cross section.

a bone's remodelling process. Ten microstructural images (1.278 mm×0.958 mm each) were captured to obtain statistical parameters of osteonal bone, including the diameter of Haversian canals, transverse and conjugate diameters of osteons, orientation of osteons (angle between the direction of transverse diameter and the x-axis, from -90° to 90°) and positions of concentric Haversian canals and osteons. All the parameters were measured with image analysis software, Image-Pro Express [10]. The obtained data were analyzed and fitted with polynomial distribution functions to statistically describe the random distributions of the microstructural constituents.

In this study, the models of microstructured cortical bone were generated with a novel MATLAB code written by the author based on the probability density functions of the analyzed images. Statistical analysis of optical measurement showed that the volumetric ratios of osteons and porosity ranged from 45% to 58% and 2.5% to 4.6%, respectively. Therefore, three different finite-element models were developed to represent the range of the obtained data, see Table 1.

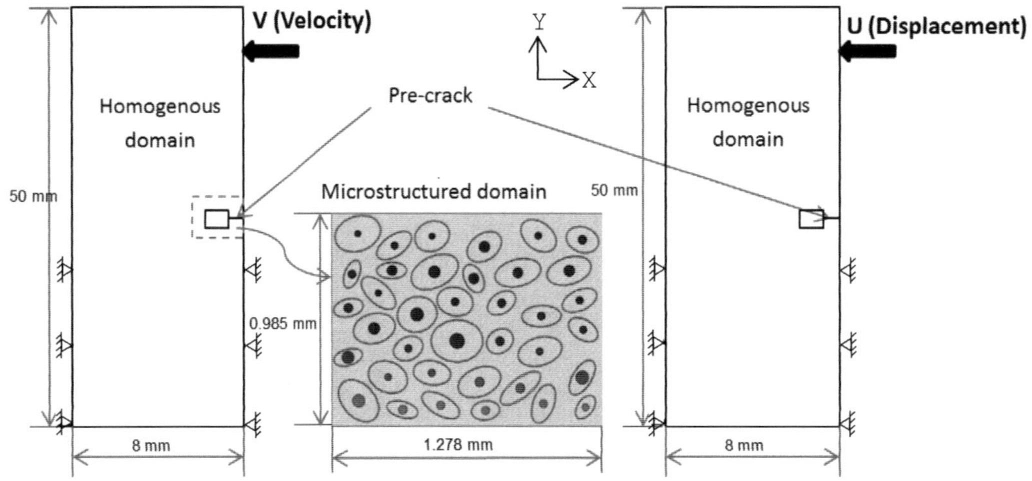

Figure 2. Geometry, boundary conditions and loading regimes of dynamic model (left) and quasi-static model (right)

Table 1. Microstructural parameters for all models

	Osteon volumetric ratio (%)	Porosity ratio (%)
Model A	46.9	4.66
Model B	51.8	3.72
Model C	57.9	4.21

2.1. Model configurations

In this study, three different two-dimensional finite-element models (FEM) were developed in ABAQUS for both dynamic and quasi-static loading regimes to investigate the effect of loading conditions on the fracture process in the cortical bone tissue at the microscale level. The models were configured based on previous Izod experimental tests performed by the same research group [11]. A full description of Izod test can be found in the previous paper by the authors [11]; the microstructured models of 1.278 mm×0.958 mm (length × width) were embedded into a homogenized material (50 mm×8 mm) (length ×width) as shown in Fig. 2.

In the simulations, 4-node bilinear plain-strain quadrilateral (CPE4R) elements were chosen for both parts of the model due to significant difference between the element size of microstructured and homogenized domains. Based on the Izod tests, an impact velocity of 3.46 m/s was chosen to provide initial energy for crack propagation [11]. In order to compare the fracture behaviour for both dynamic and quasi-static loading conditions, the maximum displacements during the contact time in the dynamic finite-element model were used as prescribed boundary condition for quasi-static models (Fig. 2). Table 2 summarizes the loading conditions for all the used models.

Table 2. Loading conditions for both dynamic and quasi-static models

Model	Applied velocity (m/s)	Model	Applied displacement (mm)
Dynamic Model A	3.46	Quasi-Static Model A	1.90
Dynamic Model B	3.46	Quasi-Static Model B	1.61
Dynamic Model C	3.46	Quasi-Static Model C	1.65

2.2. Mechanical properties

In terms of the mechanical behavior, cortical bone can be considered as transversely isotropic material in 2D FE models [7, 12]. The material properties of osteons, interstitial bone, cement line and homogenous material media were implemented based on the previous research can be found in [4]. Following Budyn and Hoc [13], the elastic modulus of cement lines were considered to be 25% lower than that of osteons. The Poisson's ratios of each phase were obtained from literature [4]. The elastic properties used in our simulations are summarized in Table 3.

Table 3: Material properties of microstructural constituents of cortical bone tissue and homogenous material

Model	Elastic modulus (GPa)	Poisson's ratio	Strain energy-release rate (N/mm)
Homogenous material	10.46	0.167	0.422
Osteons	9.13	0.17	0.86
Interstitial matrix	14.122	0.153	0.238
Cement lines	6.85	0.49	0.146

2.3. Cohesive behaviour and fracture properties for X-FEM

An X-FEM-based cohesive technique for a crack initiation and propagation was utilized to study the crack growth along an arbitrary, solution-dependent path under various loading conditions. Belytschko and Black [14] introduced an extended-finite-element method for crack-propagation analysis without a remeshing process. In our simulations, the whole bone specimen was chosen as an enrichment region; an initial pre-crack with a length of 300 μm was placed within the homogenized media near the microstructured region and parallel to the impact/displacement loading direction, see Fig. 2. Crack initiation and evolution criteria were chosen based on the critical maximum principal strain of 0.6% and strain energy release rate reported in [4, 7].

3. Results and discussion

The effect of the random distribution of cortical bone microstructural features on the path of a propagating crack under dynamic loading regime can be observed clearly (Figs. 3). Due to the random distribution of the microstructural features of the studied cortical bone tissue, a crack follows different paths affected by the local microstructure as shown in Figs. 3a-c. The crack deflection mechanism reported in [5] was also observed in the current simulations: the crack tends to propagate around cement lines and was attracted towards Haversian canals. The cement lines offered a weak path for cracks and acted as protective structures of the osteonal bone to prevent further damage to the cortical bone tissue [15]. It can be seen also that the crack continued to propagate in the homogenized media after crossing the microstructred domain with a practically straight path with no alternations, see Fig. 4 for both dynamic and quasi-static loadings.

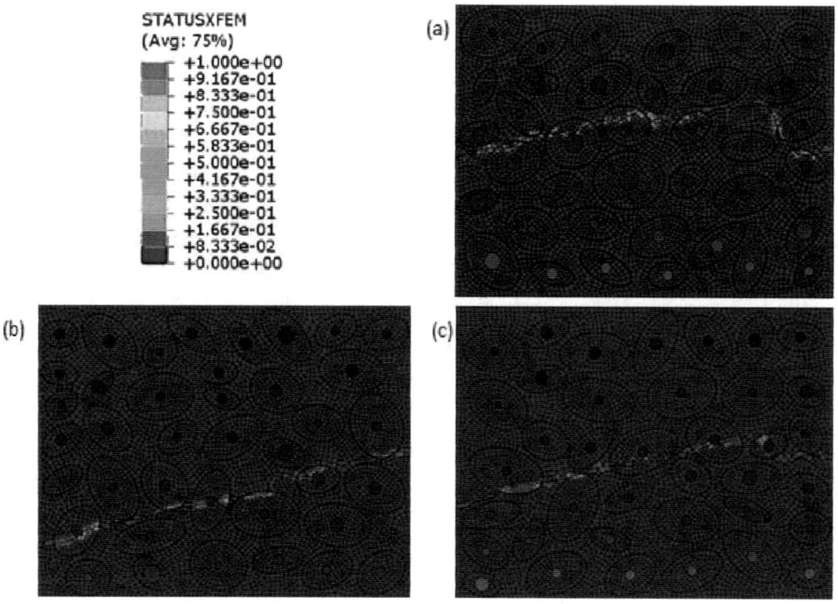

Figure 3. Final crack paths within three different random microstructures in case of dynamic loading: (a) model A; (b) model B; (c) model C

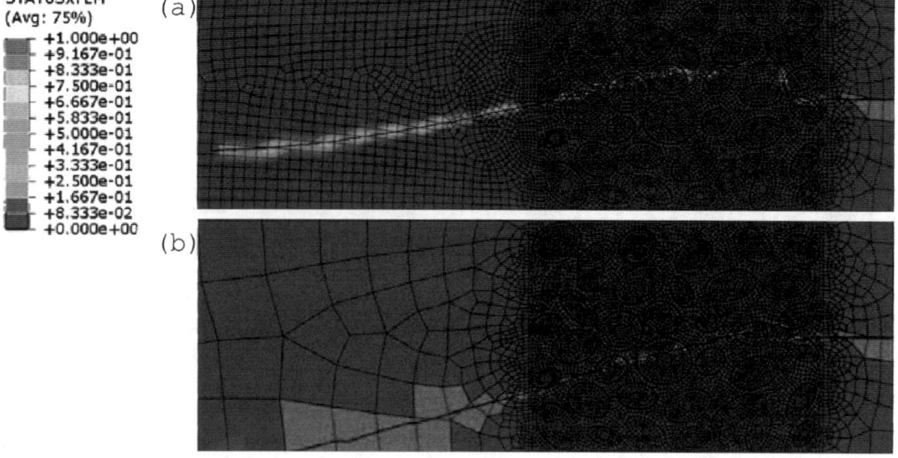

Figure 4. Final crack for Dynamic Model A (a) and Quasi-Static Model A (b)

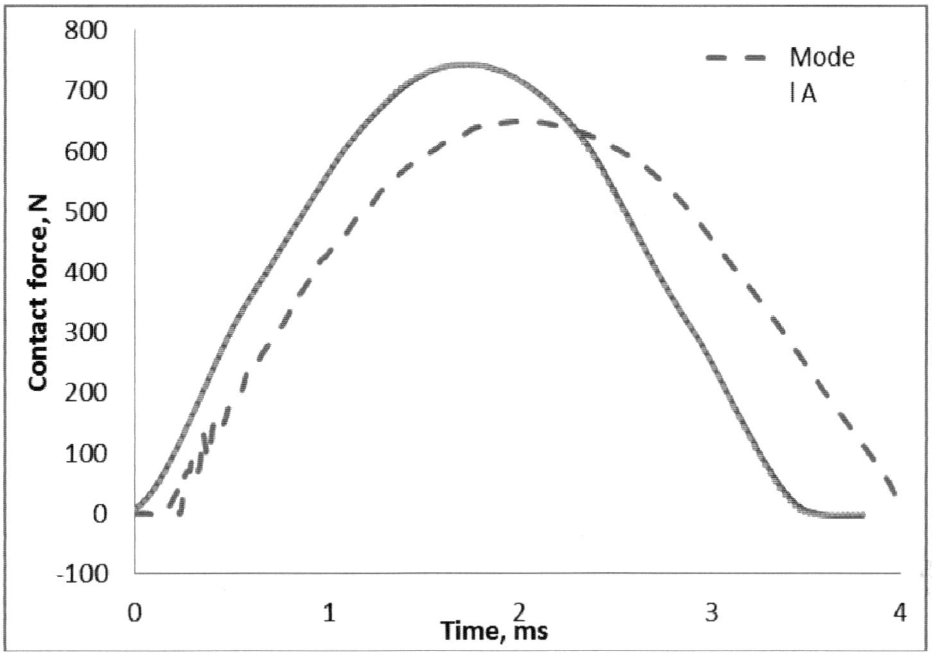

Figure 5. Contact force-time profile for dynamic loading

Although the criterion of crack initiation for all models was the same, the crack-growth rate was different for each microstructured model as shown in Figs. 6 and 7. Additionally, by comparing simulation results for the three microstructured models, Model B revealed the longest total crack length (2934.4 μm for A, 3144.5 μm for B, 2957.6 μm for C) but the shortest crack length in the microstructured part (1466.7 μm for A, 1343.7 μm for B, 1410.2 μm for C), as shown in Fig. 6. Generally, the crack propagated as the contact force increased, and it reached close to its maximum value when the force was at the peak point (671.3 N at 1.96 ms for A, 853.1 N at 1.68 ms for B, 812.1 N at 1.74 ms for C). When the force started to decrease, the crack propagated only slightly (Figs. 5 and 6). It was also observed that the evolution of contact forces for Models B and C were closer to one another compared to Model A owing to the differences in the volume fraction of the microstructural constituents. In Model A, the crack propagated towards a cement line and cut thought an osteon initially, while in Models B and C, the crack grew initially into the interstitial matrix, see Figs. 3a-c. As a result of the deflection mechanism of cement lines, the force-time response (Fig. 5) for Model A was different compared to that of two other models.

A comparison between the crack-propagation behaviors under different loading conditions shows that models subjected to quasi-static loading conditions generally had a shorter total crack length (2744.7 μm for A, 2462.4 μm for B, 2657.6 μm for C) compared to those found for dynamic loading conditions. In addition, the crack trajectories in quasi-static models were smoother than those in the models with dynamic loadings conditions. Therefore, the crack lengths within the microstructured domain (1382.6 μm for A, 1423.7 μm for B, 1410.2 μm for C) varied from those in their dynamic counterparts, see Fig. 6. It was noticed that models loaded dynamically had crack-propagation rates in the microstructured domain that tended to slow down when reaching the homogenized material as the contact force almost reached its peak point. On the other hand, variations of the crack-propagation rates in the quasi-static models extended over the whole fracture process (Figs. 5 and 6). As a whole, both the loading conditions and the variation of microstructural constituents strongly affected the crack-propagation characteristics of cortical bone.

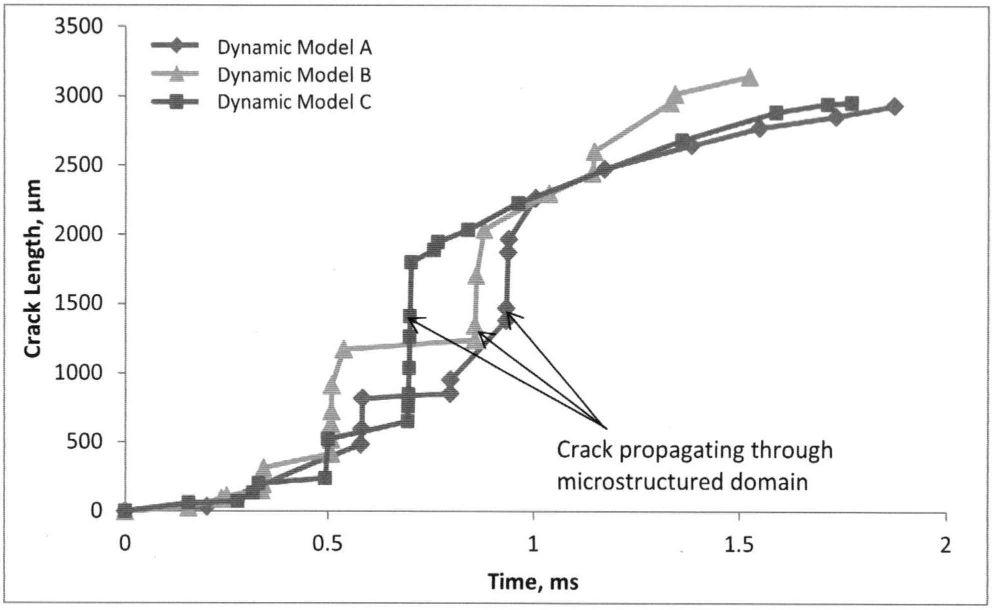

Figure 6. Evolution of crack length under dynamic loading

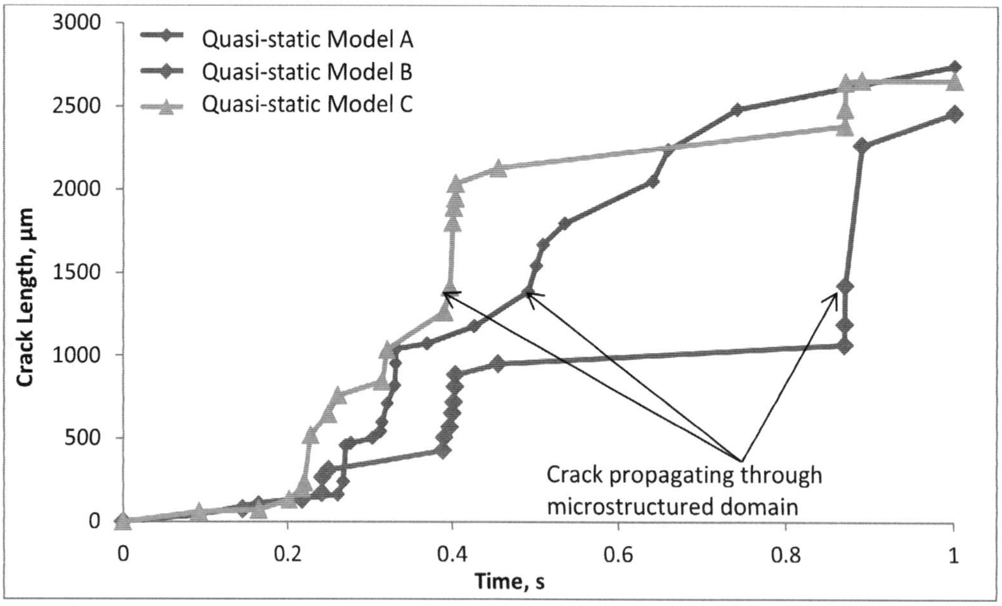

Figure 7. Evolution of crack length under quasi-static loading

4. Conclusions

The effect of the random microstructure of the cortical bone tissue on crack initiation and propagation under both dynamic and quasi-static loadings was studied using 2D FE models. Three different realizations of bone microstructure were obtained using optical measurements and used to develop the microstructured finite-element models embedded into a homogenized-material model. The obtained results demonstrated different crack-propagation behaviors for different microstructural statistical realizations. The finite-element models loaded in dynamic conditions demonstrated higher crack lengths than those loaded in quasi-static conditions. These developed models provided a new important insight for understanding the fracture processes of the cortical bone tissue subjected to impact events and can be used for future evaluations of the age-related disease associated with fracture of cortical bone tissue.

References

[1] Department for Transportation, available at: https://www.gov.uk/government/uploads/system/uploads/attachment_data/file/9286/road-accidents-and-safety-quarterly-estimates-q2-2012.pdf
[2] Kisser R and Bauer R. 2012, available at: http://www.safetyinsports.eu/upload/downloads/Report_Burden_of_Sport_Injuries.pdf
[3] Ebrahimi H, Rabinovich M, Buleta V, Zalcman D, Shah S, Dubov A, Roy K, Siddiqui F S, Schemitsch E H, Bougherara H and Zdero R, 2012 *J. Mech. Behav. Biomedi. Mater.* **16** 121-135
[4] Abdel-Wahab A A; Maligno A R and Silberschmidt VV 2012 *Comp. Mater. Sci.* **52** 128-135
[5] Li S, Abdel-Wahab A A, Demirci E and Silberschmidt V V 2013 *Int. J. Fracture* DOI:10.1007/s10704-013-9814-7
[6] Najafi A R, Arshi A R, Eslami M R, Fariborz S and Moeinzadeh M H 2007 *J. Biomech.* **40** 2788-2795
[7] Li S, Demiric E and Silberschmidt V V 2013 *J. Mech. Behav. Biomedi. Mater.* DOI:10.1016/j.jmbbm.2013.02.021
[8] Hamed E, Lee Y and Jasiuk I 2010 *Acta. Mech.* **213** 131–154
[9] Budyn E, Hoc T and Jonvaux J 2008 *Comp. Mech.***42** 579-591
[10] *V. Image-Pro Express*, Version media cybernetics, 2005
[11] Abdel-Wahab A A and Silberschmidt V V 2011 J. *Theor. Appl. Mech.* **49**(3) 599-619
[12] Abdel-Wahab A A, Alam K and Silberschmidt V V 2010 *J. Mech. Behav. Biomedi. Mater.* **4** 807-820
[13] Budyn E and Hoc T 2007 *Revue Européenne de Mécanique Numérique* (*Europ. J. Comp. Mech.*) **16** 213–236
[14] Belytschko T and Black T 1999 *Int. J. Numer. Meth. Eng.* **45**(5) 601-620
[15] Augat P and Schorlemmer S 2006 *Age Ageing* 35-52

Finite element modeling and experimentation of bone drilling forces

W A Lughmani[1], K Bouazza-Marouf [1] and I Ashcroft [2]
[1] Wolfson school of mechanical and manufacturing Engineering Loughborough University, Epinal way, LE11 3TU, UK
[2] The University of Nottingham, NG7 2RD, UK

E-Mail: W.A.Lughmani@lboro.ac.uk

Abstract. Bone drilling is an essential part of many orthopaedic surgery procedures, including those for internal fixation and for attaching prosthetics. Estimation and control of bone drilling forces are critical to prevent drill breakthrough, excessive heat generation, and mechanical damage to the bone. This paper presents a 3D finite element (FE) model for prediction of thrust forces experienced during bone drilling. The model incorporates the dynamic characteristics involved in the process along with the accurate geometrical considerations. The average critical thrust forces and torques obtained using FE analysis, for set of machining parameters are found to be in good agreement with the experimental results.

1. Introduction

In orthopaedic surgery, drilling and tapping are extensively carried out before the insertion of screws into bone. The desired outcome of bone drilling process is accurately positioned holes without mechanical and thermal damage to surrounding tissue.

Drilling into bone is a fundamental skill that can be both very simple, such as drilling through long bones, or very difficult, such as drilling through the vertebral pedicles where incorrectly drilled holes can result in nerve damage, vascular damage or fractured pedicles [1,2]. Similarly, large uncontrolled forces experienced during bone drilling may result in drill breakthrough, imparting damage to surrounding tissue [3-5] and promote crack formation which could yield bone failure [6]. Moreover, drilling forces are main source of heat generation during bone drilling [7]. Therefore, it is important to understand the effects of bone drilling conditions and material behaviour on the bone drilling forces to select favourable drilling conditions, and assist in robotic surgery procedures [3, 8].

Various studies have been carried since late 1950 to optimise the bone drilling performance based on measurable parameters such as, drilling force, drilling torque, rotational speed, feed rate, temperature and accuracy of the drilled hole [9-16]. Although these studies resulted in various conclusions, they have not provided a functional relationship that represent the effects of drilling conditions, drill bit geometry and material behaviour on drilling forces.

Finite element (FE) models of drilling can provide insight into the underlying mechanics of material behaviour at high strain and strain-rate regimes. An accurate and reliable FE simulation of drilling

[1] To whom any correspondence should be addressed.

enables good predictions of thrust forces and torque by taking into account the complex drill geometry and process parameters. In this study, a 3D finite element (FE) model of drilling in bovine cortical bone is developed. An elastic-plastic material model is used to predict the behaviour of cortical bone during drilling. The model is used to predict the thrust force and torque during the specific cutting conditions.

2. Finite element analysis of drilling in cortical bone

2.1. Methodology

A three dimensional (3D) Lagrangian FE model of drilling of cortical was developed in a commercially available FE software ABAQUS/Explicit. The model accurately characterises the dynamic characteristics of the drilling process accounting for the contact interaction between a drill bit and cortical bone surface. A dynamic failure criterion is applied to control the element removal which has undergone severe deformation. The details of this FE model are discussed in following sections.

2.1.1. Geometry and meshing.

A rectangular plate with overall dimensions of 5mm × 5mm × 5mm was modeled as cortical bone specimen. Figure 1 shows the geometrical configuration of the drill bit and the bone model. The diameter, point angle and helix angle of the drill bit used in this study are taken as 2.5 mm, 118° and 30°, respectively. The drill bit was modeled as a discrete rigid body in order to reduce the computing time and resources. In order to model the kinematics of the drill bit accurately, a lumped mass and inertia was applied at a reference point located at chisel edge of the drill bit.

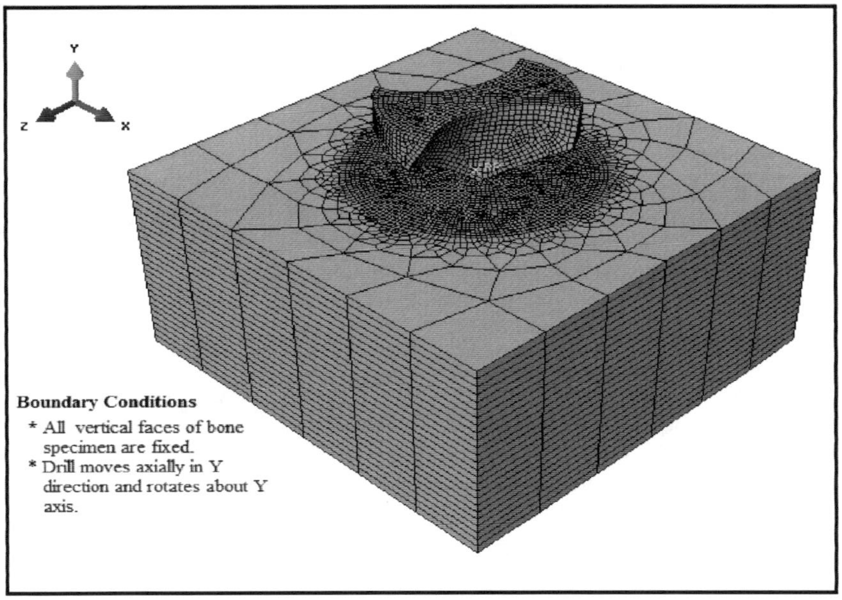

Figure1. FE model of bone drilling

An extensive mesh convergence study was carried out in this FE analysis to achieve the balance between mesh discretisation and stress distribution in cortical bone surface, for accurate prediction of drilling forces in the available computing resources. A refined mesh with a minimum size of 5 μm was used at and in the immediate vicinity of the volume to be drilled, while a coarse mesh was used to discretise the workpiece away from the path of drill bit. The cortical bone was modeled using 8-node, 3D brick elements of type C3D8R, while a twist drill was modeled with 4-node, 3D discrete rigid elements of type C3D4. The cortical bone model consists of 101320 elements while the drill bit was meshed with 4850 elements. The drilling conditions were applied on reference point of the drill bit which has single node mass and rotary inertia.

2.1.2. Boundary conditions and contact interactions
The twist drill was fed into the workpiece in the axial direction using a velocity boundary condition, which represents the feed rate during experiments. A reference point which accounts for drill mass and inertia was constrained in X and Z directions while feed was applied in the Y direction. The combination of machining parameters used in this study to predict drilling forces are listed in table 1.

Table 1. Machining parameters used in bone drilling

Drill	HSS, ⌀ 2.5 mm, point angle 118°
Spindle speed (rpm)	800, 1200, 1500
Feed (mm/rev)	0.05, 0.1, 0.1875

Contact between the twist drill and the cortical bone was defined by the general contact algorithm available in ABAQUS/Explicit. This algorithm generated the contact forces based on the penalty-enforced contact method. The friction coefficient μ is used to account for the shear stress of the surface traction τ with the contact pressure p and can be represented as $\tau = \mu p$.
In this case, the frictional contact between a drill and cortical bone was modeled with a constant coefficient of friction of 0.7 [17]. The models require on average 54 h on 36 Intel quad-core processors with 48 GB RAM each to finish the analysis using High Performance Computing (HPC) facility available at Loughborough University.

2.1.3. Material model and failure law
The cortical bone in this study was modeled as a rate dependent transversely isotropic material. The elastic-plastic law was used for the material constitutive model of finite element simulation. The high strain rate data was obtained by using Split Hopkinson Pressure Bar (SHPB) experiments. The SHPB experiments were conducted at strain rates of 5000/s, 6400/s and 7500/s. The mechanical properties used in finite element analysis are summarized in table 2.

Table 2. Mechanical properties of bone used in finite element model

Density (Kg /m^3)	2000
Young's Modulus Longitudinal (GPa)	20
Young's Modulus transverse (GPa)	18
Strain at failure (%)	5

The dynamic failure law [18] was used to initiate and propagate the failure based on the SHPB experimental data. The failure was calculated for each element and was defined by

$$D = \sum \frac{\Delta \varepsilon^{p}}{\varepsilon^{f}} \tag{1}$$

where $\Delta\varepsilon^p$ was the increment of the equivalent plastic strain, and ε^f was equivalent strain to fracture. Failure was allowed to occur when D = 1, and the concerned elements were removed from the computation.

3. Drilling Experiments

To accomplish the aims of this study, bone drilling has been conducted on custom designed electromechanical test rig with single setting of specimen as per ASTM F543-02 standard. A microcontroller PICI8F6620 from the Microchip is used for interfacing test rig with the computer, and a 12-bit, eight channels data acquisition system is used for data recording. The experimental setup is shown in figure 2.

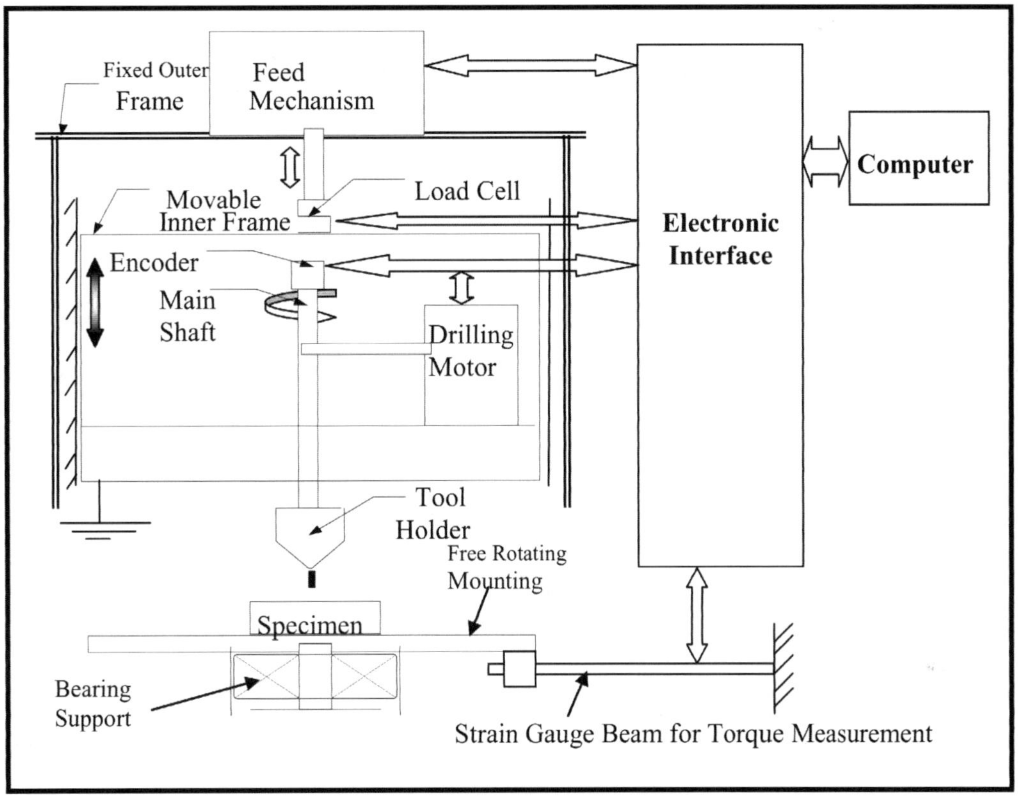

Figure 2. Schematic of Drilling Rig

Cortical bone used in this investigation was taken from mid-diaphysis of bovine femur. Bovine femur bone samples were purchased from a local butcher. Any extra tissues present around bone were cleaned using knife and scrapper before testing. Bone used in saturated condition for testing. An industrial drill bit (Model A9762.2X125 Dormer UK) was used for these experiments.

4. Results and Discussion

In order to allow a better comparison of the experimental and simulated thrust force and torque in drilling cortical bone, a feed of 150 mm/min was chosen from the experimental feed data with a spindle speed 800 rpm. The FE simulations were carried out using these process parameters and later used to predict the thrust force and torque for other feed rates. Figure 3a and b shows the experimental and simulated data for the thrust force and torque in drilling of cortical bone. The average maximum thrust force (obtained for the period of complete drill engagement) in the experimental trial was 70 N whereas FE model estimated 72. The experimentally measured torque was 1.62 N-cm compared to the torque value predicted as 1.7 N-cm by FE simulation. This shows that the FE model estimated the thrust force and torque accurately, with 2.9% and 8% deviation from the respective test results.

(a)

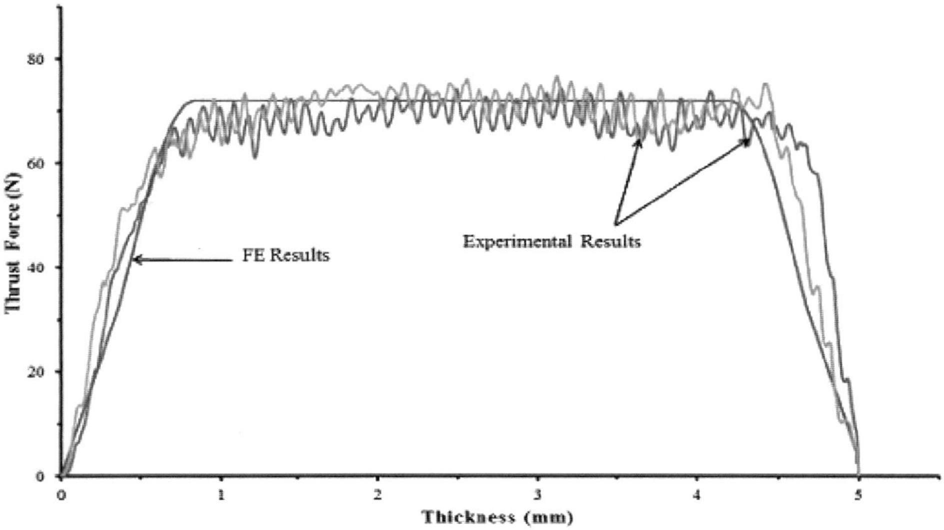

(b)

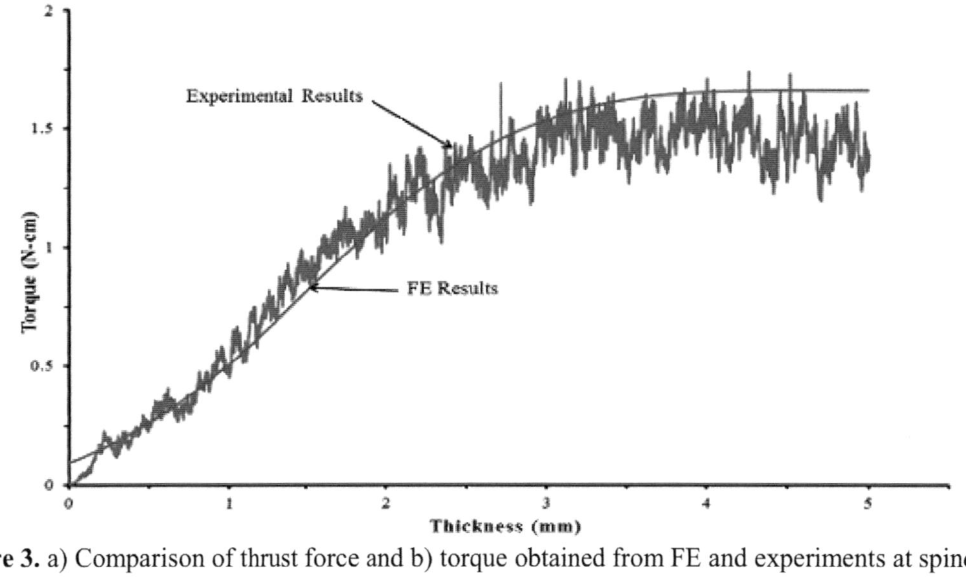

Figure 3. a) Comparison of thrust force and b) torque obtained from FE and experiments at spindle speed of 800 rpm and a feed rate of 0.1875 mm/rev.

Figure 4a shows the effect of drilling conditions on the average maximum thrust force. The FE model estimated the thrust force between 28 N and 70 N for the range of feed rates modeled. The obtained results indicate that thrust force in drilling increases with the increasing feed rate. It can be observed from Figure 4a that the average maximum thrust force was the highest at the feed rate of 0.1875 mm/rev and lowest at the feed rate of 0.05 mm/min. Figure 4b shows the drilling torque corresponding to different spindle speeds. The overall results show that the torque increased with the increase in spindle speed.

5. Conclusion

In this paper the effect of different machining parameters on thrust force and torque in drilling of a cortical bone was investigated both experimentally and numerically. A 3D FE model of drilling in cortical bone was developed. The dynamic failure criterion for damage initiation was used for hole making process in drilling. The following observations were made in this study:

- The FE model predicted the drilling thrust force and torque with reasonable accuracy when compared to experimental results.
- The validated drilling model was used to determine the thrust force, and torque for different drilling conditions. It was observed that the thrust force and torque increased with an increase in the feed rate and spindle speed. Thrust force and torque may be reduced by using the low feed which can be achieved by a combination of low feed rate and high spindle speed, while drilling in cortical bone.

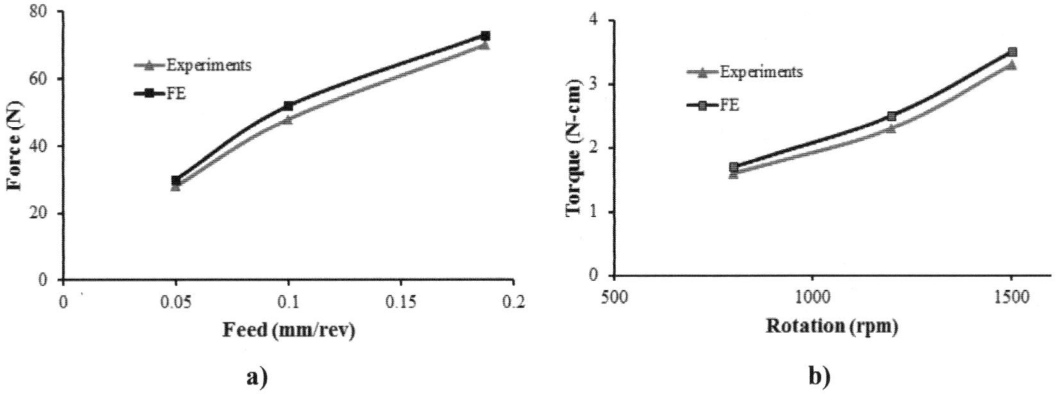

Figure 4. a) Comparison of thrust force b) torque obtained from FE and experiments at different cutting conditions

References

[1] Van K, Vander J, Van R, and Fabry G 1996 Internal fixation of the spine in traumatic and scoliotic cases. *The potential of pedicle screws. Technol. Health cares,* **4** 365-8

[2] Carmouche J J, Molinari R W, Devine J and Patience T 2005 Effects of pilot hole preparation technique on pedicle screw fixation in different regions of the osteoporotic thoracic and lumbar spine. *J. Neurosurg. Spine* **3** 364-370

[3] Brett P N, Baker D A, Taylor R and Griffiths M V 2004 Controlling the penetration of flexible bone tissue using the stapedotomy microdrill *Proc. IMechE, Part I J. Syst. Control Engng* **218** 343–351

[4] Kendoff D, Citak M and Gardner M J 2007 Improved accuracy of navigated drilling using a Drill accuracy of navigated drilling using a drill alignment device *J Orthopaedic Research* **25**, 951-957

[5] Ong F R and Bouazza-Marouf K 1999 The detection of drill-bit break-through for the enhancement safety in mechatronic assisted orthopaedic drilling *Mechatr.* **9** 565–588

[6] Kasiri S, Reilly G and Tylor D 2010. Wedge indentation fracture of cortical bone: experimental data and predictions, *J. Biomech. Engng* **132** 081009

[7] Augustin G, Davila S, Mihoci K and Udiljak T 2008 Thermal osteonecrosis and bone drilling osteonecrosis and bone drilling parameters revisited *Arch. Orthopaed. Trauma Surg.* **128** 71–77

[8] Brett P N, Fraser C A and Kamel Y 1995 Automatic surgical tools for penetrating flexible

tissues *IEEE Engng Med. Biol. Mag.* **14** 264–270
[9] Thompson HC 1958. Effect of drilling into bone. *J. Oral Surgery* **16** 22–30
[10] Jacobs C H, Pope M H, Berry J T and Hoaglund F 1974 A study of the bone machining process-Orthogonal cutting *J. Biomech.* **7** 131–136
[11] Jacobs C H, Berry J T, Pope M H and Hoaglund F T 1976 A study of the bone machining process-drilling *J. Biomech.* **9** 343–349
[12] Wiggins K L and Malkin S 1976 Drilling of bone *J. Biomech.* **9** 553–559
[13] Farnworth G H and Burton J A 1974 Optimization of drill geometry for orthopaedic surgery *Proceedings of the 14th International Conference on Machine Tool Design and Research Conference, Manchester, England*
[14] Karmani S and Lam F 2004 The design and function of surgical drills and K-wires *Curr. Orthopaed.* **18** 484–490
[15] Franssen B B G M, Van Diest P J, Schuurman A H and Kon M 2008. Keeping osteocytes alive: a comparison of drilling and hammering K-wires into bone *J. Hand Surgery* **33** 363–368
[16] Tsai M D, Hsieh M S and Tsai C 2007 Bone drilling haptic interaction for orthopaedic surgical simulator. *Comput. Biol. Med.* **37** 1709–1718
[17] Davidson S, James D. Drilling in bone: modeling heat generation and temperature distribution Transactions of ASME, *J. Biomech. Engng* 2003 **125(3)** 305–314
[18] 2010 *Abaqus User's manual*, version **6.11 RI**: Hibbitt. Karlsson & Sorensen Inc.

Analytical and experimental studies on the strain rate effects in penetration of 10wt % ballistic gelatin

L Liu, Z Jia, X L Ma and Y R Fan*

State Key Lab of Fluid Power Transmission and Control, Zhejiang University, Hangzhou, 310027, P.R. China

E-mail: yurunfan@zju.edu.cn

Abstract. This work concentrates on modeling the super-elastic behavior of 10wt% ballistic gelatin at 4 ℃ and the mechanical responses at quasi-static and high-speed penetrations. Uniaxial compression and simple shearing experiments were carried out to determine the moduli in Mooney-Rivlin model describing the elastic behavior of gelatin at low strain rates. The failure mode is determined to be elastic fracture as the tensile stretch ratio exceeds a critical value. For high compression strain rates, the available results from the split Hopkinson pressure bar (SHPB) experiments for 10wt% gelatin were carefully examined and assessed. Linear relationship between the moduli and the strain rate is established. Based on these material parameters, an analytic solution of stress for the quasi-static and quasi-dynamic expansion of spherical cavity in gelatin is derived. As a consequence, the work needed to open unit volume of cavity, P_s, which is the key parameter in studying penetration problems, is linearly increasing with the characteristic strain rate. The application of P_s to our quasi-static and high-speed penetration experiments is discussed and assessed.

1. Introduction

As a physical surrogate of the human body to evaluate penetration problems, ballistic gelatin is widely used in ballistic tests. There are several reports on the uniaxial compression or elongation experiments of gelatin [1-3]; all of them, including that done by our research group, showed that gelatin is nonlinearly elastic at large deformation. The stress-strain relation of gelatin is rate sensitive, especially at high strain rates, as demonstrated by the Split Hopkinson Pressure Bar (SHPB) experiments of e.g., Salisbury and Cronin [3]. So viscoelastic behavior of gelatin must be taken into consideration.

To analyze penetration tests, a widely used analytical method, the cavity expansion model, is adopted. This method is to calculate the wall pressure on a spherically expanding cavity in an infinite medium at a constant wall velocity [6-7].The traditional spherical cavity expansion models consider the hard materials bearing small elastic deformation (metals, rocks, concretes etc.), in which Hookean law for linear elasticity is appropriate. However, for some soft materials such as gelatin the small elastic strain assumption for elastic deformation should be abandoned. Based on Mooney-Rivlin constitutive equation, a cavity expansion model for large elastic deformation is derived and used to analyze our penetration tests.

To characterize the dynamic mechanical responses of gelatin, several experimental methods can be adopted, such as SHPB test [3], indention (punch) test [4], plate-impact test [5] and ballistic penetration test[1]. In the present investigation we examine the available experimental data by SHPB tests of [3] to determine the elastic moduli and their dependence on the strain rate, as well as our quasi-static mechanical tests. Finally quasi-static and high-speed penetration tests are carried out to characterize the mechanical responses of 10wt% ballistic gelatin.

2. Mooney-Rivlin model for ballistic gelatin

The constitutive equation of Mooney-Rivlin model ([8]) which is commonly used for large elastic deformation is

$$\boldsymbol{\sigma} = -p\mathrm{I} + 2c_1\mathbf{B} - 2c_2\mathbf{B}^{-1} \tag{1a}$$

$$\boldsymbol{\sigma} = -p\mathbf{I} + G\mathbf{B},\ 2(c_1 + c_2) = G \tag{1b}$$

where $c_1$and c_2 are material parameters (modulus), **I** is the unit tensor, **B** the Finger strain tensor and $p$the isotropic pressure determined by boundary conditions. In the limit of small strain, the model becomes the linear Neo-Hookean model (Equation (1b)), where G is the shear modulus. To determine the parameters, c_1,c_2 and G describing the elastic behaviour of ballistic gelatin, we carried out uniaxial compression and simple shearing tests at low strain rates and examined the SHPB tests of [3] for high strain rates.

2.1. Quasi-static tests

A 50g of bloom 250 collagen powder was mixed with 139g of 20℃water, after 40 minutes, 311g hot water (70℃) was added for dissolution. The solution was then sealed and put into a vacuum oven at the constant temperature of 60℃ and was stirred every 15 minutes. 30 minutes later, the solution was poured into aluminium molds with the inner diameter of 40mm and the depth of 40mm, the molds were then placed in a refrigerator for 10 to 12 hours at 4(±0.5)℃ before testing. Uniaxial compression tests were carried out on the universal materials tester of Zwick/Roell Z020. The samples were released from the molds and every test was done within 3 minutes. Each sample was compressed up to 60~75% engineering strain at the engineering strain rates from0.004/s to0.208/s, defined as the loading rate divided by the original length of the specimen.

The results show that the 10wt% ballistic gelatin is highly elastic until failure by cracking, no noticeable plastic deformation was observed, the same observation was reported by Ottone and Deiber[9]and Moy and Tusit [2] for gelatin gels of different concentrations and temperatures. So the failure criterion of gelatin can be described as elastic-crack.

In Figure 1, the uniaxial compression data (negative stress in compression) are presented, as well as the variation of failure engineering stress and strain with the strain rate. Figure 1(b) indicates that the failure stress increases considerably with the increasing strain rate, while the failure engineering strain varies moderately, especially the failure strain increases even more slowly as the strain rate becomes higher. The observation inspires us to use the strain as the failure criterion for gelatin at large deformation. One choice is the maximum tensile stretch ratio in principal directions, $\lambda_m = \max\{\lambda_1, \lambda_2, \lambda_3\}$. We denote the stretch ratio in the compression direction asλ_1, and other two expansion principal directions as λ_2,λ_3. $\lambda_1 = 1+\varepsilon_E$,$\lambda_2 = \lambda_3 = 1/(1+\varepsilon_E)^{1/2}$. Since the engineering strain at failure is in the range of -0.58~-0.76 from our compression tests, the maximum tensile stretch of the 10wt% gelatin is in the range of $\lambda_m \approx 1.5\sim2.0$. And we can infer that at higher strain rates, the value of λ_m would be close to 2.0.For uniaxial compression deformation, the incompressible Mooney-Rivlin model gives the compression stress as,

$$\sigma_1 - \sigma_2 = \sigma_1 = 2\left(c_1 + c_2/\lambda_1\right)\left(\lambda_1^2 - 1/\lambda_1\right) \tag{2}$$

All the stress-strain data of our tests were transformed into the true stress vs. stretch ratio and used to fit the two parameters c_1 and $c_2$in Equation (2) based on the least-square approach. See Table 1. For the low strain rate range, the rate dependency seems insignificant compared with that at high strain rates in the SHPB test [3]. So the mean values of the moduli (c_1, c_2 and G=2(c_1+c_2)) were calculated.

Simple steady shearing tests were carried out on the Gemini200 rheometer (Bolin) by using the cone-plate fixture of 25mm diameter. After the loaded gelatin specimen reached the target temperature of 4℃in the fixture, it was left there for 4 hours aging before the test was started.In Figure **2**(a)，the four curves of our test results are rather close to each other. Compared with the significant strain rate effects above 1000/s in the SHPB test [3], the strain rate effects can be ignored at low strain rates. For simple shear deformation Mooney-Rivlin model gives linear shear stress τ and quadratic first normal stress difference N_1 with the shear strain γ,

$$\tau = 2(c_1 + c_2)\gamma = G\gamma \tag{3a}$$

$$N_1 = 2(c_1 + c_2)\gamma^2 = G\gamma^2 \tag{3b}$$

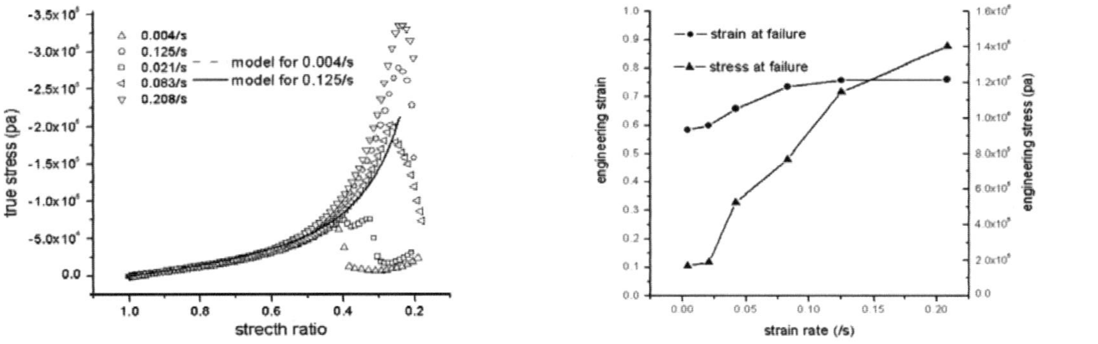

(a) (b)

Figure 1. Compression test results of 10wt% ballistic gelatin at 4℃, aging time 12h.(a) Engineering stress-strain curves; (b) Failure engineering stress and strain dependence on strain rate.

Table 1. Material parameters in Equations (1,2) fitted with uniaxial compression experimental data.

Strain rate	0.004	0.021	0.083	0.125	0.208	mean
G	19600	18100	20300	21100	23600	20000
c_1	6200	4400	6600	5700	8600	6000
c_2	3600	4700	3500	4900	3200	4000

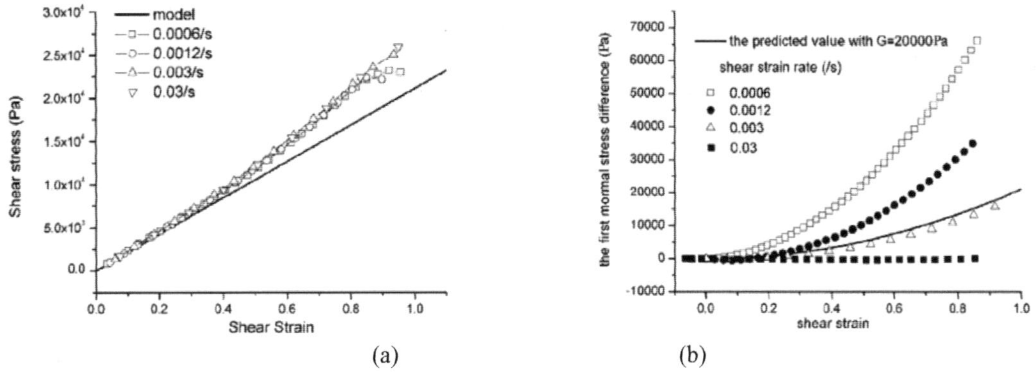

(a) (b)

Figure 2. Mooney-Rivlin model describes the simple shear deformation of gelatin with G=20000(Pa)

The fitted value of G=20000Pa from the compression tests is used in Equation (3) to predict the τ vs. γ and N_1 vs. γ, they are compared with experimental results in Figure 2. Mooney-Rivlin model predicts linear shear stress in simple shear deformation. Figure **2** shows that the curves of shear stress are very close to each other, while the first normal stress difference is very rate sensitive. Rather peculiar is that the first normal stress difference decreases as the strain rate increases, the opposite behaviour to the common viscoelastic fluids. At the present stage this peculiar behaviour is still under investigation.

2.2. High strain rate tests

Material parameters of gelatin at high loading rates are scarce and the available experimental data by SHPB show rate-sensitive stress-strain behaviour. Here we choose to let the parameters (the moduli c_1 and c_2) of Mooney-Rivlin model be the functions of strain rate to describe the strain-rate dependent response of ballistic gelatin. We read the true-stress vs. true-strain data in Figure 12 of Salisbury and Cronin [3] and transformed them into the true stress vs. stretch ratio curves as shown in Figure 3(a). The transformation from the true strain to the stretch ratio is $\varepsilon_T = \ln(\lambda)$.

In Figure 3(a), the data show modest stress increasing for the nominal (engineering) strain rates from 1000/s to 1350/s and a sudden stress increase at 1550/s. The stress strain curve at 1350/s was chosen to obtain the parameters c_1 and c_2 with Equation (2).Then the parameters at 1000/s, 1250/s and 1550/s were obtained by linear interpolation. The values of c_1 and c_2 are shown in the third to sixth columns of Table 2 and in Figure 3(b).

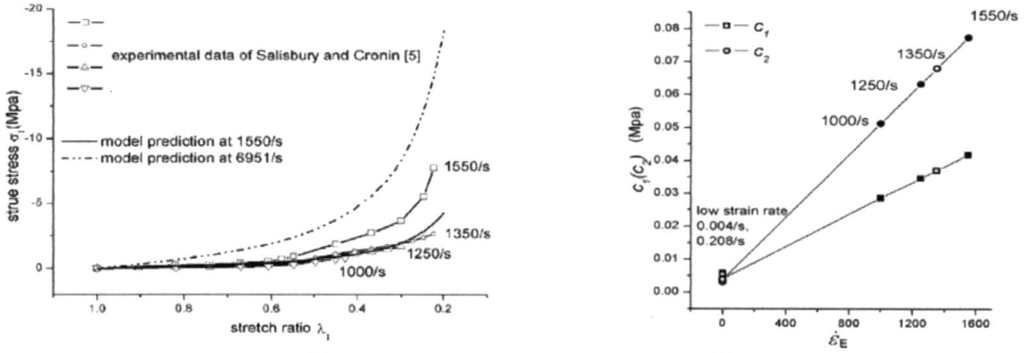

(a) (b)

Figure 3. Dependence of c_1 and c_2 on strain rate

The predictions for the true-stress vs. stretch-ratio by Mooney-Rivlin model at 1550/s are also shown in Figure 3(a). For the stretch ratios below 0.6 and at 1550/s the experimental stresses are suddenly increased and considerably larger than the model predictions. The true strain rate, defined as the compression velocity divided by the instant length of the specimen, increased as the strain increased during the compression test. Actually, for the nominal 1550/s compression test, the true strain rate could be as high as 6951/s at the end of compression. Taking this into consideration, the experimental data are reasonably between the predicted values for 1550/s and 6951/s as shown in Figure 3(a).

3. Application to penetration problems

One approach to the problem of deep penetration into semi-infinite target is to modify the analysis of the static expansion of a spherical cavity to determine the normal stress at the surface of the penetrator. We developed a cavity expansion model for soft materials like ballistic gelatin considering large elastic deformation. The elastic behaviour is modelled by the Mooney-Rivlin constitutive equation and the failure criterion by elastic fracture at the largest elastic stretch ratio λ_m. Suppose a spherically symmetric cavity expands from initial zero radius in an infinite medium, it will produce elastic and cracked response regions. The radius of the cavity is denoted by a, and the radius of the cracked region surrounding it by c. The pressure on the cavity wall derived by Liu et al. ([10] to be published) for incompressible mediums is,

$$-P_s = \sigma_r |_{r=a} = \left(c_1 \left[\left(1-(a/c)^3\right)^{1/3} \left(5-(a/c)^3\right) - 5 \right] - 2c_2 \left[\left(1+(a/c)^3\right)\left(1-(a/c)^3\right)^{-1/3} - 1 \right] \right)(c/a)^2 \quad (4a)$$

$$a/c = \left(1 - \lambda_m^{-3}\right)^{1/3} \tag{4b}$$

This solution is radial position self similar, and the pressure is equivalent to the energy needed to open unit volume of the cavity.

For the ballistic gelatin the parameter c_1 and c_2 are linear functions of the characteristic strain rate (Table 2). For low strain rates the maximum stretch ratio λ_m=1.5, while for large strain rates λ_m=2.0. The cavity wall pressure P_s is a linear function of the strain rate since the parameters c_1 and c_2 are linearly rate-dependent on the strain rate.

$$P_s = P_0 + \beta \cdot \dot{\varepsilon} \tag{5}$$

With P_0=34000 Pa and β=360 Pa/s, Equations (4) yield the pressures on the expanding cavity wall, P_s, which are listed in Table 2.

3.1. Quasi-static penetration test

The liquid gelatin at 60℃ was put into aluminium molds (120mm diameter and 50mm depth) and after solidification, the molds were sealed and put into a refrigerator for aging at 4℃ for 16~18h. The quasi-static penetration tests were carried out on the universal materials tester. Steel rods with conical head of various cone angles (20°, 30°, 40°) were stick on the upper loading surface of the tester (Figure **4**). As the cone penetrates into the gelatin block, a "cavity" is opened. Suppose the normal stress σ and frictional stress τ are uniform on the conical head, the resistance force on the rod is,

$$F = \pi z^2(\sigma \tan^2\theta + \tau \tan\theta) = Kz^2 \tag{6}$$

where θ is the semi-cone angle and z the penetration depth. So the resistance force F is proportional to the square of the penetration depth z (Figure 5(a)) and the factor K is dependent on the semi-cone angle θ (Figure 5(b)),

$$K/(\pi\tan\theta) = \sigma\tan\theta + \tau \tag{7}$$

Table 2. Material parameters of gelatin and calculated values of P_s by cavity expansion model of Equations (4)

strain rate $\dot{\varepsilon}$ (/s)	0.004~ 0.208	1000	1250	1350	1550
c_1 (MPa)	0.006	0.029	0.035	0.037	0.042
c_2 (MPa)	0.004	0.051	0.063	0.068	0.077
λ_m	1.5	2.0	2.0	2.0	2.0
P_s(MPa)	0.03	0.4	0.49	0.53	0.6

Figure 4. Quasi-static penetration test

This linear relationship between $K/(\pi\tan\theta)$ and $\tan\theta$ can be analyzed to determined the slop of σ and the intercept τ. The normal stress σ is the pressure needed to open a cavity quasi-statically. The results show that σ is in the range of 0.019~0.051 MPa at low expansion rate (0.03~1.2 mm/s) which is defined as the penetration velocity multiplied by $\tan\theta$; while the value of P_s is 0.03 MPa in the low strain rate range of Table 1. Considering the experimental errors, the agreement supports our model's estimation.

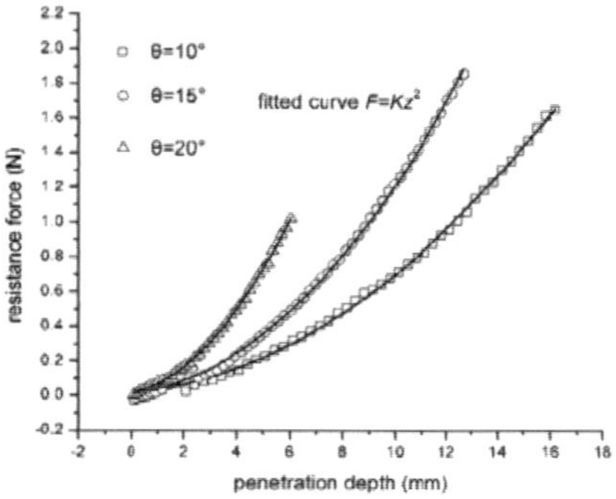

(a)

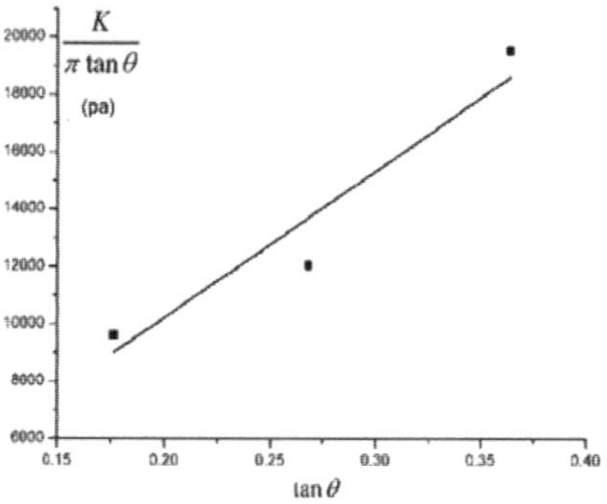

(b)

Figure 5. Fitting results of quasi-static penetration tests: (a), resistance force vs. penetration depth; (b) determination of the normal stress and frictional stress.

3.2. High-speed penetration test

For soft materials, the radial acceleration of the target material under penetration causes considerable free expansion of cavity. Since the parameter P_s characterizes the energy needed to open unit volume of cavity,we have $\Delta E=P_sV_m$, where ΔE is the total energy transferred from the projectile to the target and V_m the maximum volume of the temporary cavity. Sellier [11] proposed an empirical formula to estimate the maximum cavity volume V_m in terms of the total energy ΔE transferred to the target: $V_m=\mu\Delta E$. This is a rough estimate based on ballistic experiments. Sellier gives μ=1.26cm^3/J for 10wt% gelatin. In the present terminology, $P_s=1/\mu$=0.794MPa, this value is in the same order of magnitude as those in Table1 obtained by using the material parameters of 10wt% gelatin and Mooney-Rivlin model.

In our penetration experiments, 10wt% gelatin blocks (30cm×30cm×30cm) stored at 4℃ in a refrigerator for 24 hours were used. We carried out experiments by firing fragments of various shapes: spheres, cylinders, rhombus and a type of bullet into the gelatin blocks. The penetration velocity varied from 600m/s to 1100m/s. A high-speed video camera recorded the motion of the penetrators and the temporary cavity's wall-variations with time. The parameter P_s was estimated by $\Delta E/V_m$, where ΔE is the energy lost of the penetrators and V_m the maximum volume of the cavity produced. The experimental values of P_s are in the range of 0.2~0.6 MPa as shown in Table 2. This range is consistent with that in Table 2.

Table 2. Calculated value of P_s from penetration tests by different shaped fragments

	spheres				Rhombus	Cylinders		5.8 caliberBullet	
ρ_p(kg/m^3)	7800				7800	7800		-	
m_p(g)	0.11	0.26	0.45	0.45	0.25	0.92	0.92	4.15	4.15
v_p(m/s)	644	658	725	856	651	1032	723	820	738
V_m×10^5(m^3)	8.05	17.7	31.2	49.5	19.9	83.5	66.8	523	406
ΔE(J)	22.7	56.0	117.9	164.7	53.0	488.6	239.6	1108.0	817.7
P_s(MPa)	0.28	0.32	0.38	0.32	0.27	0.59	0.36	0.21	0.20

4. Conclusions

Uniaxial compression and simple shearing tests were carried out for 10wt% ballistic gelatin at low strain rates; the elastic moduli c_1,c_2 in Mooney-Rivlin model were obtained and the main failure mode is determined to be elastic fracture as the maximum tensile stretch ratio exceeds a critical value.

Gelatin's mechanical behavior shows rate-sensitivity, especially at high strain rates. We fit Mooney-Revlin model with the data of Salisbury and Cronin [3] and establish linear variations of the moduli with the strain rate. Consequently, the key parameter P_s in cavity expansion model, that is, the energy needed to open unit volume of cavity in 10wt% gelatin increases linearly with the strain rate. The prediction of P_s is compared favorably with the normal stress in quasi-static penetration tests and the free cavity expansion in 10wt% gelatin ballistic tests by fragments of various shapes.

References

[1] Cronin D S and Falzon C 2011 *Exp Mech* **51** 1197-1206
[2] Moy P and Tusit W *Proc. XIth Int. Congr. and Expo* June 2-5 2008 Orlando, Florida USA
[3] Salisbury C.P. and Cronin D.S.. *EXP MECH 2009*; **49** 829-840
[4] Foster M et al. *Dynamic Behavior of Materials,* 1, Conf. Proc. of the Society for Experimental Mechanics Series 99
[5] Shepherd C J et al. 2009 *AIP Conf. Proc.* **1195** 1399-1402
[6] Forrestal M J, Brar N S and Luk V K *J. Appl. Mech.-T ASME* **58** (1) 7-10
[7] Forrestal M J and Tzou D Y 1991 *Int. J.Solids Struct.* **34** (31-32) 4127-4146
[8] Macosko C W 1994 *Rheology principles, measurements, and applications* (New York:Wiley)
[9] Ottone M L and Deiber *J A* 2005 *Polymer* **46** 4928-4937
[10] Liu L et al. A spherical cavity expansion model for ballistic gelatin and its application to penetration problems. To be published in the International *J. Impact Engng*
[11] Sellier K G 1994 *Wound ballistics and the scientific background* (Amsterdam: Elsevier)

Component based modelling of piezoelectric ultrasonic actuators for machining applications

A Saleem[1], M Salah[2], N Ahmed[1], and V V Silberschmidt[3]
[1] Department of Mechanical Engineering, Taibah University, Al-Madinah Al-Munawara, Saudi Arabia
[2] Department of Mechatronics Engineering, Hashemite University, Zarqa, Jordan
[3] School of Mechanical and Manufacturing Engineering, Loughborough University, England, UK

E-mail: ashraf_saleem@yahoo.com

Abstract. Ultrasonically Assisted Machining (UAM) is an emerging technology that has been utilized to improve the surface finishing in machining processes such as turning, milling, and drilling. In this context, piezoelectric ultrasonic transducers are being used to vibrate the cutting tip while machining at predetermined amplitude and frequency. However, modelling and simulation of these transducers is a tedious and difficult task. This is due to the inherent nonlinearities associated with smart materials. Therefore, this paper presents a component-based model of ultrasonic transducers that mimics the nonlinear behaviour of such a system. The system is decomposed into components, a mathematical model of each component is created, and the whole system model is accomplished by aggregating the basic components' model. System parameters are identified using Finite Element technique which then has been used to simulate the system in Matlab/SIMULINK. Various operation conditions are tested and performed to demonstrate the system performance.

1. Introduction

Ultrasonically Assisted Turning (UAT) enhances and improves machining of hard-to-cut alloys. This technique depends on producing a vibration on the cutting tool with a frequency of 20 kHz and amplitude of 30μm peak-to-peak. Vibrating the cutting tool in this manner reduces the cutting forces up to 60% [1] and a considerable improvement in the surface finish can be obtained [2]. However, this technique has not yet been widely introduced in the industry due to instability problems in the cutting process. Instability in tool vibration causes a poor surface finish and, hence, prevents the full implementation of this process in the industry. Despite all research conducted in this area to make this machining process stable [2-5], an applicable solution is still not found. To tackle the stability issues of such machining processes, comprehensive investigation has to be conducted to find out practical solutions and all that starts from having an accurate and a comprehensive mathematical model for ultrasonic transducers.

Formulating mathematical models for ultrasonic transducers is essential to analyze and investigate the system performance in order to facilitate the controllers' development. The main challenge in modelling ultrasonic transducers is modelling the driving component in the vibrating system (*i.e.*, the piezoelectric material). Piezoelectric materials exhibit obvious nonlinear characteristics that make it more challenging when modelling the vibrating system. To overcome the nonlinearity inherent in the model's governing equations, it has become a common practice to replace it with a linearized version which provides a reasonable description of the system dynamics around an operating point. However, this approach cannot predict the global behaviour of the positioning system which considered as important as a particular solution around the operating point. Significant work has been conducted to develop a proper nonlinear model for the piezoelectric actuators in order to describe the inherent hysteresis phenomenon. The most popular models are the Preisach model [6, 7, and 8], Prandtl–Ishlinskii (PI) model [9, 10], and Krasnosel'skii Pokrovskii (KP) model [11].

Bashash and Jalili [12] introduced a nonlinear model for piezoelectric actuators combining the Prandtl–Ishlinskii hysteresis operator with a second-order linear dynamics whereas Yeh *et al.* [13], used a non-linear spring element in the hysteresis model and utilized a Maxwell slip structure. Recently, Ru *et al.* [14] proposed a new mathematical model for the hysteresis by using polynomial fitting. Salah *et. al.* [15] has utilized Coleman-hysteresis model for modelling the dynamic behaviour of the piezoelectric actuators. They proposed a nonlinear robust control strategy to actively control the displacement of the effective tip. Lyapunov based analysis tools were employed to prove that the tip tracking error is practically regulated to zero.

In this paper, a component-based modelling technique is adopted to create a nonlinear model of ultrasonic transducers. The transducer is decomposed into its basic components and each component is modelled separately. The model of the whole system is created by simply aggregating all components' models. Simulation results show that the model could predict the behaviour of the system satisfactorily under different conditions.

2. Component-based modelling for ultrasonic transducers

Component-based techniques, in general, have been used extensively in systems development due to their great potential to reduce time and cost, improve system maintainability and flexibility, and enhance system quality [16]. Therefore, adopting this technique to model and simulate ultrasonic transducers provides the following potentials: (1) component-based model is valid over the whole working range of the transducer. This is due to including the nonlinear phenomena of piezoelectric material without the need for linearization, (2) this technique provides flexibility of configuration choice and direct interpretation, (3) changes in the real system components can be reflected easily in the model, (4) the capability of modelling a wide variety of ultrasonic types and handling different materials and shapes, (5) facilitate the modelling of different complexity in order to allow for variation in the future.

Ultrasonic transducers can be decomposed into three main components as shown in Figure 1. These components are: one piezoelectric ring (PZT), and two mass-spring-dampers (MSD 1 and MSD 2). The Load component mimics the external force that acts on the transducer as a result of machining. In this model, the back section of the transducer is neglected because of the existence of a nodal point between the piezoelectric rings and the transducer's back section that prevents vibration waves to propagate back. The two mass-spring-damper components represent the concentrator part. The Kelvin–Voigt component models the effect of the resultant axial load from the machining process.

2.1. PZT component model

The piezoelectric material is represented by a component that has three inputs and two outputs as depicted in Figure 1. The inputs are the voltage, speed and displacement of the driven load (i.e. the first mass-spring-damper which is attached directly to the component). The speed and displacement of the driven load are required to calculate the load force on the piezoelectric ring while the input voltage excites the piezoelectric material to produce the required displacement and speed. The outputs of this component are the displacement and speed of the piezoelectric component. The dynamic behaviour of the piezoelectric component can be expressed as [15].

$$m_p\ddot{x}_o = \left(\frac{T_{em}}{C_C}\right)q - \left(\frac{T_{em}^2}{C_C}\right)x_o - \bar{F} \tag{1}$$

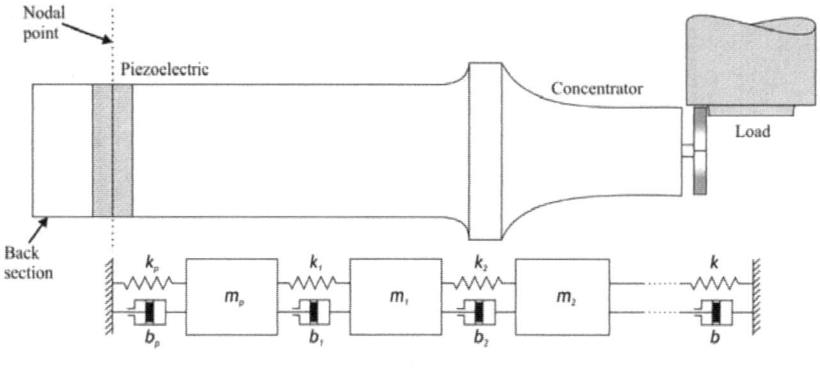

(a)

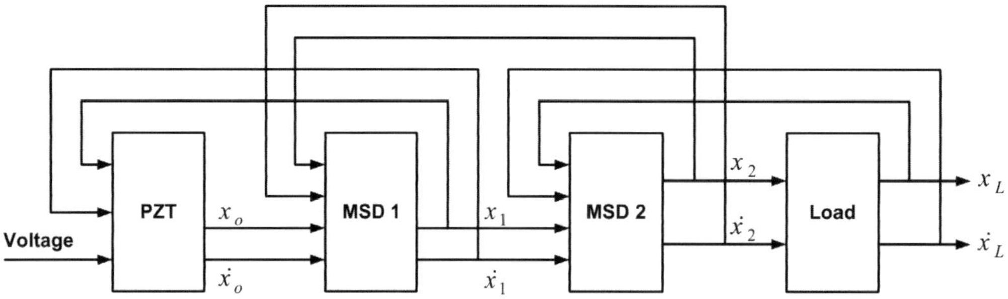

(b)

Figure 1. (a) Ultrasonic transducer; (b) proposed component-based model of ultrasonic transducers

and

$$\bar{F} = k_p x_o + b_p \dot{x}_o + k_1(x_o - x_1) + b_1(\dot{x}_o - \dot{x}_1) \tag{2}$$

where $x_o(t), \dot{x}_o(t), \ddot{x}_o(t)$ are the displacement, speed, and acceleration of the effective tip (where the turning tool is attached), respectively, $x_1(t), \dot{x}_1(t)$ are the displacement and speed of the MSD 1, respectively, $x_2(t), \dot{x}_2(t)$ are the displacement and speed of the MSD 2, respectively, $q(t)$ is the charge induced within the piezoelectric, m_p is the mass of the piezoelectric material, T_{em} is the elongation constant, C_c is the internal capacitance, k_p, b_p, k_1, b_1, are the spring constants and damping coefficients of the PZT and MSD 1, respectively, and $\bar{F}$ is the perpendicular force acting on the piezoelectric actuator.

A nonlinear hysteresis model (*i.e.*, $q = H(V_h)$ can be defined to describe the relationship between the input voltage, $V_h(t)$, and the induced charge, $q(t)$ as [17]

$$H = f + d \tag{3}$$

where the input voltage, $V_h(t)$, is defined as $V_h = V - V_c$. The signal $V_c(t)$ is the voltage across the internal capacitor of the piezoelectric and $V(t)$ is the voltage applied on the piezoelectric component and represents the control input. The function $f(V_h)$ is a subsequently defined signal and the function $d(V_h)$ is defined as

$$d(.) \triangleq \begin{cases} [q_o - f(V_{h_o})]e^{-\delta|V_h - V_{h_o}|} + e^{-\delta V_h} \int_{V_{h_o}}^{V_h} (g(\tau) - f'(\tau))e^{\delta\tau} d\tau, when\ \dot{V}_h(t) > 0 \\ [q_o - f(V_{h_o})]e^{\delta|V_h - V_{h_o}|} - e^{\delta V_h} \int_{V_{h_o}}^{V_h} (g(\tau) - f'(\tau))e^{-\delta\tau} d\tau, when\ \dot{V}_h(t) < 0 \end{cases} \quad (4)$$

where q_o is the induced charge at $t = t_o$, V_{h_o} is the input voltage at $t = t_o$, δ is a positive constant, $f'(\tau) = \frac{\partial f(V_h)}{V_h}$ and $g(\cdot)$ is a subsequently designed signal. To validate the expressions in equation (4), the signals $d(V_h)$, $f(V_h)$ and $g(V_h)$ have to satisfy certain properties that can be found in [17, 15]

2.2. Mass-Spring-Damper (MSD) component model

The MSD component represents a mass that is connected with two springs and two dampers. As shown in Figure 1, the component has four inputs and two outputs. The inputs are speed and displacement of the previous and subsequent masses that are connected to this component whereas the outputs are the resultant speed and displacement of the current mass. Parameters required for this component are stiffness and damping coefficients of the current and adjacent systems. If this component is not connected to other systems, then all corresponding parameters and inputs are set to zero.

In order to be consistent in modelling different components in the system, the transducer can be modelled with two separate MSDs, as shown in Figure 1. The equation of motion for each MSD component block can be written as follows:

$$m_n\ddot{x}_n = -k_n(x_n - x_{n-1}) - b_n(\dot{x}_n - \dot{x}_{n-1}) + k_{n+1}(x_{n+1} - x_n) + b_{n+1}(\dot{x}_{n+1} - \dot{x}_n) \quad (5)$$

where m_n, k_n, and b_n denote the mass, stiffness coefficient, and damping coefficient, respectively, of the nth component and k_{n+1} and b_{n+1} are the stiffness and damping coefficients of the subsequent mass.

2.3. Load component model

The load component is considered as the load force from the machining process (*i.e.*, milling, turning, or drilling). This component is important to study the influence of the machining process on the ultrasonic transducer behaviour. The component has two inputs and two outputs as shown in Figure 1. The inputs are the ultrasonic resultant displacement and speed whereas the outputs are the resultant displacement and speed during the machining process.

Application of the load (*i.e.*, machining process) causes nonlinearity and instability in the ultrasonic transducer behaviour [18]. Therefore, it is important to include the load model in the proposed transducer system to simulate and investigate its influence on the system behaviour. This investigation assists in the development of control systems.

The mathematical model of this component can be created based on Kelvin–Voigt model. This model describes one-dimensional contact interaction between the ultrasonic transducer and a workpiece. Equation (6) represents the load model:

$$F_L = \frac{1}{2}(1 + \operatorname{sgn}(x))(kx + b\dot{x}) \quad (6)$$

where $x = x_2 - \Delta$, k is the contact stiffness, b is the contact damping, Δ is the initial interference/gap, x_2 is the resultant displacement of the ultrasonic transducer, and sgn(·) is the standard signum function.

In order to validate the proposed system model, system parameters should be identified for each component as well as the resonant frequencies of the system in order to get maximum amplitude from the system. In this work, parameters identification is done theoretically using the simple beam model as mentioned in the previous section. The mass, damping and stiffness parameters are extracted and used to compute the natural frequencies of the reduced system. The computed natural frequencies are compared with those computed from the full finite element model. Also, the frequency response of the reduced model is compared with the frequency response of the full finite element model.

3. Numerical simulation

The component-based model of the ultrasonic transducer is created using Matlab/Simulink. All systems parameters (namely the mass, stiffness, and damping coefficient of MSD1 and MSD2) are set based on the identification results obtained from the FEM. To investigate the properties of the proposed ultrasonic transducer model, the amplitude - frequency curve has been created as shown in Figure 2 by exciting the system with a sinusoidal signal at different frequencies and measuring the MSD2 displacement. It can be noticed that the system has three resonant frequencies at 6500Hz, 18000Hz, and 20500Hz which are correspond to m_p, m_2, and m_3 respectively. Although the maximum amplitude was attained at 6500Hz, high frequencies are utilized in the subsequent analyses since it was proved that the machining at 20kHz gives the best results [1].

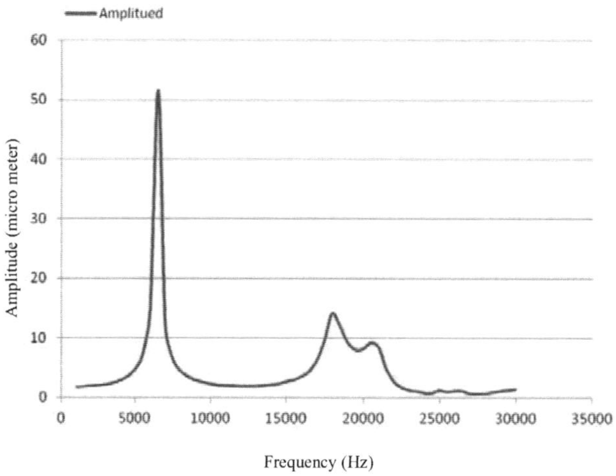

Figure 2. Frequency response of the ultrasonic transducer model

Based on the frequency response of the system, a sinusoidal input voltage with a 20.5kHz is applied to the system. The displacement of the ultrasonic transducer tip (cutting tool) is shown in Figure 3. It is clear that the amplitude of the vibration is 30μm peak-peak at a frequency of 20kHz as required for the machining. However, it can be observed that the peaks of the displacement amplitude are not consistent. This is due to the hysteresis effect between the system input (applied voltage) and output (tip displacement) which are shown in Figure (4). Piezoelectric materials are hysteretic in nature, and therefore they may exhibit oscillatory closed-loop behavior, as well as poor tracking and potential instability [19].

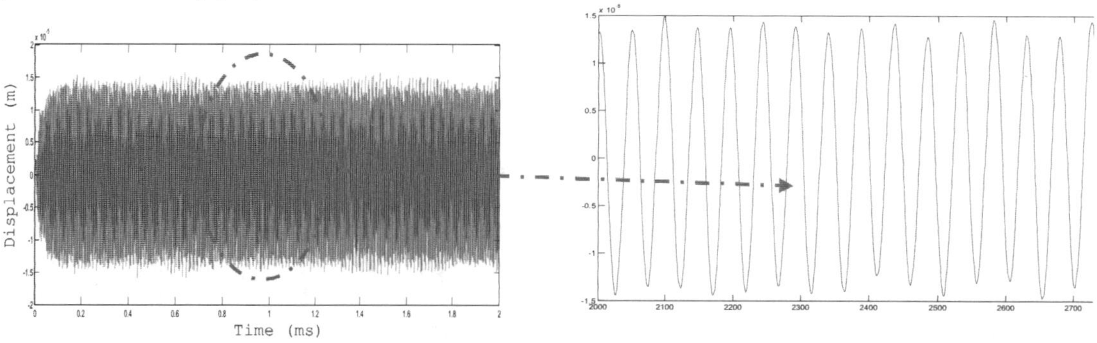

Figure 3. Displacement of the ultrasonic transducer tip

In Figure 4, the response of system components (*i.e.*, PZT, MSD 1, and MSD 2) are presented. It is clear how the hysteresis effect propagates throughout the system components and influences the their response.

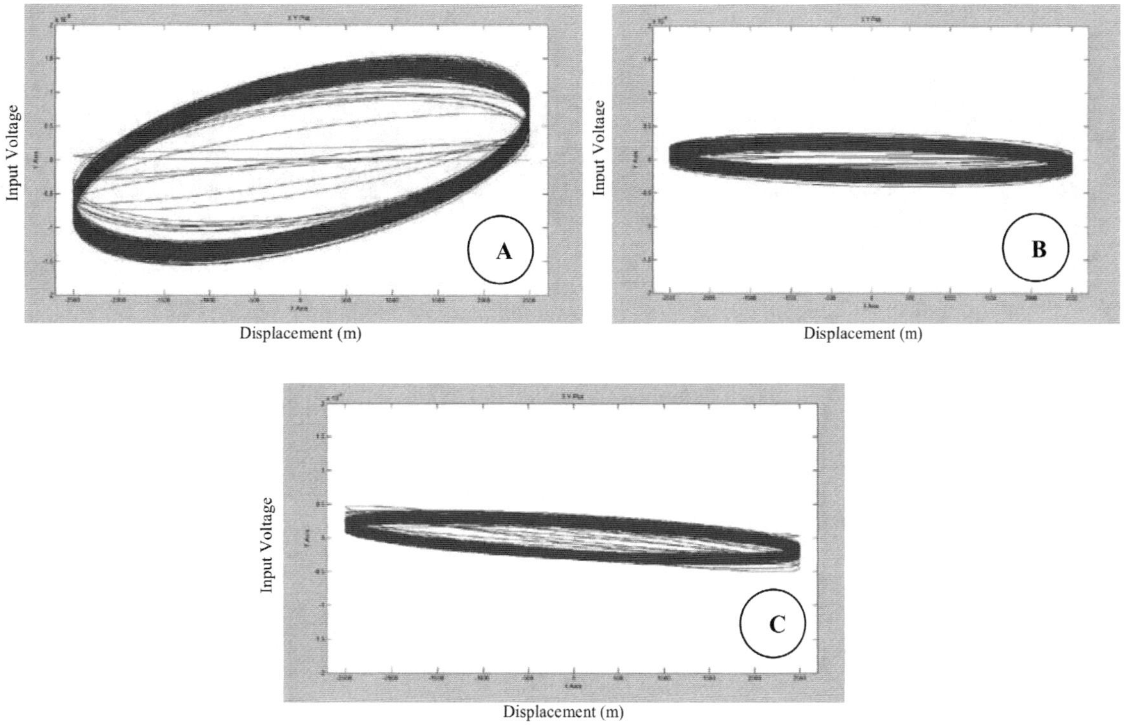

Figure 4. Hysteresis effect on the ultrasonic transducer's components: (A) MSD 2's displacement, (B) MSD 1's displacement, and (C) PZT's displacement

To investigate the influence of the nonlinearity and hysteresis of piezoelectric actuator, load contact stiffness was increased gradually on the transducer's tip (cutting tool) as depicted in Figure 5. One of the most important observations in the system's response is the attenuation of the peak-to-peak displacement (*i.e.*, amplitude) of the vibration as shown in Figure 6. Moreover, the hysteresis effect on the overall system (*i.e.*, ultrasonic transducer) is increased. This is clear from the high deficiency between the peaks of the output signal (*i.e.*, MSD 2 displacement).

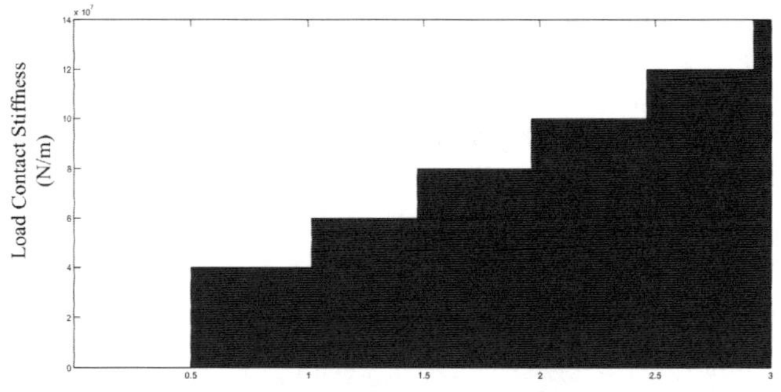

Figure 5. Load contact stiffness profile applied on the ultrasonic transducer

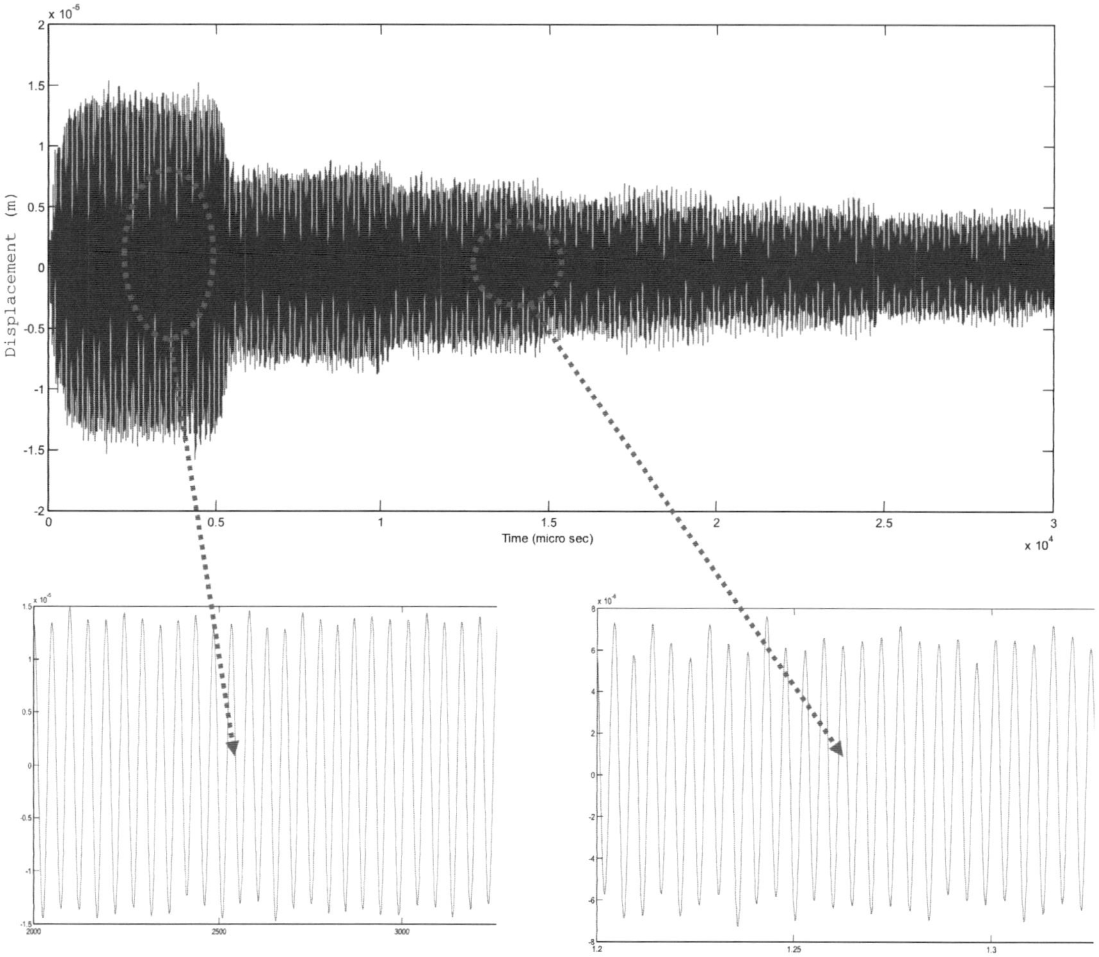

Figure 6. Ultrasonic transducer's response to the change of load contact stiffness

4. Concluding remarks

In this paper, a component-based model for piezoelectric ultrasonic transducers is introduced. The proposed technique is based on decomposing the entire system into individual components that can be used in different configuration and represented by mathematical expressions. The components are configured to be combined in series to formulate the ultrasonic transducer. To investigate the performance of such component-based modelling, the combined system was simulated in Matlab/Simulink. In the simulation, the system parameter values were identified using Finite Element technique. Various operation conditions were applied to test the performance of the system. From the performed numerical simulations, it was clear that the nonlinearity due to the inherent piezoelectric actuator's hysteresis has a big impact on the performance of the machining process and that it should be considered when control strategies are introduced.

5. Acknowledgement

This research was supported by the Strategic Technologies Program of the National Plan for Science, Technology and Innovation in the Kingdom of Saudi Arabia under the project no. 09-ADV809-05.

References

[1] Ahmed N, Mitrofanov A V, Babitsky V I and Silberschmidt V V 2007 Analysis of forces in

ultrasonically assisted turning *J. Sound Vib.* **308**(3-5) 845-854
[2] Babitsky V I, Kalashnikov A N, and Molodtsov F V 2004 Autoresonant control of ultrasonically assisted cutting. *Mechatronics* **14** 91-114
[3] Babitsky V I., Kalashnikov A and Astashev V K 2004 Autoresonant control of nonlinear mode in ultrasonic transducer for machining applications. *Ultrasonics* **42**(1-9) 29-35
[4] Babitsky V I, Mitrofanov A V and Silberschmidt VV 2004 Ultrasonically assisted turning of aviation materials: simulations and experimental study *Ultrasonics* **42** 81-86
[5] Maurotto A, Muhammad R, Roy A and Silberschmidt V V 2013 Enhanced ultrasonically assisted turning of a β-titanium alloy *Ultrasonics* **53**(7) 1242-1250
[6] Croft D, Shed G, and Devasia S. 2001 Creep, hysteresis, and vibration compensation for piezoactuators: Atomic force microscopy application *Trans. ASME J. Dyn. Syst. Meas. Contr.* **123**(1) 35–43
[7] Mayergoyz I D 1991 *Mathematical Models of Hysteresis* (New York: Springer)
[8] Natale C, Velardi F and Visone C 2001 Identification and compensation of Preisach hysteresis models for magnetostrictive actuators *Physica* B **306**(1) 161–165
[9] Brokate M and Sprekels J 1996 *Hysteresis and Phase Transitions* (New York: Springer)
[10] Visintin A 1994 *Differential models of hysteresis* (New York: Springer)
[11] Krasnoselskii M A and Pokrovskii A V 1989 *Systems with hysteresis*. (New York: Springer)
[12] Bashash S and Jalili N 2007 Robust multiple frequency trajectory tracking control of piezoelectrically driven micro/nanopositioning systems *IEEE Trans. Contr. Syst. Technol.* **15** (5) 867–878
[13] Yeh T, Lu S and Wu T 2006 Modeling and identification of hysteresis in piezoelectric actuators *ASME J. Dyn. Syst. Meas. Contr.* **128** 189–196
[14] Ru C, Chen L, Shao B, Rong W and Sun L 2009 A hysteresis compensation method of piezoelectric actuator: model, identification and control *Contr Eng. Pract.* **17** 1107–1114
[15] Salah M H, McIntyre M L, Dawson D M, Wagner J R and Tatlicioglu E 2012 Charge feedback-based robust position tracking control for piezoelectric actuators *IET Contr. Theo. Appl.* , **6**(5) 1–14
[16] Pour G 1998 Moving toward component-based software development approach *Proc. Tech. Obj.-Orient. Lang.* 296-300
[17] Coleman B and Hodgdon M 1987 On a class of constitutive relations for ferromagnetic hysteresis *Arch. Rat. Mech. and Anals.* **99**(4) 375-396
[18] Voronina S and Babitsky V 2008Autoresonant control strategies of loaded ultrasonic transducer for machining applications *J. Sound Vib.* **313**(3-5) 395-417
[19] Al-Janaideh M, Sumer D, Yan J, Amato A M D', Drincic B, Aljanaideh K, and Bernstein D S 2012 Adaptive control of uncertain linear systems with uncertain hysteretic input nonlinearities *ASME 2012 5th Annual Dynamic Systems and Control Conference joint with the JSME 2012 11th Motion and Vibration Conference October 17-19, 2012, Fort Lauderdale, Florida, USA.*

Study of ultrasonically assisted turning of stainless steel and brass alloys

S M A Mahdy[1,2,3], M A Gouda[1] and V V Silberschmidt[2]

[1]Mechanical Engineering Dept., Faculty of Engineering, The British University in Egypt (BUE), El-Shorouk city, Cairo 11837, Egypt
[2]Wolfson School of Mechanical and Manufacturing Engineering, Loughborough University, Loughborough, Leicestershire LE11 3TU, UK

Email: sameh.mahdy@bue.edu.eg

Abstract. Ultrasonically assisted turning (UAT) is a hybrid machining technique employing high-frequency small-amplitude vibration superimposed on the tool movement during turning. It is superior to conventional turning (CT) with regard to cutting forces, surface quality and machining accuracy. The aim of this study is to investigate the effect of different machining parameters on stainless steel and brass alloys, during both UAT and CT, and evaluate improvements of cutting forces, surface roughness, surface integrity, and machining accuracy. An experimental setup for UAT at Loughborough University was used to accomplish this investigation. This setup used a Picoscope data acquisition add-on with Kistler three-component dynamometer, Seco DNMG cutting inserts were utilized, a surface tester from Taylor Hobson was used to evaluate surface roughness, Alicona Infinite Focus microscope was used to evaluate surface roughness and surface integrity, while Metris CMM with Renishaw probe was employed to evaluate machining accuracy. Cylindrical workpieces of steel and brass alloys were turned under CT and UAT conditions; cutting forces, surface roughness, and machining accuracy produced with both techniques where compared. Significant improvements were noticed in the ultrasonically assisted machining when compared to the CT for both alloys.

1 Introduction

The ultrasonic vibration cutting method is an efficient cutting technique for difficult-to-machine materials [1]. The ultrasonic vibration cutting method is a more effective cutting process than conventional cutting in terms of cutting force, cutting instability, tool blunting, tool wear, chip generation, surface finish and so on, to machine difficult-to-cut materials such as Ni- and Ti-based super alloys, hardened steels, optical glasses, ceramics, tungsten carbides, etc. [1-4] . Ultrasonic Assisted Turning (UAT) is a process of turning a work piece with the help of high-frequency low-amplitude vibrations. This process makes use of the classical lathe machine to superimpose this high-frequency vibration on the tool while cutting. The vibration direction in relation to the tool movements has a great effect on the surface produced [1]. This direction could be either in lateral feed direction or in the longitudinal feed direction [5]. The first moves the tool intermittently in a hammer-wise movement in relation to the work piece's rotating surface; this movement results in an instantaneous increase in the radial force component. The second direction is linked to the tool movement in a

[3] To whom any correspondence should be addressed

direction parallel to the work piece's rotating axis; this movement results in an instantaneous increase in the axial force component each time the vibratory movement brings tool closer.

The effect of this is the multi-fold decrease of cutting forces as well as improvements of the surface integrity of the produced surface [6, 7]. Also the effect incorporates the enhancements of machining accuracy with the decrease in the cutting force [8]. Introduction of vibration promotes stability of the cutting process; this is achieved thanks to intermittent cutting contact interaction between the tool and the work piece [9]. UAT is investigated nowadays because of its unquestionable benefits when machining hard-to-machine alloys as well as tough materials, with the ability to reach mirror-like surface finish [10]. Also UAT proves to be superior to other processes when machining brittle materials such as glass and ceramics; it machines the surface to the required specifications without the need to costly post-processing operations usually associated with these materials. Aside from the obvious advantages, other effects can include the hardening of the machined surface due to the continuous hammering effect of the tool [11]. Different materials have been studied under UAT to try to evaluate the optimum machining conditions as well as to increase our knowledge in that new area [12]. These materials include super alloys as well as aviation materials, both being perfect candidates for such study because of their hard machinability under conditions of the classical conventional cutting (CT) [13]. Some of the most renowned brittle materials are glass and ceramics. Ultrasonically assisted machining process i.e. UAM, has been utilized in many of the basic metal cutting processes in an effort to improve the process. Ultrasonic Assisted Drilling (UAD) is used to minimize the cutting forces as well as improve the surface produced [2]. Ultrasonic Assisted Milling (UAM) is used with the high speed multi-point cutters to enhance both material removal as well as quality of surface [4]. Ultrasonic Assisted Turning (UAT) can be utilized with either single-direction vibration or two-direction vibration. In single-direction vibration the system vibrates in only one direction and the transducer system is fairly straight forward, while in the case of two-direction vibration (ultrasonic elliptical vibration cutting) the system is much harder to design [14], [12], and [15]. The aim of this paper is to study experimentally the effect of changing different machining parameters on machining stainless steel (SS) and brass under conditions of both CT and UAT.

2 Experimental technique

Experimental work was carried out at Wolfson School of Mechanical and Manufacturing Engineering at Loughborough University, where a Harrison M300 lathe was retrofitted with an ultrasonic vibration attachment to allow for both conventional cutting and ultrasonic assisted turning on the same machine delivering a vibration of a frequency of 20 kHz and an amplitude of 8 μm. This ultrasonic attachment is composed of a band of piezoelectric rings, a horn and a dynamometer; the system is attached to a high-frequency pulse generator as well as a data acquisition system for auto resonance feedback along with force measurements. As shown in figure 1 a cylindrical SS work piece was turned using CT and UAT; both tangential and radial forces where measured (using the dynamometer and the data acquisition with a PC). Surface roughness measurements were taken as shown in figures 2-3, while measuring roundness is presented in figure 4. An experimental plan was designed as follows for both material under investigation.

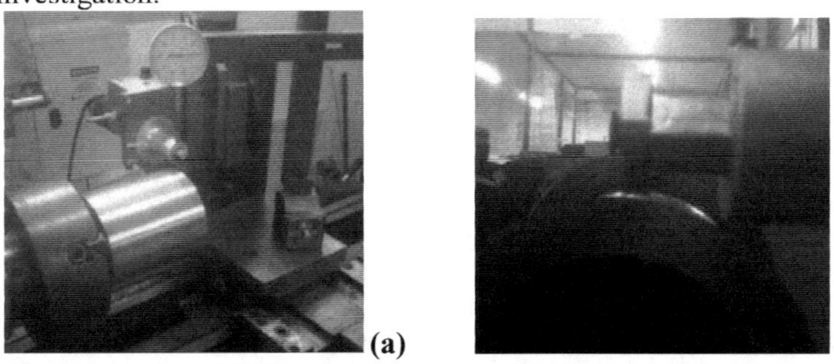

(a) **(b)**

Figure 1. Experimental setup with SS workpiece mounted, ultrasonic vibration attachment, and cutting tool positioned for cutting: (a) whole setup, (b) cutting-tool view.

Figure 2. TAYLOR HOBSON™ surface tester used to evaluate R_a for SS specimen

(a) **(b)**

Figure 3. ALICONA™ Infinite Focus® microscope used to evaluate surface roughness S_a for SS (a) and brass specimens (b)

(a) **(b)** **(c)**

Figure 4. Different workpieces during roundness measurements (R_o) on METRIS™ CMM (a) with Renishaw™ probe (b) for both CT and UAT; (c) measurement of machined surface for brass specimen

2.1 Stainless steel alloy material

In tests the austenitic standard stainless steel grade SAE 304 (BS 304S16 / DIN 1.4301) (ISO designation X2CrNi18-9) was used; this steel has the following composition: [C:0.06/Mn:0.9/Si:1/P:0.05/S:0.02/Cr:18/Ni:8] and properties: tensile Strength: 515 MPa, yield strength (0.2%): 205 MPa, density: 8000 kg/m^3, elastic modulus: 193 GPa. It was machined using the following machining parameters: cutting speed (v): 25, 30, 35, 40, and 45 m/min; depth of cut (a): 0.1, 0.2, 0.3, 0.4, and 0.5 mm; feed rate (f): 0.03, 0.05, 0.1, 0.16, and 0.2 mm/rev.

2.2 Brass alloy material

It is an alpha/beta brass alloys, which contains copper and zinc (ISO designation: CuZn37Pb2) (BS BZ131 / UNS C35300). This alloy has the following composition: [Cu:62/Pb:2] and the following mechanical properties: tensile strength: 420 MPa, yield strength: 200 MPa, density: 8500 kg/m^3, elastic modulus: 105 GPa. It was machined using the following machining parameters: cutting speed (v): 25, 35, 45, and 55 m/min; depth of cut (a): 0.1, 0.2, 0.3, 0.4, and 0.5 mm; feed rate (f): 0.1, 0.16, and 0.2 mm/rev.

2.3 Cutting tool

The cutting insert utilized during experiments is the industrial sintered carbide coated insert from SECO: DNMG 150608 MF1 CP500. This insert is suitable for intermittent metal cutting operations of high-speed steels. Insert as in figure 5 was mounted on a compatible tool holder at the end of the horn whose function was to concentrate the ultrasonic vibration movement at the tool-work piece interface.

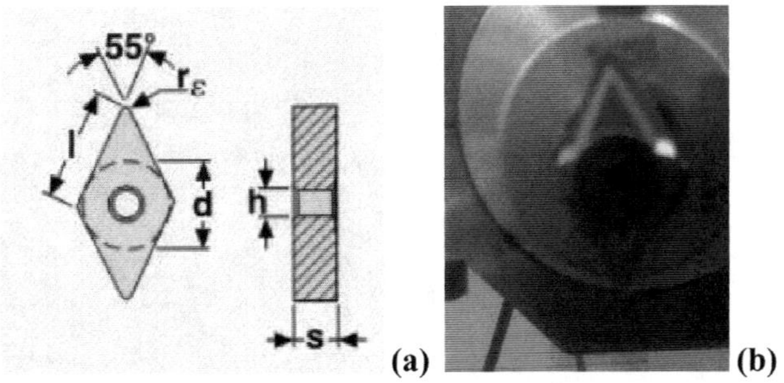

Figure 5. Schematic diagram of cutting tool insert (a) courtesy of SECO Gr [16], and real image (b)

3. Results and discussions

3.1. Cutting of stainless steel alloy material

3.1.1. Cutting forces measurement

In general, the cutting forces (tangential F_t and radial F_r) increased with the increasing cutting speed (v), depth of cut (a), or feed rate (f) for both cases of CT and UAT as shown in figures 6-8. Apparently, under different cutting conditions the cutting forces in UAT are lower than those in CT. At cutting speed v = 35 m/min, as example, the cutting forces F_t and F_r decreased from 162.2 N and 124.3 N in CT to 72.2 N and 71.0 N in UAT, i.e. by 56.2% and 42.9%, respectively. At depth of cut a = 0.3 mm, these cutting forces decreased from 177.3 N and 105.9 N in CT to 65.2 N and 42.7 N in UAT (by 63.2% and 59.6%, respectively). At feed rate f = 0.2 mm/rev, they decreased from 194.6 N and 149.2 N 126.0 N and 133.2 N (by 35.2% and 10.7%).

This decrease in cutting forces in case of UAT if compared to the CT is due to the lowering of friction levels between tool and workpiece due to the intermittent nature of tool-chip contact under UAT as stated in [9]. Another reason for that decrease might be attributed to the absorption of ultrasonic vibration energy produces an acoustic softening effect on the material as stated by Skelton et al. [17]. Also the decrease in cutting forces is expected to yield a better surface in case of UAT than in case of CT, as the system stability is enhanced by adding the superimposed vibrations, not limited to the surface improvements only but also to the improvement of the machining accuracy which is the ability of machine to produce required shape with close out of roundness values due to the decrease of the force. Another expected result from vibration action imposed on acoustic softened surface with lowered cutting forces on a much stable system is the close surface roughness values as well as surface integrity due to surface healing with each tool-workpiece hammering action.

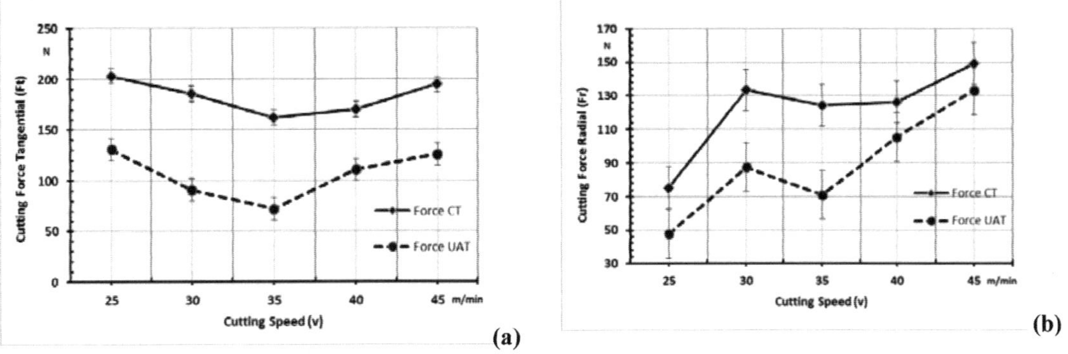

Figure 6. Effect of cutting speed (v) on tangential cutting force (F_t) (a) and radial cutting force (F_r) (b) for turning of SS under CT and UAT (f = 0.2 mm/rev, a = 0.3 mm)

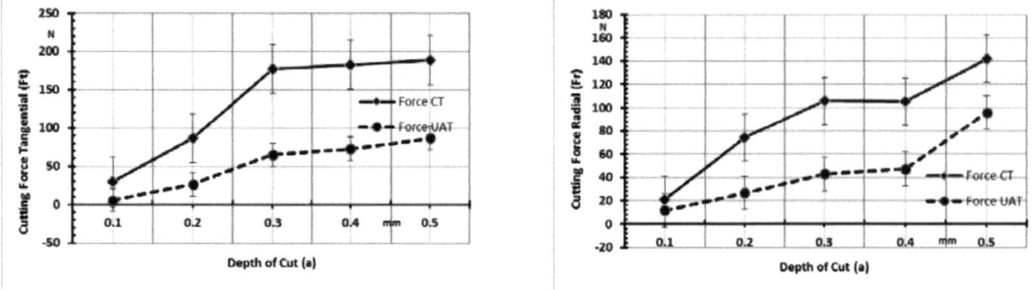

(a) **(b)**

Figure 7. Effect of depth of cut (a) on tangential cutting force (F_t) (a) and radial cutting force (F_r) (b) for turning of SS under CT and UAT (v = 35 m/min, f = 0.1 mm/rev)

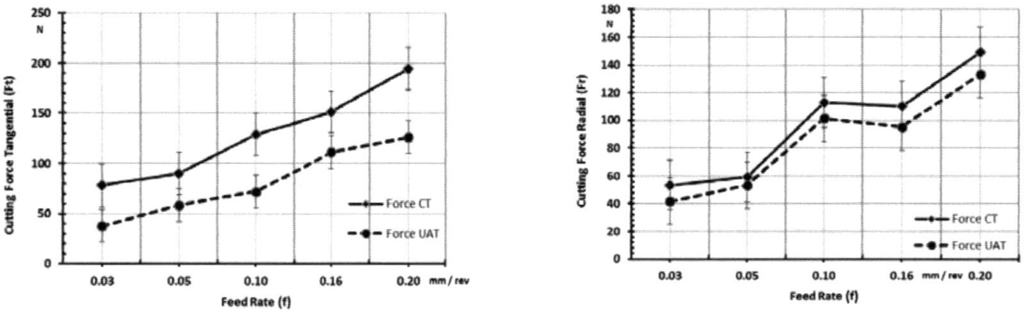

(a)

(b)

Figure 8. Effect of feed rate (*f*) on tangential cutting force (F_t) (a) and radial cutting force (F_r) (b) for turning of SS under CT and UAT (v = 45 m/min, a = 0.3 mm)

As shown in figures 9-11 the average roughness (R_a) and (S_a) in case of UAT is lower than that obtained in case of CT; it means that the vibration-assistance turning causes improved surface quality. At cutting speed v = 30 m/min, the surface roughness R_a and S_a decreased from 0.1301 μm and 0.9860 μm in CT to 0.1185 μm and 0.9140 μm in UAT, (by 8.9% and 7.3%, respectively). At depth of cut a = 0.4 mm, these parameters decreased from 0.6020 μm and 0.5360 μm in CT to 0.3330 μm and 0.4010 μm in UAT, which means that the R_a decreased by 44.6% while S_a decreased by 25.1%. At feed rate f= 0.1 mm/rev, R_a and S_a decreased from 0.1700 μm and 0.8890 μm in CT to 0.1570 μm and 0.7910 μm in UAT (by 7.6% and 11%).

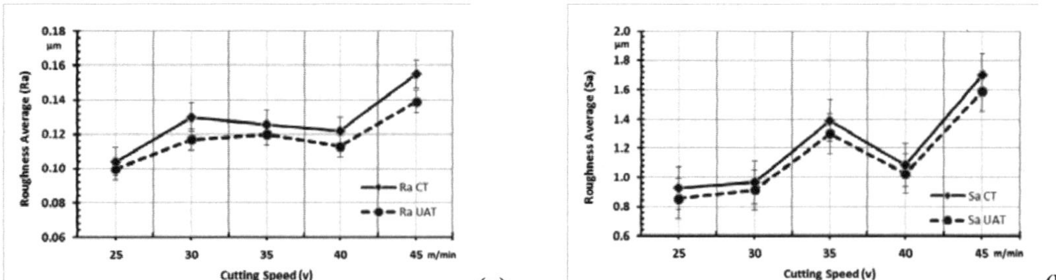

(a) **(b)**

Figure 9. Effect of cutting speed (*v*) on average roughness (R_a) (a) and average roughness (S_a) (b) for turning of SS under CT and UAT (f = 0.2 mm/rev, a = 0.3 mm)

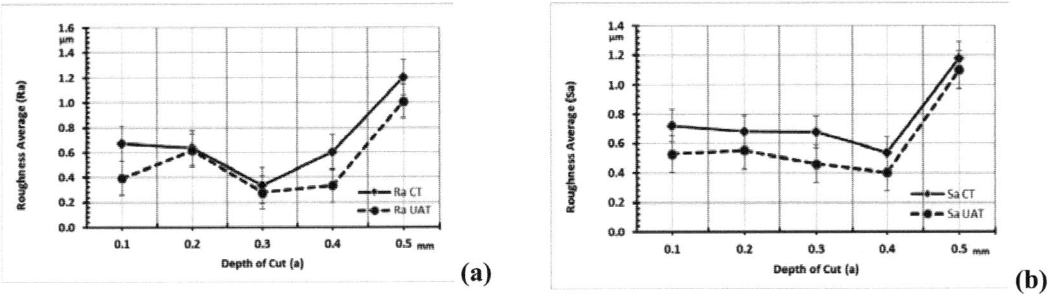

Figure 10. Effect of depth of cut (a) on the average roughness (R_a) (a) and the average roughness (S_a) (b) for turning of SS work piece under CT and UAT (v = 35 m/min, f = 0.1 mm/rev)

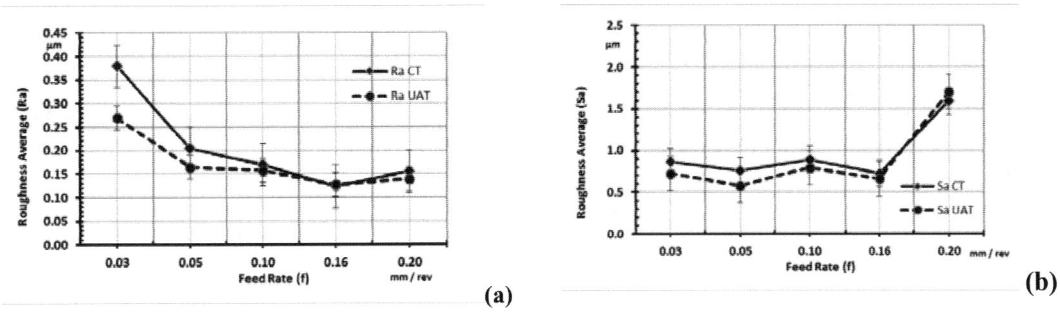

Figure 11. Effect of feed rate (f) on average roughness (R_a) (a) and average roughness (S_a) (b) for turning of SS under CT and UAT (v = 45 m/min, a = 0.3 mm)

Machining accuracy measurements

As for the machining accuracy, it was obtained by the out of roundness measurements on the CMM machine. As shown in figures 12-14, the out of roundness (R_o) in case of UAT is lower than that obtained in case of CT. It means that the use of UAT results in improved system stability as forces decrease, affecting in their turn the accuracy of the produced surface. From figure 12 at a cutting speed v = 35 m/min as example, the out of roundness (R_o) decreased from 0.0230 µm in CT to 0.0134 µm in UAT, which means a decrease by 41.7%. At depth of cut a = 0.4 mm (figure 13), R_o decreased from 0.0227 µm in CT to 0.0068 µm in UAT, decreased i.e. by 70%. At feed rate f = 0.16 mm/rev (figure 14), the out of roundness decreased from 0.0197 µm to 0.0138 µm (by 29.9%).

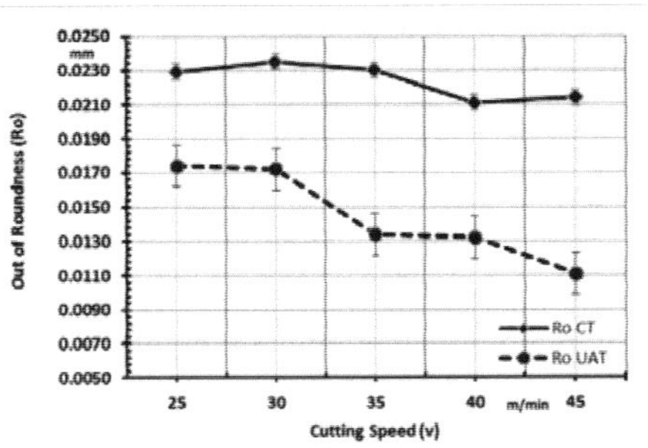

Figure 12. Effect of cutting speed (v) on out of roundness (R_o) for turning SS under CT and UAT (f = 0.2 mm/rev, a = 0.3 mm)

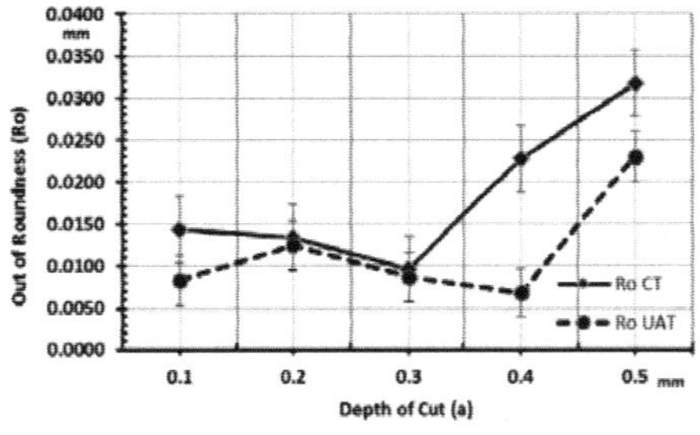

Figure 13. Effect of depth of cut (a) on out of roundness (R_o) for turning SS under CT and UAT (v = 35 m/min, f = 0.1 mm/rev)

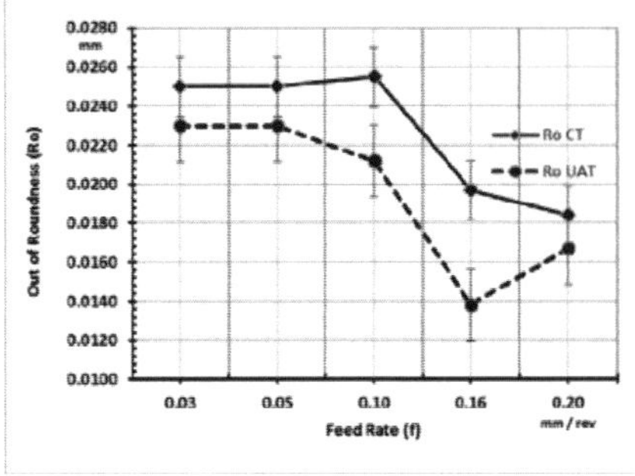

Figure 14. Effect of feed rate (*f*) on the out of roundness (R_o) for turning of SS under CT and UAT (v = 45 m/min, DOC a = 0.3 mm)

3.2 Cutting of brass alloy material

Force measurements

For brass alloy material the increase of the cutting conditions (v, a, and f) causes the increase of F_t and F_r components for both cases of CT and UAT as stainless steel alloys as shown in figures 15-17. It can be noticed that in case of UAT the cutting forces is lower than that in case of CT. At cutting speed v= 45 m/min as example for comparison between CT and UAT. It can be found that F_t decreased by 61.5% while F_r decreased by 80.3% while at depth of cut a = 0.3 mm the cutting forces F_t decreased by 85.2% while F_r decreased by 31.9%.

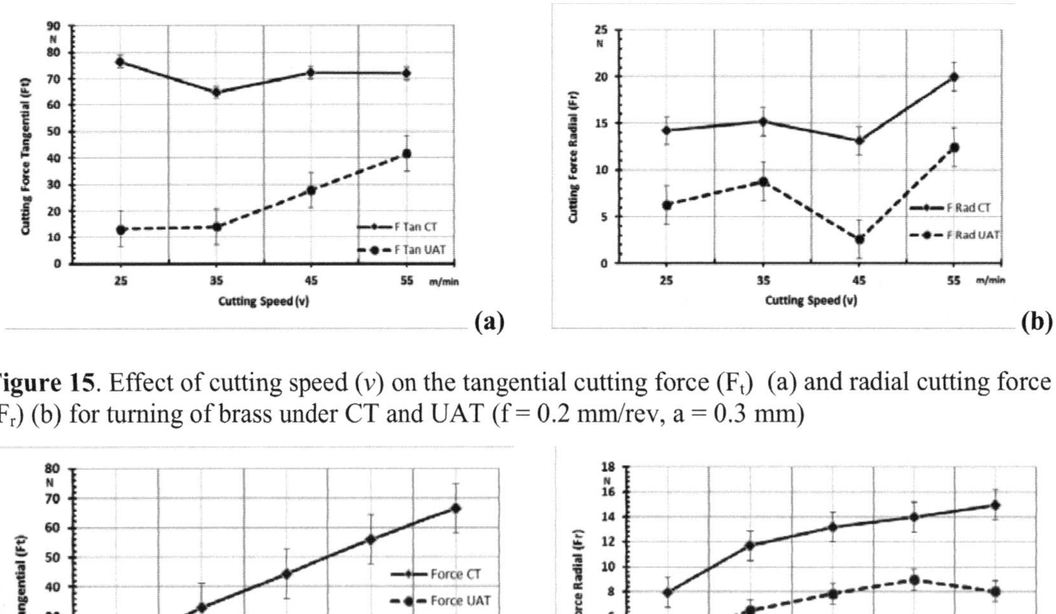

Figure 15. Effect of cutting speed (*v*) on the tangential cutting force (F_t) (a) and radial cutting force (F_r) (b) for turning of brass under CT and UAT (f = 0.2 mm/rev, a = 0.3 mm)

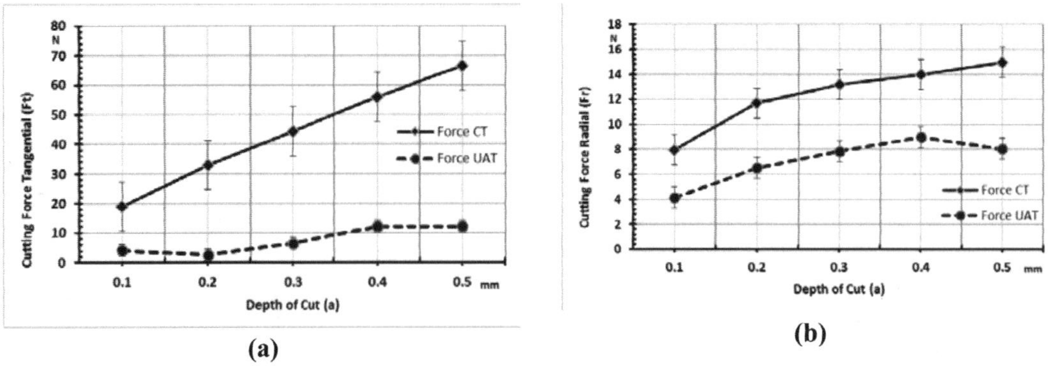

Figure 16. Effect of depth of cut (*a*) on the tangential cutting force (F_t) (a) and radial cutting force (F_r) (b) for turning of brass under CT and UAT (v = 35 m/min, f = 0.1 mm/rev)

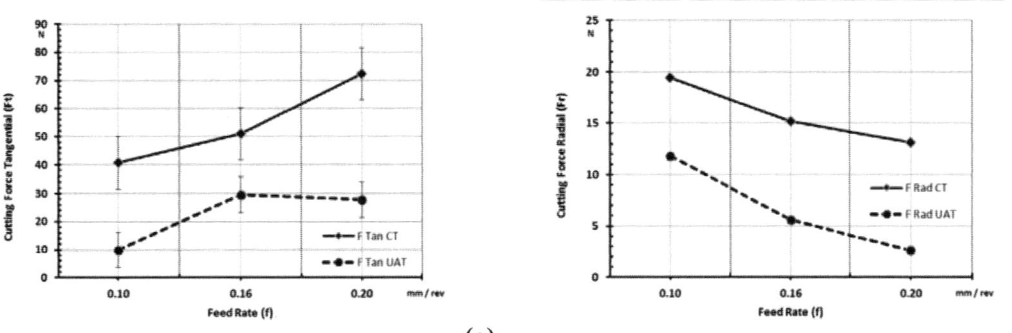

(a) (b)

Figure 17. Effect of feed rate (*f*) on the tangential cutting force (F_t) (a) and radial cutting force (F_r) (b) for turning of brass under CT and UAT (v = 45 m/min, a = 0.3 mm)

As can be detected from figures *18-20.* UAT causes improvements of surface quality where R_a and S_a at v = 45 m/min decreased by 23.1% and 17.5% respectively. While at a = 0.4 mm, R_a and S_a decreased by 33.6% and 9.6%. But for f = 0.2 mm/rev causes the decrease of R_a and S_a by 23.1% and 17.5%.

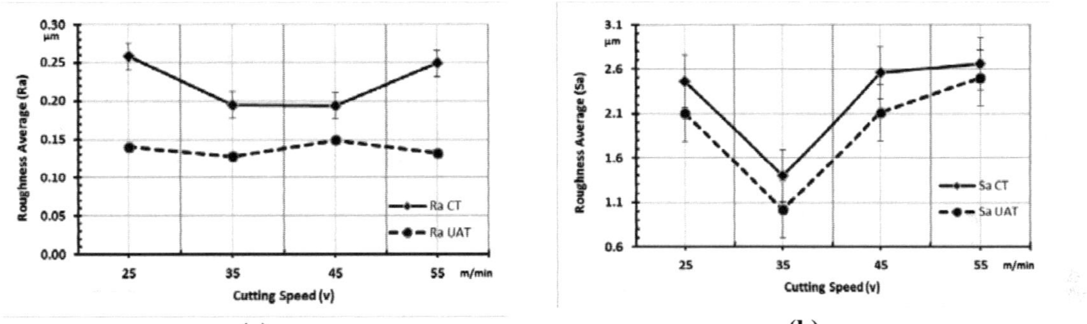

(a) (b)

Figure 18. Effect of cutting speed (*v*) on the average roughness (R_a) (a) and average roughness (S_a) (b) for turning of brass under CT and UAT (f = 0.2 mm/rev, a = 0.3 mm)

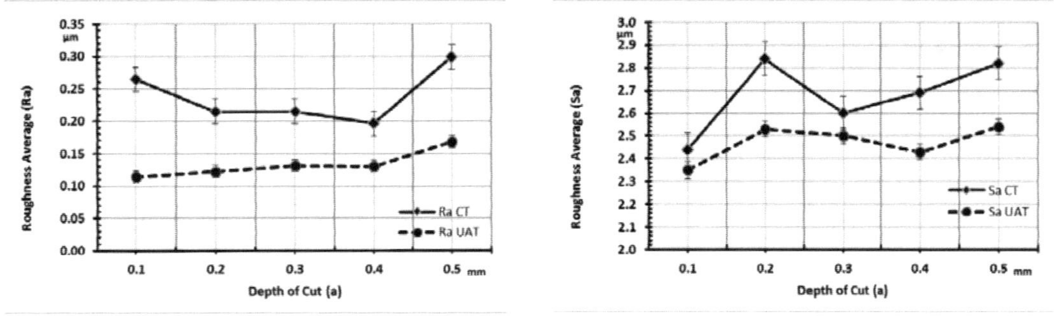

(a) (b)

Figure 19. Effect of depth of cut (*a*) on the average roughness (R_a) (a) and average roughness (S_a) (b) for turning of brass under CT and UAT (v = 35 m/min, f = 0.1 mm/rev)

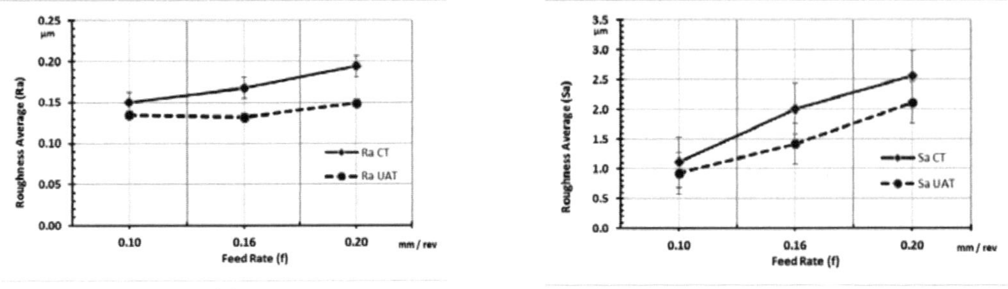

(a) (b)

Figure 20. Effect of feed rate (*f*) on average roughness (R_a) and average roughness (S_a) for turning of brass under CT and UAT (v = 45 m/min, a = 0.3 mm)

It can be deduced that UAT causes also the decrease of out of roundness as a measure of machining accuracy for brass material as in figures 21-23; at v = 45 m/min, R_o decreased by 30.6%, while at a = 0.3 mm, R_o decreased by 30.4%, but at f = 0.1 mm/rev, R_o decreased by 44.3%.

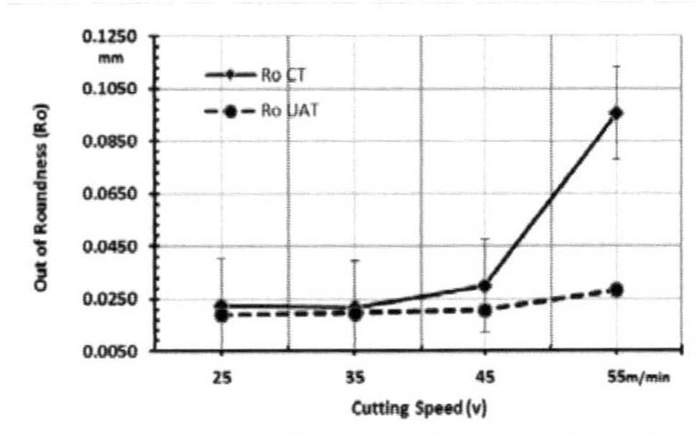

Figure 21. Effect of cutting speed (*v*) on out of roundness (R_o) for turning brass under CT and UAT (f = 0.2 mm/rev, a = 0.3 mm)

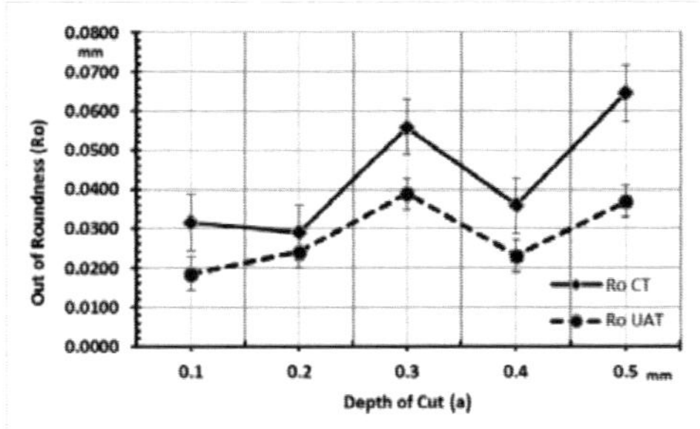

Figure 22. Effect of depth of cut (*a*) on out of roundness (R_o) for turning brass under CT and UAT (v = 35 m/min, f = 0.1 mm/rev)

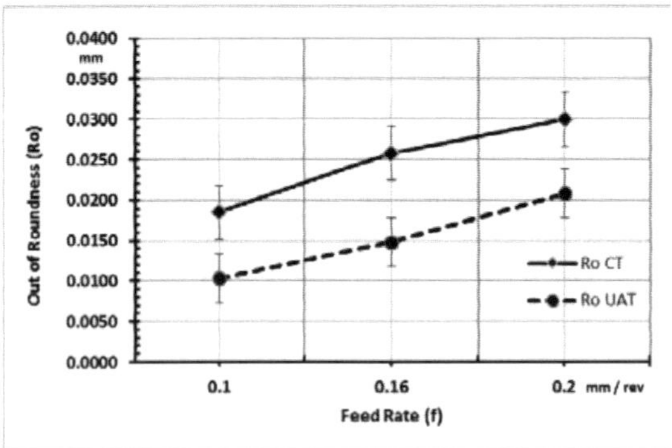

Figure 23. Effect of feed rate (*f*) on the out of roundness (R_o) for turning of brass under CT and UAT (v = 45 m/min, a = 0.3 mm)

3.3. Empirical relationships

For machining the SS work piece some empirical formulas as in equations 1-4 can be deduced by curve fitting of the obtained experimental data, for CT to deduce the tangential F_t and radial F_r cutting force components as follows:

$$F_t = 2461.6 \times v^{-0.1482} \times a^{1.2015} \times f^{0.4636} \quad (1)$$

$$F_r = 42.59 \times v^{0.9323} \times a^{1.1398} \times f^{0.541} \quad (2)$$

The respective expressions for UAT have the following form:

$$F_t = 2737.1 \times v^{-0.1629} \times a^{1.7271} \times f^{0.6072} \quad (3)$$

$$F_r = 3.365 \times v^{1.5305} \times a^{1.2054} \times f^{0.5848} \quad (4)$$

Here (*v*) is the cutting speed in m/min, (*a*) the depth of cut in mm, and (*f*) is the feed rate mm/rev, where the above formula is valid in the cutting conditions ranges obtained in the present research. A comparison of the calculated values using the above formula with the obtained experimental data is shown in figures 24-25.

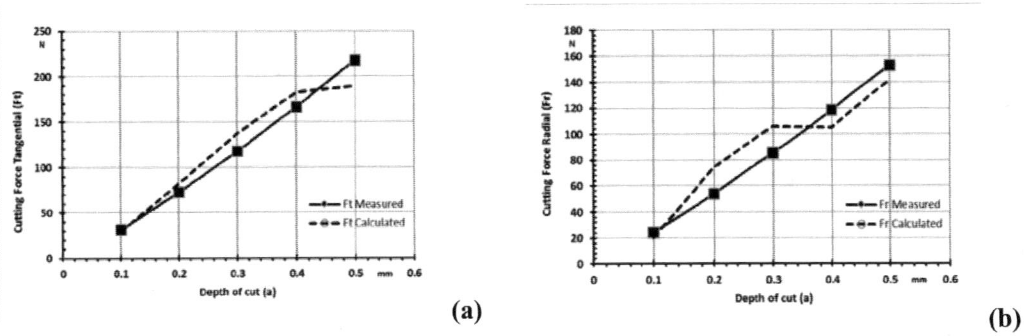

Figure 24. Calculated and measured tangential (F_t) (a) and radial (F_r) (b) force components under CT for SS specimen

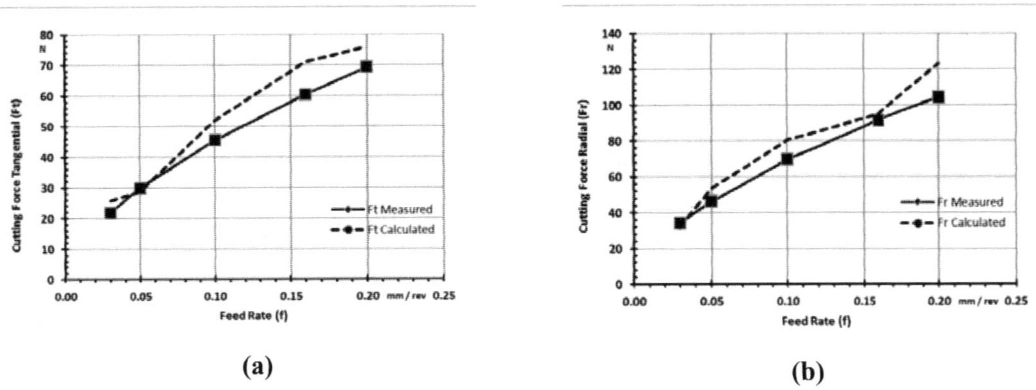

Figure 25. Calculated and measured tangential force(F_t) **(a)** and radial (F_r) **(b)** components under UAT for SS specimens

4. Conclusions

Based on the undertaken experimental analysis of two turning techniques, it can be concluded that:

a. The tangential and radial components of the cutting force in case of UAT are lower than those in

CT for both studied materials: stainless steel and brass.

b. The average surface roughness in case of UAT is lower than that in case of CT in both materials.

c. The out of roundness as a measure of machining accuracy in case of UAT is lower than that in case of CT in both materials.

d. Empirical relationships were suggested for the tangential and radial components under UAT and CT for stainless steel material.

References

[1] Nath C and Rahman M 2008 Effect of machining parameters in ultrasonic vibration cutting. *Int. J. Machine Tools Manuf.* **48**(9) 965-974

[2] Pujana J, Rivero A, Celaya A and López de Lacalle L 2009 Analysis of ultrasonic-assisted drilling of Ti6Al4V *Int. J. Machine Tools Manuf.* **49**(6) 500-508

[3] Nath C, Rahman M and Neo K S 2009 Machinability study of tungsten carbide using PCD tools under ultrasonic elliptical vibration cutting *Int. J. Machine Tools Manuf.* **49**(14) 1089-1095

[4] Shamoto E, Suzuki N and Hino R 2008 Analysis of 3D elliptical vibration cutting with thin shear plane model *CIRP Annals – Manuf. Technol.* **57**(1) 57-60

[5] Mitrofanov A, Ahmed N, Babitsky V and Silberschmidt V V 2005 Effect of lubrication and cutting parameters on ultrasonically assisted turning of Inconel 718 *J. Mater. Process. Technol.* **162–163** 649-654

[6] Voronina S and Babitsky V 2008 Autoresonant control strategies of loaded ultrasonic transducer for machining applications J. Sound Vibrat. **313**(3–5) 395-417

[7] Amini S, Soleimanimehr, H, Nategh M, Abudollah A and Sadeghi M 2008 FEM analysis of ultrasonic-vibration-assisted turning and the vibratory tool *J. Mater. Process. Technol.* **201**(1–3) 43-47

[8] Nath C, Rahman M and Neo K S 2009 A study on the effect of tool nose radius in ultrasonic elliptical vibration cutting of tungsten carbide *J. Mater. Process. Technol.* **209**(17) 5830-5836

[9] Mitrofanov A, Babitsky V and Silberschmidt V V 2003 Finite element simulations of ultrasonically assisted turning *Comput. Mater. Sci.* **28**(3-4) 645-653

[10] Li X and Zhang D 2006 Ultrasonic elliptical vibration transducer driven by single actuator and its application in precision cutting Journal of Materials Processing Technology, **180**(1-3) 91-95

[11] Brehl D and Dow T 2008 Review of vibration-assisted machining Precision Engng **32**(3) 153-172

[12] Babitsky V, Kalashnikov A, Meadows A and Wijesundara A 2003 Ultrasonically assisted turning of aviation materials *J. Mater. Process. Technol.* **132**(1-3) 157-167

[13] Babitsky V, Mitrofanov A, and Silberschmidt V V 2004 Ultrasonically assisted turning of aviation materials: simulations and experimental study *Ultrasonics.* **42**(1-9) 81-86

[14] Moriwaki T and Shamoto E 1995 Ultrasonic elliptical vibration cutting. *CIRP Annals – Manuf. Technol.* **44**(1) 31-34

[15] Shamoto E, Suzuki N, Moriwaki T and Naoi Y 2002 Development of ultrasonic elliptical vibration controller for elliptical vibration cutting. *CIRP Annals – Manuf. Technol.* **51**(1) 327-330

[16] *SECO Group, 2012, SECO DNMG insert, SECO tools media catalogue* p 19, Retrieved September 2012, www.secotools.com/gb

[17] S.R.C. Effect of ultrasonic vibration on the turning process 1969 *Int. J. Machin. Tool Design Res.* **9**(4) 363-374

Vibration-assisted machining of single crystal

S A Zahedi, A Roy and VV Silberschmidt
Wolfson School of Mechanical and Manufacturing Engineering, Loughborough University, LE11 3TU, UK

E-mail: S.Zahedi@lboro.ac.uk

Abstract. Vibration-assisted machining offers a solution to expanding needs for improved machining, especially where accuracy and precision are of importance, such as in micromachining of single crystals of metals and alloys. Crystallographic anisotropy plays a crucial role in determining on overall response to machining. In this study, we intend to address the matter of ultra-precision machining of material at the micron scale using computational modelling. A hybrid modelling approach is implemented that combines two discrete schemes: smoothed particle hydrodynamics and continuum finite elements. The model is implemented in a commercial software ABAQUS/Explicit employing a user-defined subroutine (VUMAT) and used to elucidate the effect of crystallographic anisotropy on a response of face centred cubic (f.c.c.) metals to machining.

1. Introduction

Vibration-assisted machining (VAM) combines precision machining with small-amplitude tool vibration to improve the machining process, while in conventional machining a tool generally moves towards the workpiece with a constant velocity. VAM can potentially offer improved machinability for difficult-to-cut materials [1]. Single crystals are extensively used in the production of electric, piezoelectric, semiconductors, magnetic and optical components for high-performance systems such as transducers, resonators, actuators and sensors [2].

A cutting force is one of the critical governing parameters defining efficiency of the machining process. In micro-machining, a tool-workpiece interaction occurs within either a single grain of the workpiece material or a few grains. Thus, crystallographic orientation of the workpiece material significantly influences the machining characteristics. The vibro-impact interaction between the tool and workpiece in VAM during the chip formation leads to a dynamically changing cutting force in the process zone [3-5].

In the last few decades, the finite element (FE) technique has been extensively used to gain insight into the underlying mechanisms that drive a plastic response in high-deformation problems. Smooth Particle Hydrodynamics (SPH) [6,7] has several advantages compared to FE approaches, especially in the study of large-deformation processes.

In this study, we intend to address the matter of ultra-precision machining of a material at the micron scale using computational modelling. A hybrid modelling approach is suggested for ultrasonically assisted-micro machining of crystalline materials that combines two discrete schemes: smoothed particle hydrodynamics and continuum finite elements. The model is implemented in a commercial software ABAQUS/Explicit using a user-defined subroutine (VUMAT) and used to elucidate the effect of crystallographic anisotropy on a response of face centred cubic (f.c.c.) metals to machining. The current model enables us to study the influence of crystal orientation on the process variables.

2. Crystal plasticity

The crystal-plasticity (CP) framework used for description of a deformation behaviour of a single-crystal of copper is summarized below. The evolution of plastic strain rate is related to the slipping rate $\dot{\gamma}^{\alpha}$ in a slip system α :

$$\dot{\varepsilon}_{ij}^{p} = \sum_{\alpha=1}^{N} \mu_{ij}^{\alpha} \dot{\gamma}^{\alpha} , \tag{1}$$

where N is the total number of available slip systems, μ_{ij}^{α} is the Schmid tensor and equals to a dyadic product of the slip direction s_i^{α} and the slip plane normal n_j^{α} in the initial unloaded configuration. The plastic slip rate in a slip system α was presented in the form of a viscoplastic power-law expression:

$$\dot{\gamma}^{\alpha} = \dot{\gamma}_0{}^{\alpha} \operatorname{sgn}(\tau^{\alpha}) \left| \frac{\tau^{\alpha}}{g^{\alpha}} \right|^{n} . \tag{2}$$

Here, $\dot{\gamma}_o$ is the reference strain rate, τ^{α} is the resolved shear stress on the slip system α , obtained with the Schmid law ($\tau^{\alpha} = \sigma_{ij}\mu_{ij}^{\alpha}$), n is the material's rate sensitivity, g^{α} is the strength of the slip system α that characterizes the strain hardening, and sgn(∗) is the signum function of ∗. The strength of a slip system β with shear starin on it is represented by

$$g^{\alpha} = \sum_{\beta=1}^{N} h_{\alpha\beta} \gamma^{\beta} . \tag{3}$$

The hardening moduli $h_{\alpha\beta}$ in (3) are evaluated using the hardening model as follows:

$$\begin{aligned} h_{\alpha\alpha} &= h(\gamma) = h_0 \sec h^2 \left| \frac{h_0 \gamma}{g_T^{\alpha}\big|_{sat} - g_T^{\alpha}\big|_{t=0}} \right|, \\ h_{\alpha\beta} &= q h_{\alpha\alpha} (\alpha \neq \beta), \gamma = \sum_{\alpha} \int_0^t \left| \dot{\gamma}^{\alpha} \right| dt \end{aligned} \tag{4}$$

where h_0 is the initial hardening parameter, q is the latent hardening ratio and assumed to be 1, γ is the Taylor cumulative shear strain on all slip systems and $g_T^{\alpha}\big|_{t=0}$ and $g_T^{\alpha}\big|_{sat}$ are the shear stresses at the onset of yield and the saturation of hardening, respectively.

In a f.c.c. single crystal of copper, slips may occur on 12 individual slip systems, composed of three directions [110] on four distinct planes (111). Material parameters used in the FE simulations are listed in Table 1, where each slip system assumed to have the same values of plastic parameters.

Table 1. Material parameters of single-crystal copper [8]

Elastic parameters	Parameters of slipping rate	Hardening parameters
$C_{11} = 168$ GPa	$\gamma_0 = 0.00001$ s^{-1}	$h_0 = 180$ MPa
$C_{12} = 121.4$ GPa	$n = 20$	$\tau_s = 148$ MPa
$C_{44} = 75.4$ GPa		$\tau_0 = 16$ MPa

3. FE model description

A 3D domain with dimensions of 250 μm × 100 μm × 10 μm was selected as an appropriate representation of the workpiece material. The workpiece was divided into two regions, one representing the SPH domain (250 μm × 50 μm × 10 μm) and the remaining part being a continuum

FE domain (Figure 1). The continuum domain was meshed with eight-node brick elements with reduced integration (C3D8R) while the PC3D element type was used to represent the SPH area. In the continuum FE domain, a mesh with a minimum element size of 1.25 μm was sufficient to characterize accurately the machining process. A fixed boundary condition was imposed on the bottom and left-hand face of the workpiece. A cutting tool was modelled as a rigid body; its rake angle was 0° and a clearance angle 5°. A depth of cut was set at 5 μm, and the tool moved in the negative x direction. The constant cutting speed of 1200 m/s was used in our simulation for conventional machining, and a varation of velocity with time in vibration-assisted machining is shown in Figure 2. The workpiece was modelled as a deformable body. A frictionless condition was used to model the interaction between the tool and the workpiece. The crystal-plasticity model was implemented in a FE software package ABAQUS/Explicit using the user-defined material subroutine VUMAT.

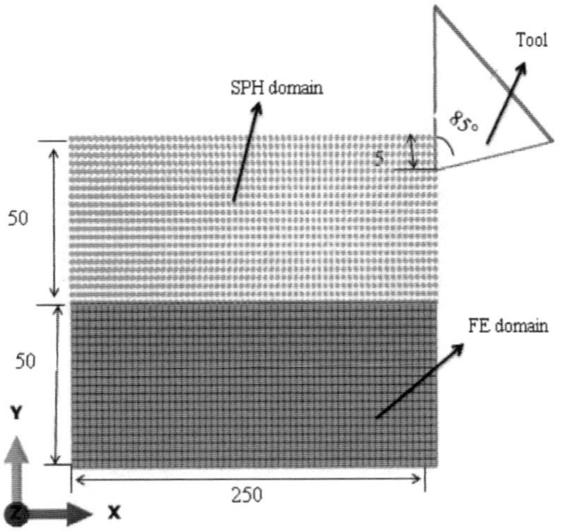

Figure 1. 3D finite-element model of micro-machining (dimensions in μm)

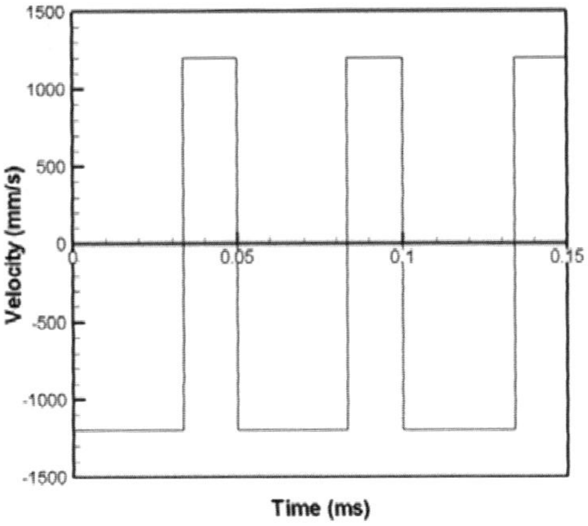

Figure 2. Vartiaion of velocity in vibration assisted machining

4. Results and discussion

The 3D finite-element model developed allows us to calculate the effect of cutting direction and crystallographic orientation on the cutting forces. To study it, two cutting directions for two crystal orientations were chosen. In our analysis, a planar configuration was used, in which the tool cuts along a particular crystal direction [hkl] on a chosen crystal plane [abc] (Figure 3). All numerical (finite-element) simulations were performed for vibration-assisted machining and convention machining with identical cutting parameters with the crystalline model for different orientations. In our simulations frequency of tool vibration was 15 kHz and vibration amplitude 0.01 mm.

The cutting force for vibration-assisted machining and convention machining is shown in Figure 4 for 0° direction for two crystal's orientation planes. In simulations of conventional machining of the crystalline workpiece the force acting on the cutting tool remains practically unchanged, as expected, whilst in vibration-assisted machining the force changes in each cycle. The cutting force grows steadily from the moment of the first contact between the vibrating tool and formed chip until it reaches the maximum level at the point of maximum penetration. The force magnitude then starts to

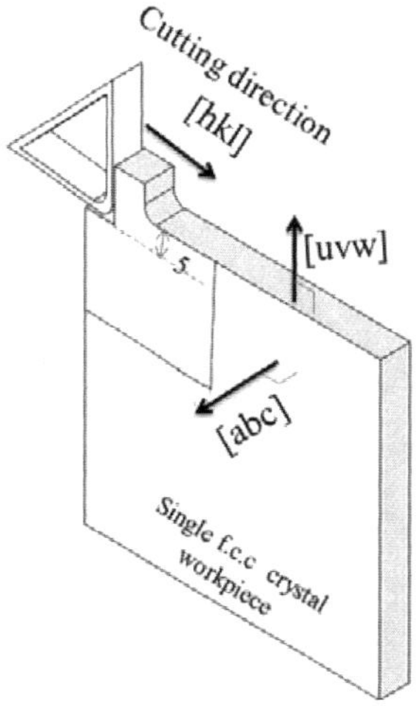

Figure 3. Orientations in modelling of machining

decline at the unloading stage until it vanishes when the tool separates from the workpiece at the process repeats with the next cycle of ultrasonic vibration.

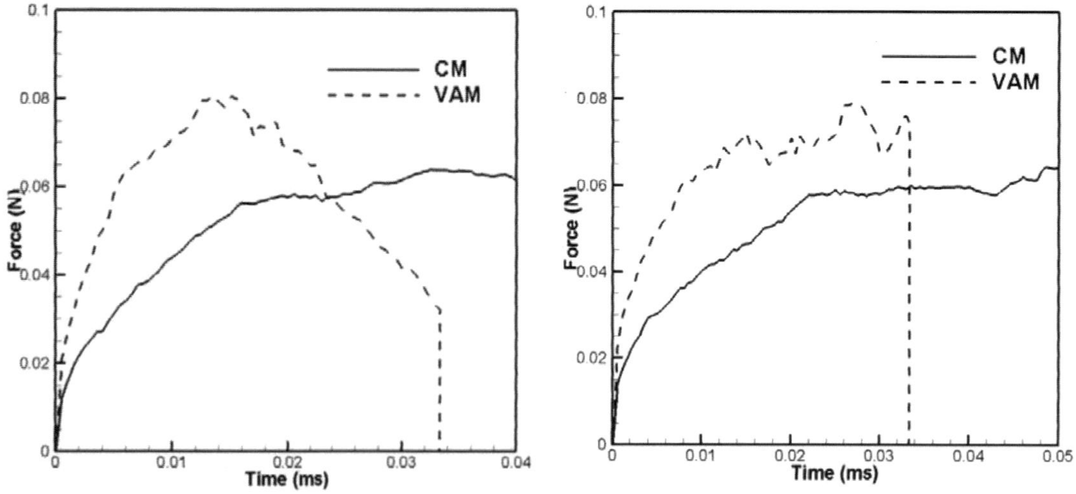

(a) (b)

Figure 4. Cutting force in conventional machining and vibration-assisted machining for (100) (a) and (101) (b) (frequency 15k Hz and amplitude 0.01mm)

The averaged value of cutting force over the vibration cycle in vibration assisted machining model for each crystalline plane was calculated. For the (100) plane, the cutting-force reduction from the level of the conventional machining is 35% and for the (101) plane it reaches 42%. This reduction in the cutting force would result in the increases in the tool life, improved surface finish and roundness of machined workpieces, it can also allow for a potential increase in the material removal rate [9].

5. Conclusion

In the present work, a 3D hybrid finite-element model of micro-machining of single-crystal copper was developed for vibration-assisted machining and conventional machining to study the level of cutting forces for machining in different orientations. The described model is a viable and numerically robust way for solving large-deformation problems. A noticeable decline in the average cutting forces was obtained for vibration-assisted machining in comparison to the conventional technique. This reduction is valid for different crystalline orientations.

Refrences

[1] Shamoto E and Moriwaki T 1999 Ultraprecision diamond cutting of hardened steel by applying elliptical vibration cutting *CIRP Ann.* **48** 441–44
[2] Rai-Choudhury P *MEMS and MOEMS Technology and Applications* 2000
[3] Babitsky V, Mitrofanov A and Silberschmidt V V 2004 Ultrasonically-assisted turning of aviation materials: simulations and experimental study *Ultrasound* **42** 81–6
[4] Ma C, Shamoto E and Moriwaki T Study on the thrust cutting force in elliptical vibration cutting. 2004 *Mat. Sci. Forum* 396-400
[5] Negishi N *Elliptical vibration-assisted machining with single crystal diamond tools* 2003 MS thesis. North Carolina State University
[6] Monaghan J J Smoothed particle hydrodynamics 2005, *Rep. on Prog. in Phy.* 68 **8** 1703-59
[7] Zahedi S A, Li S, Roy A, Babitsky V and Silberschmidt V V 2012 Application of Smooth-Particle Hydrodynamics in metal machining *J. of Phys.*: Conference Series **382** (1)
[8] Wang Y, Raabe D and Kluber C 2004 Orientation dependence of nanoindentation pile-up patterns and of nanoindentation microtextures in copper single crystals *Act. Mate.* **52** 2229–38
[9] Muhammad R, Ahmed N, Roy A and Silberschmidt V V 2012 Numerical modelling of vibration-assisted turning of Ti-15333 *Proc. CIRP* **1** 347–52

D2FAM 2013 IOP Publishing
Journal of Physics: Conference Series **451** (2013) 012039 doi:10.1088/1742-6596/451/1/012039

An investigation into magnetic electrolytic abrasive turning

M A M Mahdy[1,3], A L Ismaeial[2] and F F Aly[2]

[1]Design and production engineering department, faculty of engineering, Ain Shams University, Cairo, Egypt
[2]Reactor department, nuclear research center, atomic energy authority, cairo, EGYPT.

[3] E-mail: mahdymohsen@hotmail.com

Abstract. The magnetic electrolytic abrasive turning (MEAT) process as a non-traditional machining is used to obtain surface finishing like mirror. MEAT provides one of the best alternatives for producing complex shapes with good finish in advanced materials used in aircraft and aerospace industries. The improvement of machining accuracy of MEAT continues to be a major challenge for modern industry. MEAT is a hybrid machining which combines two or more processes to remove material. The present research focuses on the development of precision electrochemical turning (ECT) under the effects of magnetic field and abrasives. The effect of magnetic flux density, electrochemical conditions and abrasive parameters on finishing efficiency and surface roughness are investigated. An empirical relationship is deduced.

1. Introduction

The electrolytic magnetic abrasive turning (EMAT) process is used for metal removal without burrs from complex shapes and gives supper finishing surface for hard and difficult to machine workpieces. Many previous studies have reported that EMAT is a very efficient finishing method for inner and outer finishing of tubes, the edge finishing and the finishing of flat surface [1, 2]. Many researchers have reported their experimental results and confirmed the high efficiency of the magnetic finishing methods for various materials and shapes of workpiece [3, 4]. The magnetic field can accelerate the mobility of the charged particles, accelerate the electrochemical reaction and reduce the surface roughness of the workpiece more quickly, which improves the lapping efficiency [3]. Electrolytic magnetic abrasive finishing is a compound process, involving magnetic field, abrasives and an electrolyte for finishing workpieces. The electrolytic process and abrasive pad is the much easier removal of the produce passive film (or oxide film layer on the original metal surface during electrolytic processing). Moreover, in the presence of both electric and magnetic field, the negatively charged ions move toward the anode surface along a cycloid curve by the action of the Lorentz force. Under appropriate operating conditions, this phenomenon build electrolytic effects, which produces increase in finishing efficiency, yielding a superior surface and excellent finishing, better than those obtained by magnetic abrasive finishing only [5].

During abrasive electrolytic process, electrochemical first produces a passive film on the work surface, the passive film was then immediately removed using an abrasive pad, which was made of a nonwoven cloth that contained abrasives. But magnetic electrolytic abrasive process can produce a fine surface finish due to the effect of magnetic field [6, 7]. However, many operating conditions with

[3] To whom any correspondence should be addressed

wide ranges are applicable to magnetic electrolytic abrasive process within the range, short circuits will not occur in electrolysis. The finishing efficiency of magnetic electrolytic abrasive finishing is much higher than that of magnetic abrasive finishing only because of the contributions of both the electrolytic process and the effects of the magnetic field. Hence, the magnetic electrolytic abrasive finishing is a process that matches the current demand closely and can be used to accelerate the removal of material and simultaneously obtain a superior finishing surface.

Finally magnetic electrolyte abrasive finishing has better finishing characteristics than magnetic abrasive finishing, especially at a high electrolytic current. The parameters of electrode gap, magnetic flux density and electrolytic current must be appropriately fitted to produce passive film quickly. The softening of the passive film makes it easier to remove from the original metal surface [8]. The rate of workpiece revolution must also be matched to remove the passive film rapidly and thereby efficiently obtain a superior surface finish. A $NaNO_3$ electrolyte with concentration of 20 % with pad of Al_2O_3 (mesh 8000) are used to produce a good surface quality [5]. A theoretical study of electrochemical, abrasive process and magnetic is presented [9]. The aim of the present paper are the construction of a prototype of the electrolytic magnetic abrasive turning (EMAT) process and the investigation of the effect of the main parameters on the EMAT process. Magnetic fields with electrolytic machining: Magnetic field affects the move path of electrolytic ions, from a linear path to a curved (cycloid) path, and also accelerates the ions. This results in the improvement of finishing efficiency [2, 9]. Therefore in the magnetic abrasive polishing process, surface roughness and metal removal improve at the same time compared with the traditional electrolytic process.

2. Experimental set-up

To carry out the experimental investigation, an electrochemical turning (ECT) system is developed, having the provisions of controlled electrolyte flow and automatic tool electrode feed as well as electrical circuit. The electrical circuitry of the ECT set-up includes a DC power supply with electrical elements for short-circuit prevention, spark detection and operation of the ECT system. Variation of the tool feed rate is obtained with the help of DC motor driving system. Figure 1 shows the parts of the abrasive pad device. The abrasive pad causes the removal of metal layers (oxide film) due to electrochemical operation. The removal of the oxide film causes the renewal of the electrochemical process and tacking sludge from workpiece resulting in good surface finishing. Figure 2 shows a photo of the abrasive device with mechanical vibration.

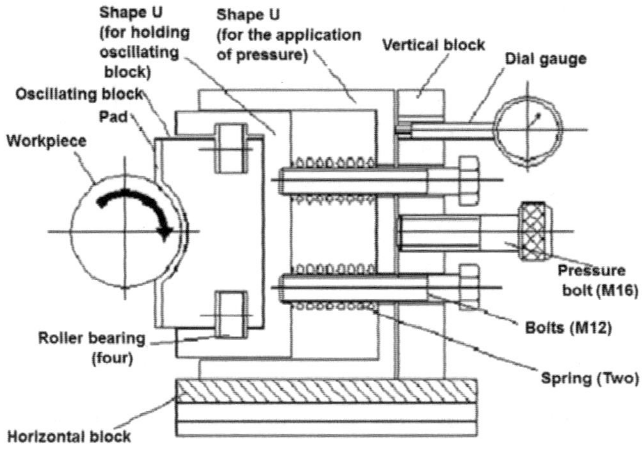

Figure 1. Parts of abrasive pad device

Figure 2. Photo of the abrasive device with mechanical vibration

The specifications of MEAT prototype are : I_{max} = 60 A, v steps = (15, 18, 21, 24, 27, 30volt), tank capacity for electrolyte Q = 60 L, pump motor power = 1.5 HP, swing over bed = 80 mm, distance between center = 300 mm and magnetic field (H = 0.06 Tesla). Figure 3 shows a photo for magneto-electrolytic-abrasive turning (MEAT) process on lathe. While Figure 4 shows the schematic diagram of the MEAT process. During electrolyte supply with magnetic field the magnetic lines causes change of pass of ions movement, which add forces. The viscoelastic abrasive pad is set on the lathe and pressurized to produce the finishing pressure. This pressure removes the passive layer easily and allows the electrolytic process to arise continuously.

Figure 3. Photo for magneto-electrolytic-abrasive turning process on lathe

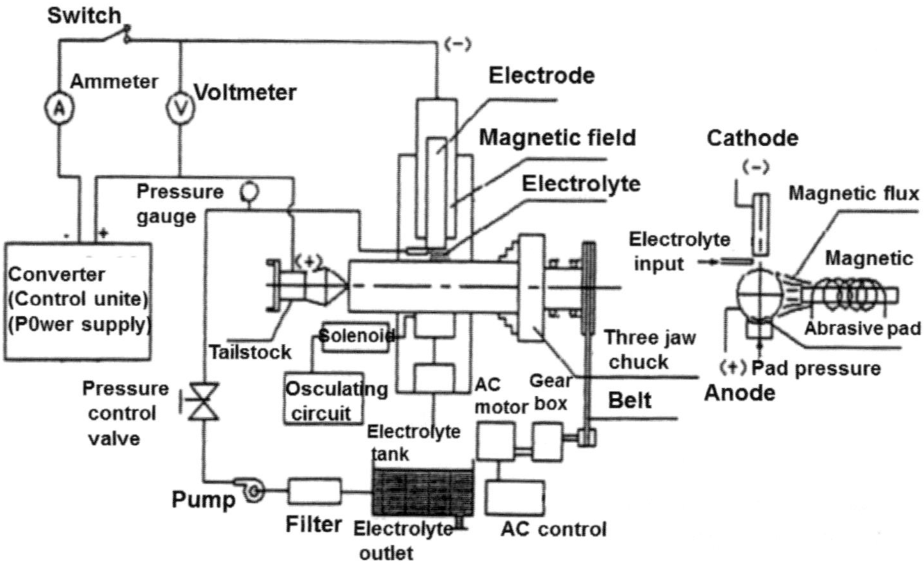

Figure 4. Diagram for the developed magneto-electrolytic abrasive turning (MEAT) process

This was investigated to establish the processes conditions of MEAT for finishing to reduce the surface roughness value to reach high quality product for important industry parts. The MEAT process involved are required to be more precise and efficient to obtain a product whilst maintaining high productivity. A recent trend for quality by the MAET machining is alternate manufacturing process for traditional machining, which improves the surface finishing at low cost and reducing the surface roughness value. The MAET is a high polishing machine with compounded more of processes in simultaneous machining. These hybrid effect are run on simple lathe model, with additional variables: speed power unit, electrolyte system, mechanical device abrasive and magnetic field.

A calibration curve is obtained for the abrasive pad devic which can be used to know the value of the exerted force on the abrasive pad and then the required pressure can be calculated. Surface roughness measurements are taken on the specimen ring workpiece of ∅ 50 x 15 mm. The roughness values recorded are average of seven readings measured along the width of the specimen. The surface roughness value for the machined surface is measured using the average roughness (R_a) in μm and peak to valley height (R_z) in μm.

3. Electrolytic-magnetic -abrasive turning (MEAT)

To investigate the effect of variables of MEAT process on the obtained roughness during the variation of machining time, rotational speed, pad pressure and material of workpiece. Tests were carried out according to the variable levels given in table 1.

Table 1. Process variable levels for MEAT process

MEAT Variables	Variable Levels							Units
	1	2	3	4	5	6	7	
Machining time t_m	4	8	12	16	20	24	28	**min**
Rotational speed n	25	100	225	400	600			**rpm**
Pad pressure	0.5	1	2					**N**
Abrasive mesh	36	100	280					**# mesh**
Magnetic field	0.06							**Tesla**

4. Results and discussion

4.1. Comparison between MEAT and EAT processes

It is required to study the further effect of magnetic field (for magnetic gap G_m = 3 mm) on the obtained surface roughness as another effect in case of EAT process. The EAT process parameters are carred out using SiC as abrasive of mesh 280 concentration 15 % of $NaNO_3$ as electrolyte with 0.3 mm intial gap between the workpiece and stainless steel electrode while the flow rate 8 L/min as well as a voltage difference of 21 volt. The effect of EAT process only on the obtained surface roughness (R_a , R_z) is shown in Figure 5. It was noticed that the average roughness is decrease from R_a = 0.13 µm to R_a = 0.07 µm only, Which means that the intial R_{ai} is decreased by 45 % under the effect of EAT while under the effect of magnetic field as MEAT process a further decrease on surface roughness is occurred. The magnetic field is normal to the electrolyte gap.

The surface roughness value is decrease in the presence of a magnetic flux density compared with the case without the magnetic field. The relationships (t_m – R_a) and (t_m – R_z) during two processes MEAT and EAT is shown in Figure 5. A comparison between EAT and MEAT processes using Na No_3 electrolyte on the obtained surface roughness (R_a, R_z) is shown in Figure 6. Under the effect of magnetic field average roughness is decreased from R_a = 0.13 µm to 0.033 µm, which means that the initial R_{ai} is decreased by 74 % which means that the magnetic field shear in decrease of initial R_{ai} by 29 % over EAT process. It is important to introduce relative improvement scales in the present investigation as the reduction % of the initial roughness. The reduction percent of the initial roughness during MEAT process is the percent ratio between improvement in initial roughness after MEAT process and initial roughness before MEAT process. Therefore the reduction % in initial roughness can be mathematically evaluated as:

$$U_a = \frac{\Delta R_a}{R_{ai}} = \frac{\mathrm{R_{ai} - R_a}}{\mathrm{R_{ai}}} = \mathrm{redu.\% \ in \ R_{ai}} \quad (1)$$

$$U_z = \frac{\Delta R_z}{R_{zi}} = \frac{\mathrm{R_{zi} - R_z}}{\mathrm{R_{zi}}} = \mathrm{redu.\% \ in \ R_{zi}} \quad (2)$$

The effect of the type of processes as ECT or MEAT on both relationships (u_a – t_m) and (u_z – t_m) is shown in Figure 7 and Figure 8. It can be found out that the effect of magnetic field causes the decrease of the initial surface roughness (R_{ai} , R_{zi}) or (U_a , U_z) by (74 % , 65 %) respectively.

4.2. MEAT process for different workpiece materials

The effect of MEAT process for different workpiece materials in case of rotating workpiece on relationships (R_a - t_m) and (R_z - t_m) is shown in Figure 9. A magnetic field created by a magnet is applied on the region of the electrolyte in a normal direction. The magnetic flux is 0.06 Tesla (T), this effect is used for further improvements of surface roughness. A magnetic gap of (G_m = 3 mm) is used between the magnetic pole and the workpiece. Through ECT process with addition of both abrasive pad and magnetic field the increase of (t_m) causes the decrease of peak to valley height (R_z) and average roughness (R_a) of the machined workpiece with $NaNO_3$ electrolyte are shown in Figures 9 and 10 for different workpiece material. It was found out that R_a decrease by the increase of (t_m) for different workpiece materials of St 70, Stainless Steel and Cr-coated steel.

The decrease of roughness is due to the both effects of abrasive pad and magnetic flux density of 0.06 Tesla (T). Where the easier removal of the produced oxide film on the cycloid pass causes more removal rate in peak roughness as shown in Figure 9. During MEAT process using SiC particles of mesh 280 as a pad and a magnetic field of 0.06 T, the surface roughness was decrease sharply in the first 16 min but from 16 to 24 min there was no big different between surface roughness. the workpiece surface by abrasive pad results in the continuity of ECT process. While magnetic field causes addition force on ions and change its moving pass to cycloid pass.

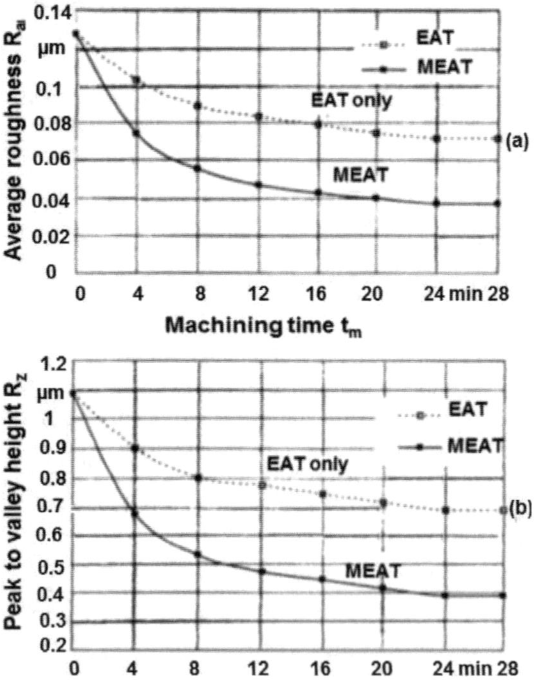

Figure 5. A comparison between EAT process and MEAT process using $NaNO_3$ electrolyte on relationships: (a) (t_m –R_a) (b) (t_m –R_z). Workpiece mat.: Cr-coated steel, Flux density = 0.06 T, abrasive type: SiC, n = 100 rpm, grain size = 280, C = 15% $NaNO_3$, v = 21 volt, Q = 8 L/min, G_i = 0.3 mm and G_m = 3 mm, t_m = 4 min

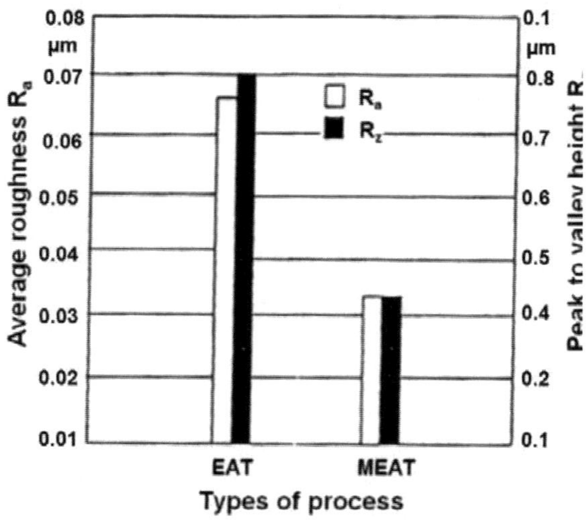

Figure 6. A comparison between EAT process and MEAT process using $NaNO_3$ electrolyte on the obtained surface roughness (R_a , R_z). Workpiece mat.: Cr-coated steel, Flux density = 0.06 T, Abrasive type: SiC, n = 100rpm, grain size = 280, C = 15 % $NaNO_3$, v = 21 volt, Q = 8 L/min, G_i = 0.3 mm and G_m = 3 mm, t_m = 28 min, R_{ai} = 0.13 µm, R_{zi} = 1.1 µm

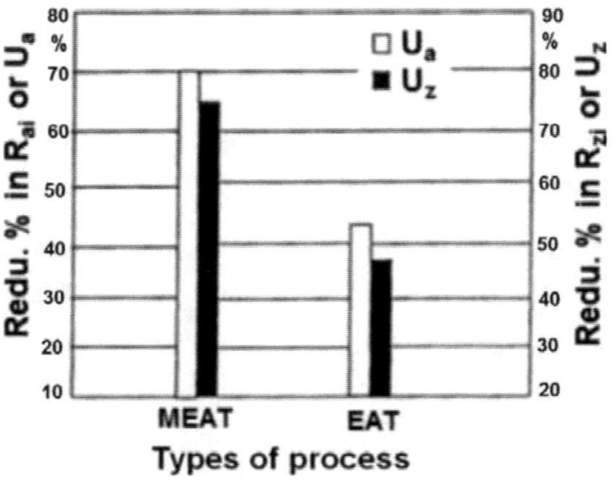

Figure 8. Effect of type of process (ECT or MEAT) on the reduction percent of initial surface roughness (U_a, U_z) using $NaNO_3$ electrolyte. Workpiece mat.: Cr-coated steel, Flux density = 0.06 T, abrasive type: SiC, n = 100 rpm, grain size = 280, C = 15 % $NaNO_3$, v = 21 volt, Q = 8 L/min, G_i = 0.3 mm and G_m = 3 mm, t_m = 28 min, R_{ai} = 0.13 μm, R_{zi} = 1.1 μm

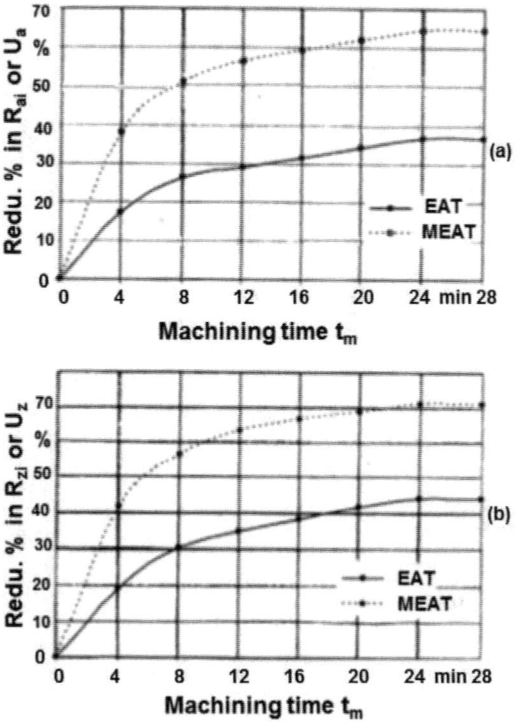

Figure 7. Effect of type of process using $NaNO_3$ electrolyte on relationships:(a)(U_{ai}–t_m) (b) (U_{zi}-t_m). Workpiece mat.: Cr-coated steel, Flux density = 0.06 T, abrasive type: SiC, n = 100 rpm, grain size = 280, C = 15 % $NaNO_3$, v = 21 volt, Q = 8 L/min, G_i = 0.3 mm and G_m =3 mm, t_m = 4min, R_{ai} = 0.13 μm, R_{zi} = 1.1 μm

By MEAT process the final surface roughness of R_a = 0.058 μm and R_z = 0.54 μm was obtained for a workpiece material of Cr-Coated steel while for stainless steel workpiece R_a = 0.064 μm and R_z = 0.6 μm was obtained. For Steel 70 workpiece R_a = 0.072 μm and R_z = 0.67 μm was obtained for the conditions given in Figure 10.

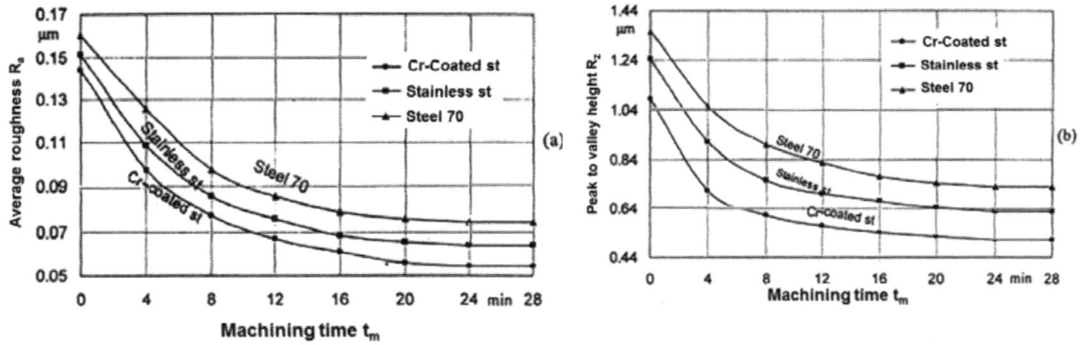

Figure 9. Effect of magnetic electrolytic abrasive turning (MEAT) process for different workpiece materials for $NaNO_3$ electrolyte on the relation: (a) (R_a - t_m) (b) (R_z - t_m)
Tool material: stainless steel, abrasive type: SiC, mesh = 280, n = 100 rpm, v = 21 volt, C = 15% $NaNO_3$, Q = 8 Lit/min, G_i = 0.3 mm, G_m = 3 mm, flux density = 0.06 T

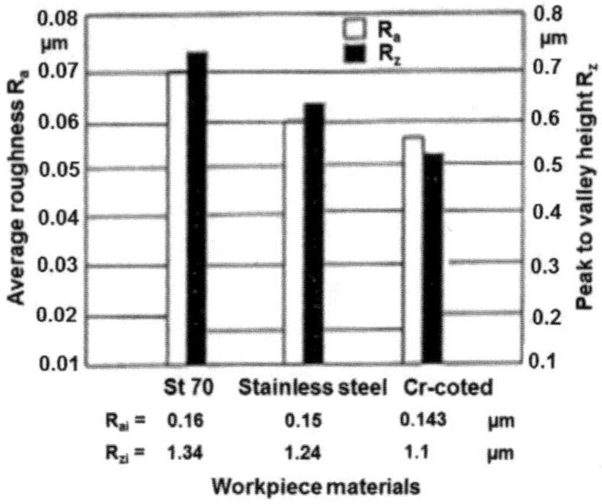

Figure 10. Effect of different workpiece material in case of (MEAT) process using $NaNO_3$ electrolyte on surface roughness (R_a, R_z).
Tool material: stainless steel, Abrasive type: SiC, Grain size: 280, n = 100 rpm, v = 21 volt, C = 15% $NaNO_3$, Q = 8 L/min, G_i = 0.3 mm, G_m = 3 mm, flux density = 0.06 T, t_m = 28 min

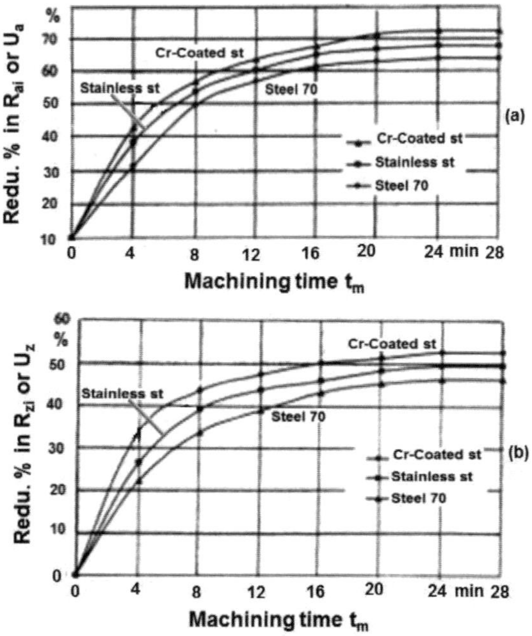

Figure 11. Effect of different workpiece material in case of (MEAT) process in case of rotating workpiece for $NaNO_3$ electrolyte on relationships:

(a) (t_m – U_a) (b) (t_m - U_z)

Tool material: stainless steel, Abrasive type: SiC, Grain size: 280, n = 100 rpm, C = 15% $NaNO_3$, Q = 8 L/min, G_i = 0.3 mm, G_m = 3 mm, flux density = 0.06 T, t_m = 28 min, v = 21 volt, (R_{ai} = 0.16 μm, R_{zi} = 1.34 μm for St 70), (R_{ai} = 0.15 μm, R_{zi} = 1.24 μm for Stainless steel) and (R_{ai} = 0.143 μm, R_{zi} = 1.1 μm for Cr-Coated steel)

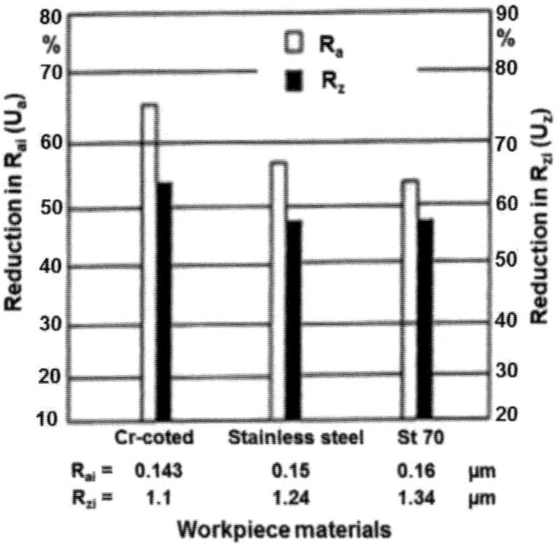

Figure 12. Effect workpiece material on U_a and U_z during MEAT after $t_m = 28\ min$.
Tool material: stainless steel, Abrasive type: SiC, Grain size: 280, n = 100 rpm, C = 15% $NaNO_3$, Q = 8 Lit/min, G_i = 0.3 mm, G_m = 3 mm, flux density = 0.06 T, v = 21 volt

The reduction percent in R_{ai} and R_{zi} or U_a and U_z is shown in Figure 11 and Figure 12 for the used materials. It can be noticed that the increase of the machining time causes the increase of U_a and U_z or the increase in the improvements of surface quality. The higher improvements occurred for Cr- coated steel due to its fine grains.

4.3. Effect of magnetic gap (G_m)
The effect of magnetic gap (G_m) on the relationships of (R_a – t_m) and (R_z – t_m) is shown in Figure 13 using $NaNO_3$ (15 %) electrolyte.

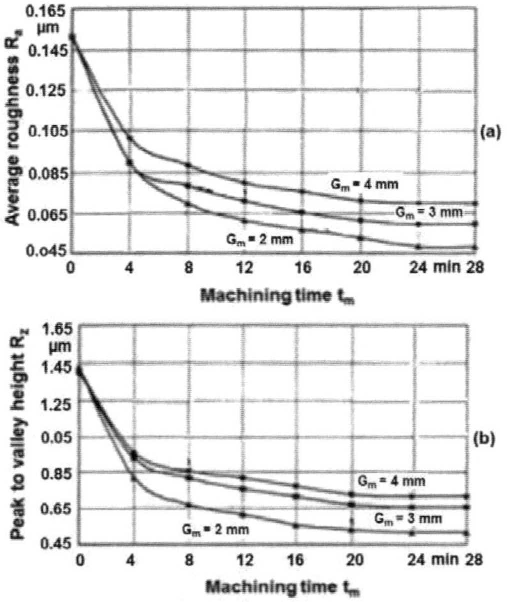

Figure 13. Effect of magnetic gap (G_m) in case of MEAT process using $NaNO_3$ electrolyte on relationship: (a) (R_a – t_m) (b) (R_z – t_m)

Workpiece mat.: Cr-coated steel, flux density = 0.06 T, Abrasive type: SiC, Grain size = 280, n = 100 rpm, C = 15% $NaNO_3$, v = 21 volt, Q = 8 L/min, and G_i = 0.3 mm

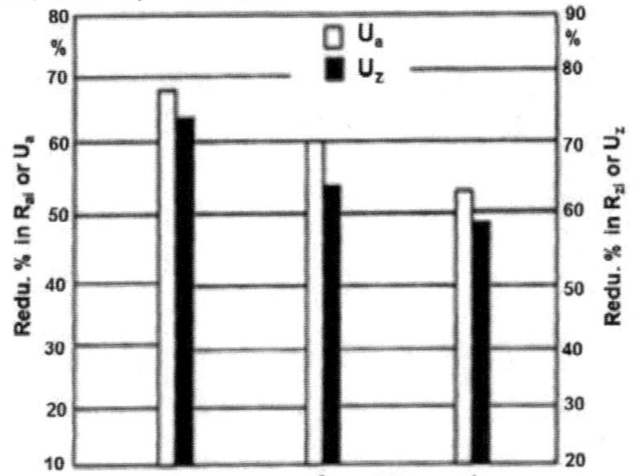

Figure 16. Effect of magnetic gap (G_m) in case of MEAT process using $NaNO_3$ electrolyte on the reduction of initial roughness (U_a, U_z).

Workpiece mat.: Cr-coated steel, flux density = 0.06 T, Abrasive type: SiC, Grain size = 280, n = 100 rpm, C = 15% $NaNO_3$, v = 21 volt, Q = 8 L/min, G_i = 0.3 mm, t_m = 28 min, R_{ai} = 0.15 μm, and R_{zi} =1.45 μm

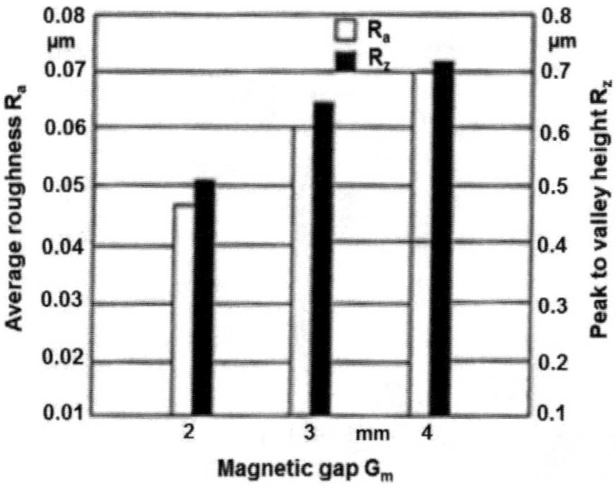

Figure 14. Effect of magnetic gap (G_m) in case of MEAT process using $NaNO_3$ electrolyte on surface roughness (R_a, R_z).

Workpiece mat.: Cr-coated steel, flux density = 0.06 T, Abrasive type: SiC, Grain size = 280, n = 100 rpm, C = 15% $NaNO_3$, v = 21 volts, Q = 8 L/min, G_i = 0.3 mm, t_m = 28 min, R_{ai} = 0.15 μm, and R_{zi} = 1.45 μm

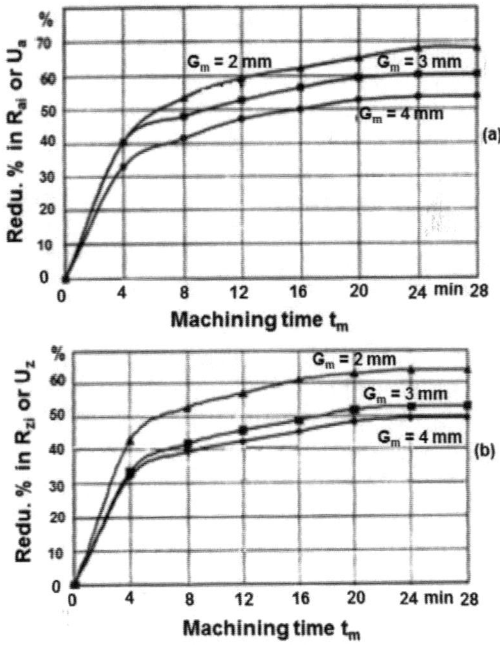

Figure 15. Effect of magnetic gap (G_m) in case of MEAT process using $NaNO_3$ electrolyte on relationship: (a) (U_a – t_m) (b) (U_z – t_m)

Workpiece mat.: Cr-coated steel, flux density = 0.06 T, Abrasive type: SiC, Grain size = 280, n = 100 rpm, C = 15% $NaNO_3$, v = 21 volt, Q = 8 L/min, G_i = 0.3 mm, R_{ai} = 0.15 μm, and R_{zi} = 1.45 μm

It can be found out that the decrease of the magnetic gap (G_m) caused the decrease of the surface roughness (R_a, R_z). The decrease of surface roughness is due to the increase of the effect of changing path of moving electrolytic ions by the decrease of (G_m). As shown in Figure 14 the decrease of magnetic gap from G_m = 4 mm to G_m = 2 mm causes the decrease of the final R_a from 0.07 µm to 0.047 µm, which means that R_a decreased by 33 %. It can be shown in Figure 15 and Figure 16 that the decrease of G_m causes the increase of both U_a and U_z the decrease of G_m from 4 to 2 mm causes the increase of U_a from 53 % to 68 % and the increase of U_z from 49 % to 64%.

4.4. Effect of rotational speed of the workpiece (n)

The effect of the rotational speed (n) on the relationships of both ($R_a - t_m$) and ($R_z - t_m$) is shown in Figure 17. It can be deduced that the increase of (n) causes the decrease of the surface roughness. The increase of (n) causes the increase of the both effects of electrochemical metal removal and the mechanical abrasion by abrasives from the peaks of roughness. As shown in Figure 18 the increase of (n) from 100 to 300 rpm cause the decrease of roughness (R_a, R_z) from (0.06, 0.65 µm) to (0.035, 0.34 µm) respectively. Which means that the increase of rotational speed from 100 to 300 rpm causes the decrease of roughness (R_a, R_z) by (42 %, 52 %) respectively. The effect of (n) on both relationships ($U_a - t_m$) and ($U_z - t_m$) is shown in Figure 19.

It was found that the increase of (t_m) or (n) causes the increase of the reduction percent of initial roughness (U_a ,U_z) as a relative measure of roughness. As shown in Figure 20 the increase of the rotational speed from 100 to 300 rpm and after MEAT process of a machining time of t_m = 28 min causes the increase of (U_a ,U_z) from (50 %, 35 %) to (70 %, 73%) respectively. The magnetic field causes the improvement in surface roughness as explained before [22]. Figure 21 shows photo of mirror-like finishing workpiece, its surface roughness was improved to R_a = 0.03 µm and R_z = 0.38 µm for both stainless steel and Cr coated workpieces.

4.5. Governing equation for the results

The reduction percent in initial average roughness value (U_a) and main working parameters (t_m , n and G_m) can be represented by the following empirical formula:

$$U_a = 0.0798 * n^{0.1923} * t_m^{0.31} / (G_m^{0.0275}) \quad (3)$$

The above empirical formula is used to deduce the reduction percent in initial average roughness in the following ranges. The magnetic field 0.06 T in the magnetic gap (G_m) from 2 to 4 mm. The rotational speed (n) from 100 to 300 rpm. The machining time (t_m) from 4 to 28 min. the MEAT conditions are C = 15 % $NaNO_3$, volt v = 21 volts, Q = 8 L/min, G_i = 0.3 mm, abrasive type SiC and grain mesh = 280. The deduced equation can be used for finishing Cr Coated steel while a correction factor may be used for other workpiece material.

4.6. Correlation between experimental and calculated (U_a):

Figure 22 shows the experimental results of reduction percent in initial average roughness (U_a) using MEAT parameters. (U_a) is estimated from the experimental data as (U_a) experimental while it can also calculated using the above empirical formulas as (U_a) calculated. Using least square method it was found a good correlation exists between (U_a) Experimental and (U_a) calculated.

5. Conclusions

1- The use of amagnetic field normal to the working gap in case of electrochemical abrasive turning process causes a further improvements in surface roughness (mirror-like) due to the effect of cycloid pass of ions.
2- The decrease of the magnetic gap (G_m) while the increase of the rotational speed (n) causes improvements in the surface roughness during MEAT.
3- Hard workpiece like Cr-coated can be easily polishing by MEAT process (R_a = 0.03µm).

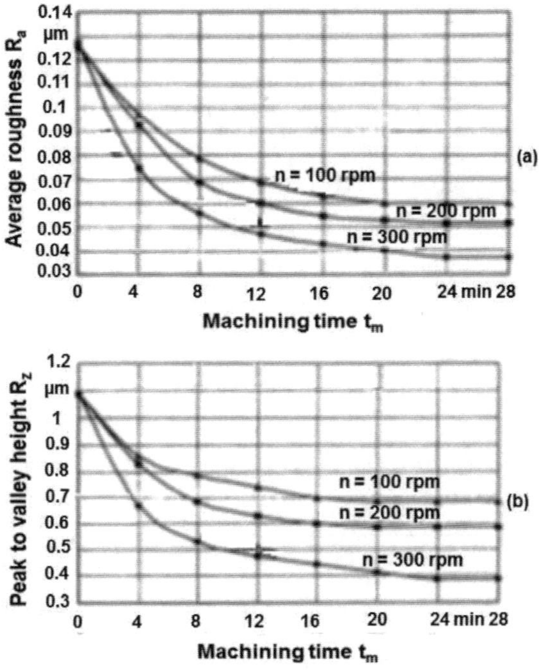

Figure 17. Effect of rotational speed (n) in case of MEAT process using $NaNO_3$ electrolyte on relationships: (a) (R_a - t_m) (b) (R_z - t_m)
Workpiece mat.: Cr-coated, flux density = 0.06 T, Abrasive type: SiC, Grain size = 280, C = 15% $NaNO_3$, v = 21 volt, Q = 8 L/min, G_i = 0.3 mm, G_m = 3 mm

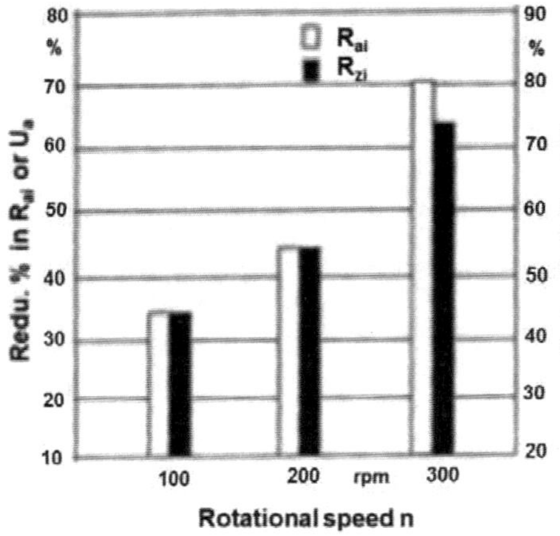

Figure 20. Effect of rotational speed (n) in case of MEAT process using $NaNO_3$ electrolyte on the reduction percent of the initial surface roughness (U_a, U_z).
Workpiece mat.; Cr-coated steel, Flux density = 0.06 T, abrasive type: SiC, grain size = 280, C = 15% $NaNO_3$, v = 21 volt, Q = 8 L/min, G_i = 0.3 mm, G_m = 3 mm, t_m = 28 min, R_{ai} = 0.125µm, and R_{zi} = 1.1 mm

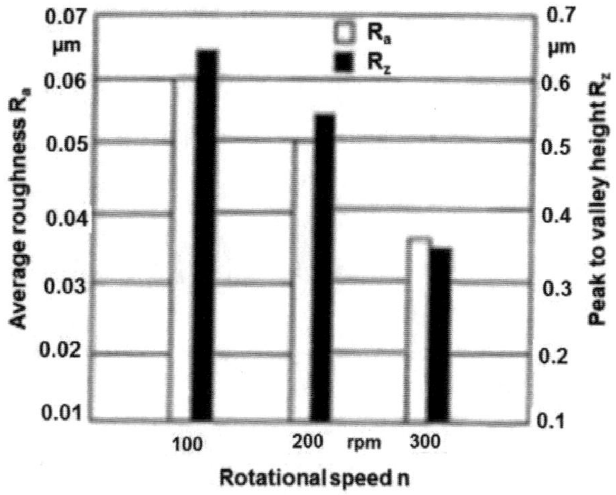

Figure 18. Effect of rotational speed (n) in case of MEAT process using $NaNO_3$ electrolyte on surface roughness (R_a, R_z).
Workpiece mat.; Cr-coated steel, flux density = 0.06 T, Abrasive type: SiC, Grain size = 280, C = 15% $NaNO_3$, v = 21 volt, Q = 8 L/min, G_i = 0.3 mm, t_m = 28 min, R_{ai} = 0.125µm, and R_{zi} = 1.1 mm

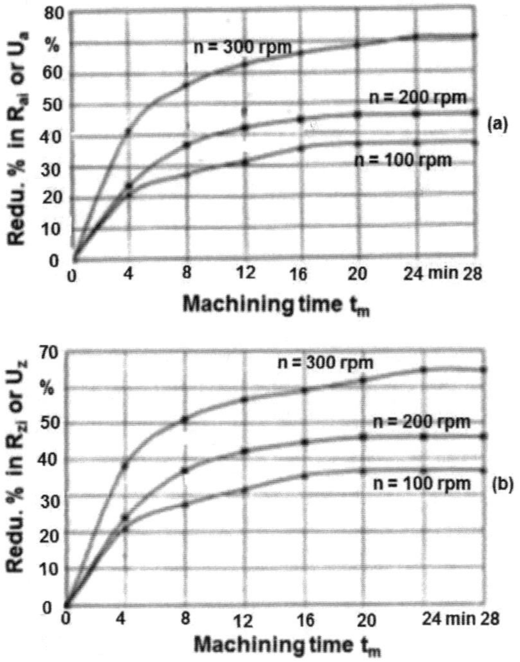

Figure 19. Effect of rotational speed (n) in case of MEAT process using $NaNO_3$ electrolyte on relationships: (a) (U_a – t_m) (b) (U_z – t_m)
Workpiece mat.; Cr-coated steel, Flux density = 0.06 T, abrasive type: SiC, grain size = 280, C = 15% $NaNO_3$, v = 21 volt, Q = 8 L/min, G_i = 0.3 mm, R_{ai} = 0.125 µm, and R_{zi} = 1.1 µm

(a) Stainless steel

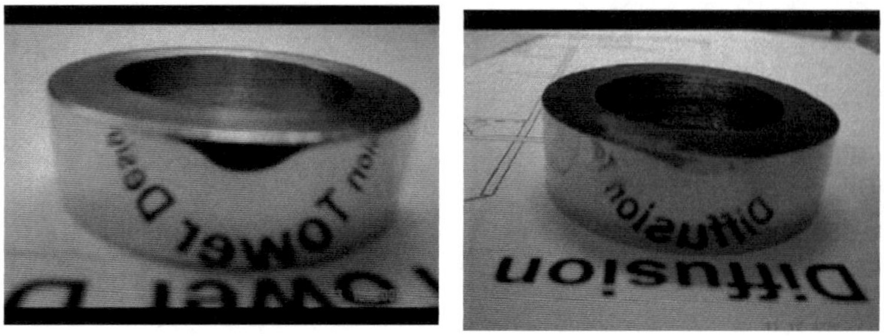

(b) Cr- coated steel

Figure 21. Photo showing mirror-like finishing two sampling workpiece, its surface roughness was improve to $R_a = 0.03$ µm and $R_z = 0.38$ µm

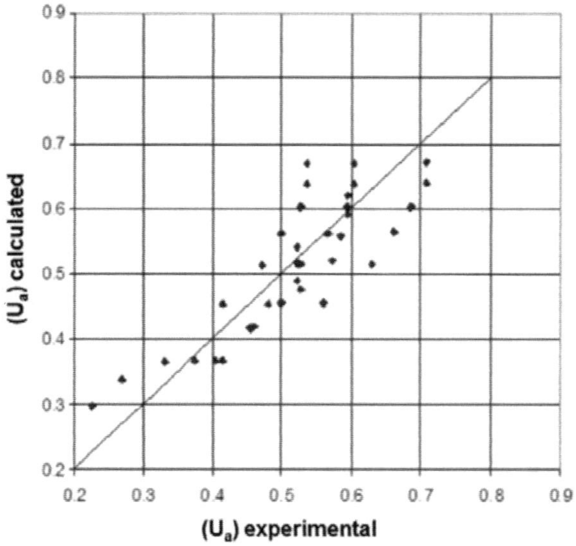

Figure 22. A good correlation between (U_a) experimental and (U_a) calculated

4- An imperical equations is deduced to obtain a relationship between the reductin percent in initial roughness (U_a) and the govering parameter machining time (t_m), rotating speed (n) and magnetic gap (G_m) as follow:

$$U_a = 0.0798 * n\wedge 0.1923 * t_m\wedge 0.31 / (G_m \wedge 0.0275) \quad (4)$$

6. Appendices

Table A1. Nomenclature; Symbols, definitions, and units used

Symbols	Definition	Units
A_i	current densities	**A/cm²**
B	the magnetic flux density	**Tesla**
C	electrolyte concentration	**weight / weight**
D	workpiece diameter before ECT	**mm**
d_e	workpiece diameter after ECT	**mm**
G	gap between tool electrode and workpiece	**mm**
G_i	initial gap between tool electrode and workpiece	**mm**

H	magnetic strength in the location of magnetic abrasive	
I	machining current	**A**
n	workpiece rotational speed	**rpm**
Q	flow rate of electrolyte	**L/min**
$\mathbf{R_a}$	average roughness	**µm**
$\mathbf{R_z}$	peak to valley height roughness	**µm**
t	electrochemical machining time	**min**
$\mathbf{t_m}$	turning machining time	**min**
v	applied voltage	**Volt**
ECT	electrochemical turning	
MEAT	magnetic-electrolytic-abrasive turning	

References

[1] Jayakumar P, Ray S and Radhakrishnan V 1997 Optimizing progress parameters of magnetic abrasive machining to reduce the surface roughness value", *J. Spacecraft Techn.* **7**(1)

[2] Kim J D and and Choi M S 1995 Simulation of an internal finishing of rectangular tube using a magnetic field", *Wear* **184** 67-71

[3] Yuquau C and Zhenjia L 1999 Research on a magnetic field in electrolytic lapping *Chinese J. Mech. Engng* **12**(1) 63–69

[4] Kim J D and and Choi M S 1996 Development and finite element analysis of the finishing system using rotating magnetic field *Int. J. Machin Tools & Manuf.* **36**(2) 245-253

[5] BIING-HWA YAN, GEENG-WEI CHANG, TSUNG-JEN CHENG and RONG-TZONG HSU, "Electrolytic magnetic abrasive finishing", *Int. J. Machin Tools & Manuf.* 43, pp.1355-1366, (2003).

[6] Kim J D, XuY M and Kang Y H 1998 Study on the characteristics of magneto-electrolytic-abrasive polishing by using the newly developed nonwoven-abrasive pads *Int. J. Machin Tools & Manuf.* **38** 1031-1043

[7] Kim J D and Choi M S 1997 Development of magneto-electrolytic-abrasive polishing system (MEAPS) and finishing characteristics of a Cr-coated roller *Int. J. Machin Tools & Manuf.* **37**(7) 997-1006

[8] Chang G W, Yan B H and Hsu R T 2002 Study on cylindrical magnetic abrasive finishing using unbounded magnetic abrasives **42** 575-583

[9] Kim J D, Xin D and Choi M S 1997 Study on the effect of a magnetic field on an electrolytic finishing process *Int. J. Machin Tools & Manuf.* **37**(4) 401-408

Ultrasonically assisted drilling: A finite-element model incorporating acoustic softening effects

V A Phadnis[1], A Roy and V V Silberschmidt

Wolfson School of Mechanical and Manufacturing Engineering, Loughborough University, Loughborough, LE11 3TU, UK

E-mail: v.a.phadnis@lboro.ac.uk

Abstract. Ultrasonically assisted drilling (UAD) is a novel machining technique suitable for drilling in hard-to-machine quasi-brittle materials such as carbon fibre reinforced polymer composites (CFRP). UAD has been shown to possess several advantages compared to conventional drilling (CD), including reduced thrust forces, diminished burr formation at drill exit and an overall improvement in roundness and surface finish of the drilled hole. Recently, our in-house experiments of UAD in CFRP composites demonstrated remarkable reductions in thrust-force and torque measurements (average force reductions in excess of 80%) when compared to CD with the same machining parameters. In this study, a 3D finite-element model of drilling in CFRP is developed. In order to model acoustic (ultrasonic) softening effects, a phenomenological model, which accounts for ultrasonically induced plastic strain, was implemented in ABAQUS/Explicit. The model also accounts for dynamic frictional effects, which also contribute to the overall improved machining characteristics in UAD. The model is validated with experimental findings, where an excellent correlation between the reduced thrust force and torque magnitude was achieved.

1. Introduction

Carbon fibre-reinforced polymer (CFRP) composites are widely used in aerospace and automobile industries and several other structural applications owing to their superior mechanical and physical properties. CFRPs outperform structural metals in durability, high strength-to-weight ratio, stiffness and density [1-4]. In spite of these advantages, machining CFRPs is cumbersome due to highly abrasive carbon fibres, low thermal conductivity of epoxy and a weak interfacial bond between reinforced fibres and matrix.

Typically, parts made of CFRP composite are manufactured to the required final shape, though machining cannot be avoided to enable assembly of individual components. Holes need to be drilled to facilitate riveting and bolting of machined components; to serve this purpose; conventional drilling (CD) methods are often adopted. Apart from poor geometric tolerance and surface finish, CD can initiate various damage modes in CFRPs such as matrix cracking, burr formation, interfacial de-bonding, delamination and fibre pull-out [1, 2, 5-7]. Consequently, the load-carrying capacity of

[1] Vaibhav A. Phadnis
Wolfson School of Mechanical and Manufacturing Engineering, Loughborough University, UK
Tel :+44 (0) 1509 227634; Email : v.a.phadnis@lboro.ac.uk

composite structure is affected. Additionally, rapid tool wear is caused by abrasive carbon fibres resulting in frequent tool replacement. To overcome these issues, a less intrusive drilling technique is needed.

Ultrasonically Assisted Drilling (UAD) is a hybrid machining technique, which has been used to improve machining of conventional metals and advanced composites in the recent past [5-12]. In UAD, high-frequency (typically in excess of 20 kHz) vibrations generated by a piezoelectric transducer are superimposed on a rotating standard twist drill bit in the axial direction, to enhance the cutting process [8, 11]. There are several advantages of UAD over CD such as reduction in drilling forces and torque, better surface finish, low tool wear and elimination/reduction in burr formation [11-13]. In CFRP, drilling forces have a direct effect on damage induced in composites. Hence it is considered to be the primary parameter affecting quality of a drilled hole [4, 7, 9]. Several studies performed over the last few years [10-12] indicate that vibrating a drill bit in the axial direction yields the maximum reduction in drilling-induced damage. In the experimental study conducted by Hochen and Hsu [13] it was concluded that ultrasonic machining have viable advantages over conventional machining processes, though, a slurry-based method was used instead of external exciting the conventional drill bit for hole making in CFRP. Wang [14] performed several experiments involving UAD in CFRP. Using a vibration frequency of 300 Hz to excite a high-speed-steel (HSS) and carbide drill bits of 0.5 mm diameter; it was observed that thrust forces were reduced relative to those in CD. Babitsky and Astashev [8, 11] reported that in UAD, hardly any axial thrust force reduction was seen whilst exciting drill bits in torsional mode, though considerable reduction in the drilling force was observed when the drill was vibrated in axial direction.

Very few researchers have attempted to model a UAD process using a finite-element (FE) technique due to the complexity involved in modelling of such processes and the extent of computational resources required, though analytical model [see e.g.14] to calculate thrust forces in UAD can be found. This paper presents development of a FE model of UAD in CFRP composite using general purpose software - Abaqus 6.11. The experimental results from our previous work are presented for validation purposes.

2. FE model of UAD in CFRP

2.1. Background and assumptions

UAD includes superposition of ultrasonic vibrations onto the relative cutting motion between a drill bit and the workpiece drilled. This collective effect results in the substantial reduction in the drilling forces and torque [5-11]. Based on the preliminary observations from our UAD experiments [5-7], a CFRP workpiece undergoes plastic deformation under the influence of ultrasonic vibrations, the extent of which depends on the intensity of vibrations, resulting in softening the workpiece in the process zone. Additionally, surface temperature of the CFRP workpiece often exceeded glass- transition temperature of thermoset epoxy polymer matrix (170 - 190^0 C), resulting in matrix burning. The reduced thrust force and torque observed in UAD is claimed to be due to the combined influence of ultrasonic vibrations giving rise to vibro-impact and localised heating phenomena. In this context, we define ultrasonic softening as, the phenomenon in which ultrasonic vibrations reduce the apparent static stress necessary for material to undergo permanent inelastic deformation.

In this work, a parametric modelling approach was used to account for acoustic (ultrasonic) and thermal softening phenomena during inelastic deformation of the CFRP workpiece. The behaviour of CFRP laminate in elastic regime was defined using the Hill's [16, 17] potential theory for anisotropic materials together with the rate-independent plasticity criterion incorporating effects of ultrasonic softening. An element-removal scheme was used based on the shear damage-initiation criterion to replicate the hole-making process. The temperature rise due to plastic straining was modelled assuming adiabatic conditions. It should be noted that the temperature is a plastic strain work conjugate in the adiabatic system; hence an element was also removed from the mesh when instantaneous temperature reaches the glass transition limit of the epoxy matrix (irrespective of instantaneous plastic strain reaching the fracture strain), thus effectively softening the material at macroscale.

The primary assumptions of this material model are:

a. Instantaneous intensity of ultrasonic vibrations is proportional to plastic strain induced in CFRP laminate. This instantaneous ultrasonic intensity is defined in terms of ultrasonic energy per unit time transferred to the workpiece and is a function of instantaneous contact pressure between the drill bit and the workpiece, the frequency and amplitude of vibration.
b. The strain-rate dependence of CFRP laminate, especially in transverse and through-thickness directions, is small enough to be neglected owing to a low loading rate [6-8].
c. The temperature rise in CFRP laminate during its interaction with the vibrating drill tip is mainly due to plastic strain developed locally.

2.2. Constitutive material model

The workpiece material used in this study was commerciality available M21/T700 quasi-isotropic laminate of thickness 10 mm with individual ply thickness of 0.25 mm. Details of the material properties used in this simulations can be found Phadnis *et al.* [7]. The thermo-mechanical plasticity model proposed in this work is based on the quadratic yield criterion for anisotropic materials by Hill [16, 17] and non-linear isotropic hardening rule for rate-independent plasticity. The constitutive equations of this model for uniaxial loading are as follows:

The total strain tensor during deformation is the sum of elastic strain tensor and plastic strain tensor, given by

$$\boldsymbol{\varepsilon} = \boldsymbol{\varepsilon}^{el} + \boldsymbol{\varepsilon}^{pl}. \qquad (1.1)$$

The elastic part of stresses is computed using a constitutive stress-strain relationship;

$$\boldsymbol{\sigma} = \mathbf{C}\,.\,\boldsymbol{\varepsilon} = (\boldsymbol{\varepsilon}^{el} - \boldsymbol{\varepsilon}^{pl}), \qquad (1.2)$$

where, $\boldsymbol{\sigma}$ and $\boldsymbol{\varepsilon}$ are the second-order elastic stress and strain tensors while $\mathbf{C}$ is the fourth-order stiffness tensor. The yield function [16] for orthotropic CFRP material in rectangular Cartesian stress components is given by

$$\begin{aligned} f(\boldsymbol{\sigma}) &= \sqrt{F(\bar{\sigma}_{22}-\bar{\sigma}_{33})^2 + G(\bar{\sigma}_{33}-\bar{\sigma}_{11})^2 + H(\bar{\sigma}_{11}-\bar{\sigma}_{22})^2 + 2L\bar{\sigma}_{23}{}^2 + 2M\bar{\sigma}_{31}{}^2 + 2N\bar{\sigma}_{12}} \\ &= \left|\boldsymbol{\sigma}^{y}\right| + R, \end{aligned} \qquad (1.3)$$

where, *F, G, L, M, and N* are constants obtained from a set of the following equations.

$$\begin{aligned} F &= \frac{(\sigma^0)^2}{2}\left(\frac{1}{\bar{\sigma}_{33}^{2}} + \frac{1}{\bar{\sigma}_{11}^{2}} + \frac{1}{\bar{\sigma}_{22}^{2}}\right) = \frac{1}{2}\left(\frac{1}{R^2_{33}} + \frac{1}{R^2_{11}} + \frac{1}{R^2_{22}}\right), & L &= \frac{3}{2}\left(\frac{\tau^0}{\bar{\sigma}_{23}}\right)^2 = \frac{3}{2R_{23}{}^2} \\ G &= \frac{(\sigma^0)^2}{2}\left(\frac{1}{\bar{\sigma}_{22}^{2}} + \frac{1}{\bar{\sigma}_{33}^{2}} + \frac{1}{\bar{\sigma}_{11}^{2}}\right) = \frac{1}{2}\left(\frac{1}{R^2_{22}} + \frac{1}{R^2_{33}} + \frac{1}{R^2_{11}}\right) & M &= \frac{3}{2}\left(\frac{\tau^0}{\bar{\sigma}_{13}}\right)^2 = \frac{3}{2R_{13}{}^2} \\ H &= \frac{(\sigma^0)^2}{2}\left(\frac{1}{\bar{\sigma}_{11}^{2}} + \frac{1}{\bar{\sigma}_{22}^{2}} + \frac{1}{\bar{\sigma}_{33}^{2}}\right) = \frac{1}{2}\left(\frac{1}{R^2_{11}} + \frac{1}{R^2_{22}} + \frac{1}{R^2_{33}}\right) & N &= \frac{3}{2}\left(\frac{\tau^0}{\bar{\sigma}_{12}}\right)^2 = \frac{3}{2R_{12}{}^2} \end{aligned} \qquad (1.4)$$

Here $\bar{\sigma}$ is the measured yield stress when applied as the only non-zero stress component, σ^0 is the initial yield stress and is τ^0 the shear stress at yield such that, $\tau^0 = \sigma^0 / \sqrt{3}$. R_{ij} are anisotropic yield ratios and can be calculated from the CFRP yield strengths. σ^y is the size of an initial yield surface, while R is the isotropic hardening term:

$$R = \sigma^0(\boldsymbol{\varepsilon}^{\mathbf{pl}}, \dot{\theta}). \tag{1.5}$$

Plastic strain during deformation is defined as

$$\boldsymbol{\varepsilon}^{\mathbf{pl}} = D\left(\frac{\bar{\sigma}}{\sigma^0} - 1\right), \text{ for } \bar{\sigma} \geq \sigma^0 \tag{1.6}$$

where $\bar{\sigma}$, σ^0, $\boldsymbol{\varepsilon}^{\mathbf{pl}}$, and D are yield stress at non-zero plastic stress, static yield stress, plastic strain and softening parameter respectively. D is defined as, $D = D(\psi, \dot{\theta})$ where ψ is ultrasonic energy per unit time and $\dot{\theta}$ is the temperature gradient due to work done towards plastic straining in an adiabatic system. It should be noted that ultrasonic energy per unit time, ψ transmitted to the CFRP surface in the immediate contact with the drill tip is defined as

$$\psi = p_c.f_u.a_u.A_c \tag{1.7}$$

where p_c is the instantaneous contact pressure exerted by the drill tip on the CFRP surface, f_u and a_u are the frequency and amplitude of ultrasonic vibration respectively. The magnitude of ψ was calculated using (1.8) for the feed rate of 8 mm/min and spindle speed of 40 rpm. The area of contact and corresponding contact pressure was obtained through a number of simulations. The effect of a range of frequency and amplitude of ultrasonic vibration on the magnitude of ultrasonic energy per unit time is shown in Fig. 1a and b respectively.

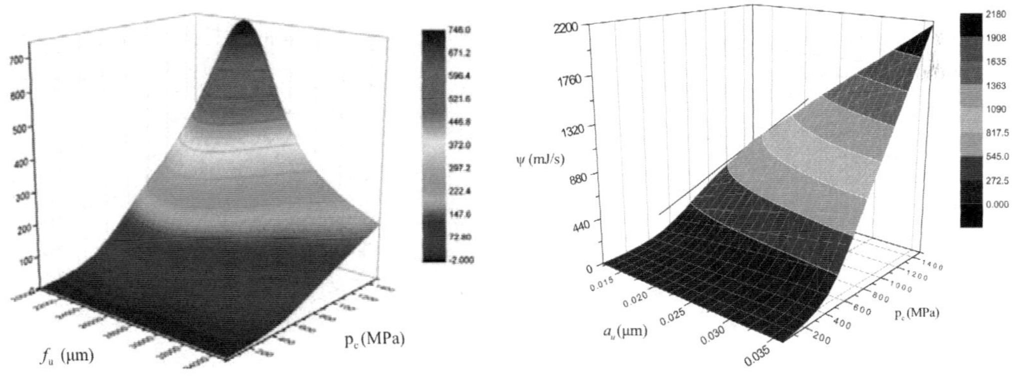

(a) (b)

Figure 1. (a) Effect of f_u on ψ (Drill feed 8 mm/min, spindle speed 40 rpm, amplitude of vibration 12 μm peak-to-peak). (b) Effect of a_u on ψ (Drill feed 8 mm/min, spindle speed 40 rpm, frequency of vibration 25 kHz)

2.3. Adiabatic heat transfer

Heat generation caused by mechanical work associated with plastic straining was modelled in an adiabatic system. The ultrasonic frequency used in these simulations was 27.8 kHz resulting in rapid accumulation of ultrasonic energy in a highly localised regime in a very short period of time. Hence it was assumed that inelastic deformation due to plastic straining occurs so rapidly that heat had hardly any time to diffuse through, and, thus, adiabatic analysis was sufficient to model the temperature rise in this local regime. The model assumed that plastic straining gives rise to a heat flux per unit volume, r^{pl} such that;

$$r^{pl} = \eta\sigma : \dot{\varepsilon}^{pl}, \tag{1.8}$$

where η is the constant inelastic heat coefficient, σ is the instantaneous stress state and $\dot{\varepsilon}^{pl}$ is the plastic strain rate.

2.4. Element removal scheme

A hole-generation process in simulations of UAD was accomplished with the help of the element removal scheme in Abaqus/Explicit [19]. This scheme was based on the fracture mechanism initiated by shear damage and damage evolution linked to the energy dissipation during degradation of material's stiffness. The in-built Abaqus [19] shear damage-criterion utilised here is a phenomenological model for predicting the onset of damage due to shear-band localisation. It assumes that the equivalent plastic strain at the onset of damage $\bar{\varepsilon}_s^{pl}$ is a function of the shear stress ratio and instantaneous plastic strain such that:

$$\bar{\varepsilon}_s^{pl} = \bar{\varepsilon}_s^{pl}(\chi_S, \bar{\varepsilon}^{pl}), \tag{1.9}$$

where χ_S is a shear stress ratio:

$$\chi_S = \frac{\tau_S}{\tau_{\max}}. \tag{1.10}$$

Here τ_S is the total accumulated shear stress at the integration point of an element and $\tau_{\max}$ is the corresponding maximum shear stress. The criterion for damage initiation is met when the following condition is satisfied:

$$\omega_s = \int \left(\frac{1}{\boldsymbol{\varepsilon}_{\mathbf{S}}^{\mathbf{pl}}(\chi_S, \boldsymbol{\varepsilon}^{\mathbf{pl}})} \right) . d\left(\boldsymbol{\varepsilon}^{\mathbf{pl}}\right) = 1 , \tag{1.11}$$

where ω_s is a state variable that increases monotonically with plastic deformation proportional to the incremental variation in the equivalent plastic strain.

The characteristic stress-strain behaviour of a material under uni-axial loading that undergoes progressive damage is shown in Fig. 2. In case of the elastic-plastic material this damage can be decomposed into two parts; softening of the yield stress and degradation of the elastic modulus. The solid curve in Fig. 3 represents the damaged stress-strain response, whereas the dashed line represents the undamaged behaviour. σ_{y0} and $\bar{\varepsilon}_0^{pl}$ are yield stress and equivalent plastic strain at the onset of damage, while $\bar{\varepsilon}_f^{pl}$ is the equivalent plastic strain at failure, also known as *fracture strain*.

When material undergoes damage, the stress-strain relationship fails to accurately present its behaviour because of a strong mesh dependency linked to the strain localisation. Hence a different approach is required to trace the strain softening branch of the stress-strain curve.

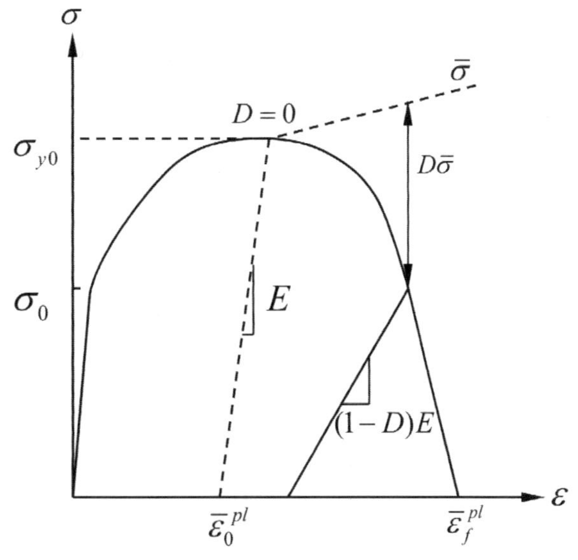

Figure 2. Stress-strain curve with progressive damage behaviour

Thus, Hillerborg's fracture energy approach [18] was employed in this model, which eventually helped to reduce mesh dependency by formulating a stress-displacement response after damage initiation. The fracture energy was idealised as work required to open a unit area of a crack, and expressed as:

$$G_f = \int_0^{u_{pl}^f} \sigma_y \, . \, d\left(\bar{u}^{pl}\right) , \tag{1.12}$$

where $\bar{u}^{pl}$ is the equivalent plastic displacement and can be considered as fracture energy conjugate of yield stress after the damage initiation: $\bar{u}^{pl} = 0$ at damage initiation and $\bar{u}^{pl} = L\bar{\varepsilon}^{pl}$ after it. Here L is the characteristic length of an element in a meshed body that depends on its geometry and formulation. D is the overall damage parameter : at damage initiation $D = 0$, while at complete damage $D = 1$. The residual elastic modulus E_r, after damage initiation can be calculated as,

$$E_r = (1 - D). \, E \, . \tag{1.13}$$

Evolution of the damage variable $\dot{d}$ associated with a relative plastic displacement was specified in a linear form as,

$$\dot{d} = \frac{\bar{u}^{pl}}{\bar{u}_f^{pl}} , \tag{1.14}$$

where $\bar{u}_f^{pl}$ is the equivalent plastic displacement at failure and can be calculated as

$$\bar{u}_f^{pl} = \frac{2G_f}{\sigma_{y0}} \tag{1.15}$$

2.5. Contact and friction

The contact and friction parameters used in the simulations were based on a number of experimental factors such as cutting speed, feed rate, drill geometry and surface properties. Contact between the twist drill and the CFRP laminate was defined by the general contact algorithm available in Abaqus/Explicit. This algorithm generated the contact forces based on the penalty-enforced contact method. A coefficient of dry friction at the interface between drill's cutting edge and laminate was assumed to be 0.03 based on the work by Lucas *et al.* [15].

2.6. FE model setup

2.6.1. Geometric modelling and boundary conditions

A 3D geometry of 6 mm twist drill with a point angle of 118° and helix angle of 31° was modelled in Abaqus. The drill was modelled as a rigid body, and the inertia effect was accounted for. The mesh was optimized to reduce the computational cost without compromising accuracy. A mesh size of CFRP work piece was refined in the immediate vicinity of the drilled area to capture high stress gradients during the drilling process. Elements in the refined cylindrical zone were removed when the failure criterion was met during simulations using element deletion discussed in section 2.4.

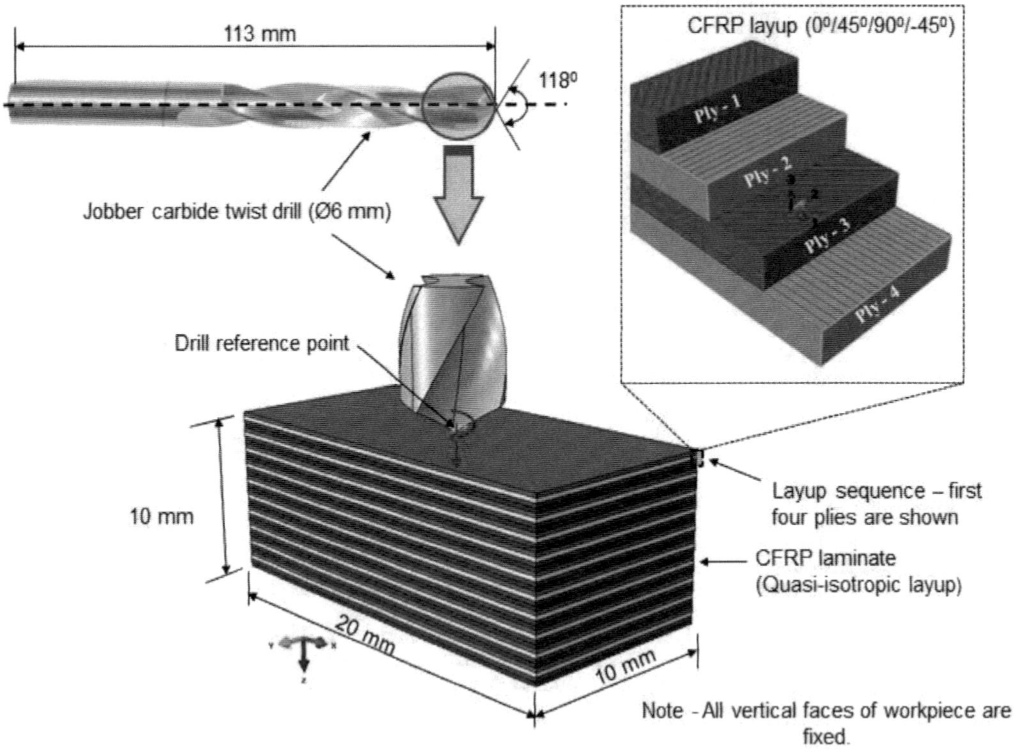

Figure 3. Finite-element model of UAD in M21/T700 composite laminate

The CFRP laminate was fixed at all four vertical faces, while the drill was constrained to rotate only about its own axis with a specified speed and fed vertically downwards into the work piece. The FE analysis was performed with experimentally optimised drilling parameters: spindle speed of 40 rpm, feed rate of 8 mm/min, vibration frequency of 27.8 kHz, peak-to-peak vibration amplitude of 12 μm. In order to incorporate the vibrating boundary condition, a time-dependent sinusoidal movement representing a cycle of ultrasonic vibration was imposed on the drill in the axial direction. Each wave consisted of 27,800 periodic sampling points for one second of simulated period of drilling. The drill is thus vibrated along the vertical direction as well as rotated about its own axis independently.

2.6.2. Choice of finite-elements

The Hill's potentential mainly depends on the deviatoric stress components and the plastic response becomes almost incompressible [16, 17]. Thus, finite-elements were chosen so that they could accommodate incompressibility, where plastic flow dominates the response. The fully integrated first-order continuum elements in Abaqus/Explicit were used with selectively reduced integration, whereby the volumetric strain was calculated at the centroid of the element only. In an adiabatic heat transfer analysis, the heat equation solved at each integration point is:

$$\rho c(\theta)\dot{\theta} = r^{pl} \quad , \tag{1.16}$$

where r^{pl} the heat flux per unit volume due to plastic strain is, ρ is the material density $c(\theta)$ is specific heat and $\dot{\theta}$ is the instantaneous temperature gradient. Since adiabatic heat is calculated using material properties and no external heat flux is added into the system, temperature degrees of freedom were not required. Hence eight-node, first-order, reduced-integration hexahedral elements of type C3D8R were used in this simulation. The CFRP model consisted of 820000 elements with the smallest element size of 5 μm. The drill was meshed with 35000 elements, with the smallest element size of 125 μm. The drill was modelled with four-node, 3D discrete rigid elements of type C3D4.

3. FE Results

The results obtained from FE simulations comprise the average of peak thrust force and torque for UAD in CFRP laminate. The thrust force and torque obtained under similar machining parameters and material properties with conventional drilling technique [5-7] are also presented for the comparison purpose and are shown in Fig. 4a and b.

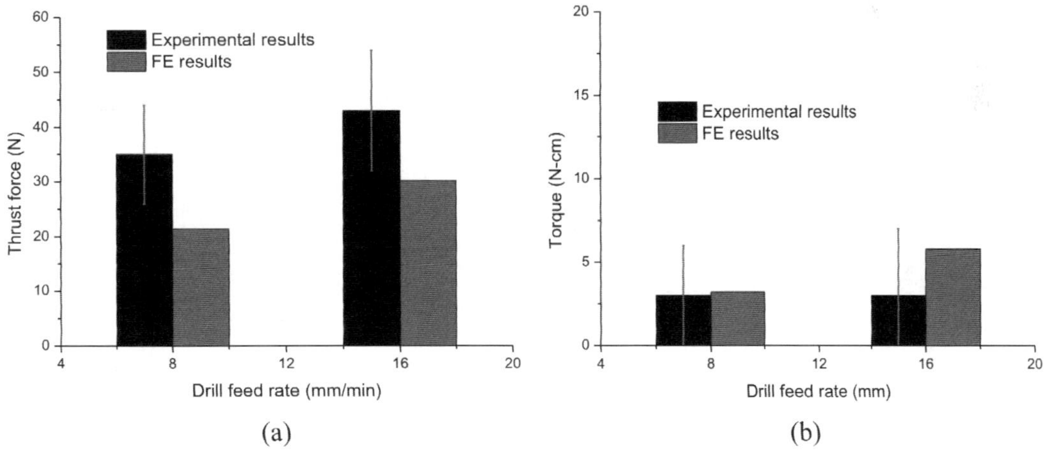

Figure 4. Comparison of peak thrust force (a) and peak torque (b) using experiments and FEA ($a = 12$ μm, $f = 27.8$ kHz)

4. Conclusions

A finite-element model of UAD in CFRP is developed using Abaqus 6.11. A constitutive material model suitable to model both volumetric and thermal softening in CFRP laminate under ultrasonic vibrations is proposed. The FE model replicated UAD process effectively. The FE results were found to be in good agreement with experimental findings. Additionally, the relative thrust force and torque reduction indicates good co-relation with the experimental results, demonstrating potential capabilities of the developed model to predict features of UAD.

References

[1] Rodriguez J P, Ashcroft I A and Silberschmidt V V 2008 *Compos. Sci. Technol.* **68** 2401-9
[2] Xia Z H and Curtin W A 2008 *Compos. Sci. Technol.* **68** 2526
[3] Lee C K 2009 *Mater. Chem. Phys.* **114** 125-33
[4] Zitoune R, Krishnaraj V, Almabouacif B S and Collombet F 2012 *Compos. Part B.* **43** 1480
[5] Makhdum F, Jennings L T, Roy A and Silberschmidt V V 2012 *J. Physics: Conf. Series.* **382** 12019
[6] Makhdum F, Nurdiya D K, Roy A and Silberschmidt V V 2012 *Solid State Phenom.* 188:170-5
[7] Phadnis V A, Makhdum F, Roy A and Silberschmidt V V 2012 *Proc. CIRP.* **1** 455-9
[8] Astashev V K and Babitsky V I 2007 *Ultrasonic processes and machines: dynamics, control and applications* (Berlin: Springer)
[9] Thomas P and Babitsky V I 2007 *J. Sound. Vib.* **308** 815
[10] Alam K, Mitrofanov A V and Silberschmidt V V 2011 *Med. Eng. Phys.* **33** 234
[11] Babitsky V I, Astashev V K and Meadows A 2007 J. *Sound Vib.* **308** 805-14
[12] Wiercigroch M, Wojewoda J and Krivtsov A 2005 *J. Sound Vib.* **280** 739-57
[13] Hocheng H, Taib N H and Liu C S 2000 *Compos. Part A.* **31** 133-42
[14] Wang X and Wang L 2004 *J. Mat. Process Tech.* **148** (2) 239-44
[15] Lucas M, MacBeath A and Cardoni E 2006 *Ultrasonics* **44** 503-509
[16] Hill R 1990 *J. Mech. Phys. Solids.* **38**(3) 405-17
[17] Hill R 1998 *The Mathematical Theory of Plasticity.* Clarendon Press
[18] Hillerborg A 1985 *Mat. Struct.* **18**(4) 291-6
[19] Abaqus 6.11 2011 *User manual* Dassault System Rhode Island US

Numerical study of strain-rate effect in cold rolls forming of steel

J Falsafi, E Demirci[1] and V V Silberschmidt
Wolfson School of Mechanical and Manufacturing Engineering,Loughborough University, LE11 3TU, Loughborough, UK

E-mail: e.demirci@lboro.ac.uk

Abstract. Cold roll forming (CRF) is a well-known continuous manufacturing process, in which a flat strip is deformed by successive rotating pairs of tools, without changing the material thickness. In the past decades, to lessen the process-development efforts, finite-element simulations have been increasingly employed to improve the process design and predict the manufacturing-induced defects. One of the important aspects in design of the CRF process is consideration of resulting strains in the final product as the material passes through several complex forming stands. Sufficient knowledge of longitudinal strain in the workpiece is required to set various process parameters. Increasing a process speed in a roll forming operation can bring cost advantages, but the influence of the forming speed on the strain distribution should be explored.
This study is focussed on a strain-rate effect in the CRF process of steel sheets. The strain-rate dependency of a plastic behaviour observed in most metals can affect the finished product's quality as well as process parameters. This paper investigates the influence of the strain rate on longitudinal strains induced in the roll forming operation by incorporating a phenomenological Johnson–Cook constitutive model, which allows studying the impact of the process speed on the output product. Taking advantage of 3D finite element analysis, a roll forming process was simulated using MCS.Marc, comprising a complete set of forming stations. Through the changing of the process speed, the strain rate impact on longitudinal peak strains and forming length was investigated. The results highlight the effect of the strain rate on edge thinning and subsequent undesirable distortions in the product.

1. Introduction

Cold roll forming is a highly productive method for the production of long profile. In this forming process, the material undergoes progressive forming by a series of rolls into a certain cross-section product. The sequence of rolls breaks the forming process down to several stages. Each stage comprises two stands between which the sheet metal is deformed.

The distance between one forming stages to the next is called the forming length. In roll-forming process elongation occurs in the edges because they travel a longer path than the centre line. If the lead-in length is not sufficiently long, the edges will elongate plastically and in the finished product the edges are

[1] To whom any correspondence should be addressed

longer than the mid line, causing the edge wrinkles, warp or twist in the product. Figure 1 depicts how further elongation occur around flange edge compared to the web of a U channel.

Since the deformation in cold-roll forming is quite complex and the curved surfaces are difficult to describe, the process design mainly used to be developed from experimental knowledge on an empirical basis. However; in the last decade the finite element analysis has been shown to provide reasonably good prediction of what takes place in the roll forming process. Based on the level of complexity of the model and the expected accuracy, the solution time can be from a few hours to several days.

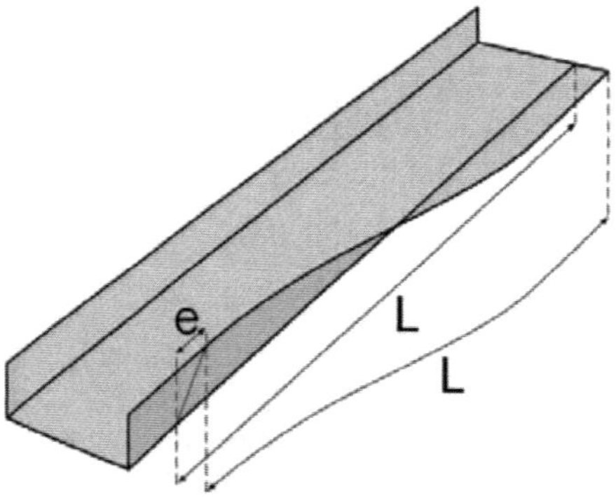

Figure 1. The material at the edge travels a longer distance than the rest of the material which makes it prone to buckle

Determination of strain distribution based on the process and materials product have been main objective of many recent researches. Lindgren Lindgren [1] studied the influence of yield strength on the longitudinal peak membrane strain in the flange and the deformation length. It was shown that when the yield strength is increased, the longitudinal peak membrane strain decreases and the deformation length increases. Han et al. [2] declared the same dependency. Zeng et.al [3]investigated the product quality with respect to material properties. They observed significant impacts of the yield point as well as the work-hardening exponent on the final product quality, whereas forming speed and friction at roll-sheet interface appear to play a minor role. Bui and Ponthot [4] demonstrated a combined experimental and numerical study sensitivity of the springback behaviour with respect to material properties. The effect of the yield limit and the strain hardening exponent on springback and edge waviness was highlighted. Wiebenga et al. [5] presented a robust optimization techniques to obtain settings of adjustable tools in the final roll forming stand. The results highlighted significant effect of tool adjustment on the dimensional quality of the product by compensating for scattering material properties.

Additional to the material properties, several studies also focused on the relation between redundant straining and the process design. Li et al. [6] carried out extensive experimental studies on the resultant residual stresses in roll formed square hollow sections. It was highlighted that the forming process has a significant effect on the distribution of longitudinal residual stresses on the outside surface of the product, while the distribution of transversal residual stresses is similar for the same dimension. Jeong et al. [7] investigated the longitudinal strain in U channel using FEM as a result of two types of flower patters for forming process of the same product. Salmani Tehrani et al. [8] studied the roll forming process design on local edge buckling, demonstrating a critical value of profile angle of the rolls to prevent local edge wave in the product. Optimization on the basis of different influential parameters Paralikas et al. [9]optimised the inter-distance between roll stations to minimize the elastic longitudinal and shear strains as well as the strip edge wave. A similar numerical optimization presented by Zeng et al. [10] to obtain the optimal settings for the forming angle increment and roll radius in order to reduce the springback and peak longitudinal strains.

All the above mentioned researches studied upon influence of material and process parameters in the longitudinal edge straining. The material model used does not include the implications of strain rate, though some of them tried to considerate speed parameter. This paper investigates the influence of the strain rate on longitudinal strains induced in the roll forming operation by incorporating a phenomenological Johnson–Cook constitutive model, which allows studying the impact of the process speed on the output product.

2. Theory, Johnson-Cook model

In order to predict the response of a material at various strain rates, the constitutive material model must include strain rate effect. The J-C model is simple approach to correlate flow stress to strain rate and primarily intended for use in computer codes[11]. The equivalent Von Mises flow stress according to Johnson-Cook model is given in generic form of equation(1):

$$\sigma = [A + B\varepsilon^n][1 + C.\ln\dot{\varepsilon}^*][1 - T^{*m}] \quad (1)$$

Where ε is the equivalent plastic strain, $\dot{\varepsilon}^* = \dot{\varepsilon}/\dot{\varepsilon}_0$is the dimensionless plastic strain rate for $\dot{\varepsilon}_0 = 1\,s^{-1}$.$T^{*m} = (T - T_r)/(T_m - T_r)$, in which T^* is the homogeneous temperature, T_r, T_m are room and meltingtemperature, respectively. Five material constants are A, B, C, n and m. A is the yield stress; B and exponent n represent the strain hardening effects of material. The expression in the second set of brackets represents the strain rate effect through constant C. the third bracket is to incorporate the temperature effect into the material model.

3. Modelling and Simulation

3.1 Finite Element Modelling of the roll forming process

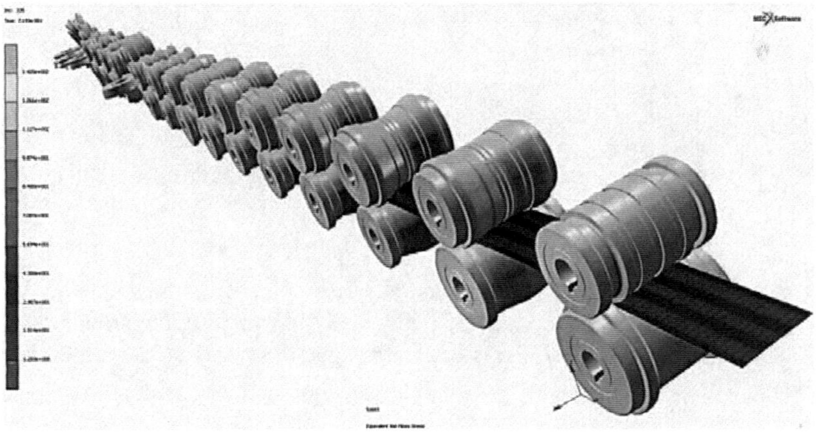

Figure 2. 3D finite element model of the roll forming process

The three-dimensional FE simulation has been carried out using MSC.Marc 2012. Complete geometry of real roll forming process has been created as shown in Figure 3. This complete roll forming process comprises 25 forming stands. There are two options available to simulate such process, first is to rotate the rollers with friction accounted, which move the strip forward and form it. Second is to keep the strip in place and move the non-rotating rollers with constant speed over the material without friction. This speaks to the idea of rolling without slipping between rollers and the surface of material. As for second technique Sheu [12] simulated a six stand CRF, in which the tools were moved over a fixed end blank at a defined speed. It was found that the resulted motion is the same, but boundary conditions are easier to specify. In this study the second approach has been opted.

To describe the contact condition and the interaction between the tools and the sheet, the analytical surface of rollers are defined as rigid bodies, and the element-based surface as the deformable body. In a sheet material, thickness is significantly smaller than other dimensions and the through thickness stress is negligible; hence shell and membrane elements can be used to model the sheet. In this simulation a bilinear thick shell element (Element 75) [13] has been used to model the sheet with nodes in the centre plane and 13 integration points in the thickness direction. Figure 4 depicts mesh in the width direction is distributed in a way to have finer mesh around bending regions. Mesh size in longitudinal direction is quite coarse and there is no remeshing involved in the simulation. In total the model comprise 6700 elements 67 transverse and 100 longitudinal elements. The simulation is performed at different speeds, 300, 1000, 5000, 15000 mm/sec.

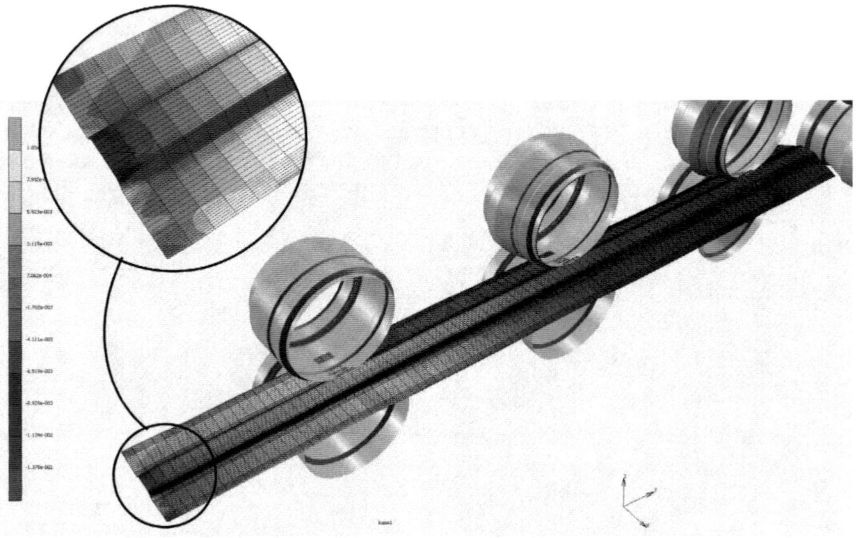

Figure 5. 3D finite element model of the strip and forming stands

3.2 Material:

The material is modelled as an elasto-plastic material. Isotropic hardening and the Von Mises yield surface and the Johnson –cook flow rule are used. For the cold roll forming simulation the influence of temperature is neglected. The set of J-C material constants for mild steel modelling is taken form literature, Vedantam[14] as equation.(2).

$$\sigma = \left[217 + (234)\varepsilon^{(0.643)}\right]\left[1 + (0.0756)ln\dot{\varepsilon}^{*}\right] \qquad (2)$$

Summary of the material and geometry parameters employed for the sheet are listed in Table 1.

Table 1. Summary of simulation parameters

Material	Mild Steel
Material Model	Isotropic Elasto Plastic, Johnson-Cook
E	210 GPa
ν	0.3

A	217 MPa
B	243 MPa
C	0.0756
n	0.643
Sheet Thickness	1 mm
Strip Length	850 mm
Velocity	300, 1000, 5000, 15000 mm/s
Friction Coefficient	0.0
Roll Material	Rigid

3.3 Boundary Condition

Two different boundary conditions have been used in the simulation, the first is defined to suppressed the front edge in longitudinal direction, and second; suppression of translation in out of plate direction of a line of nodes which do not move during the forming process. This choice of boundary conditions represents the situation in which a plane remains a plane during the deformation.

4. Results and analysis

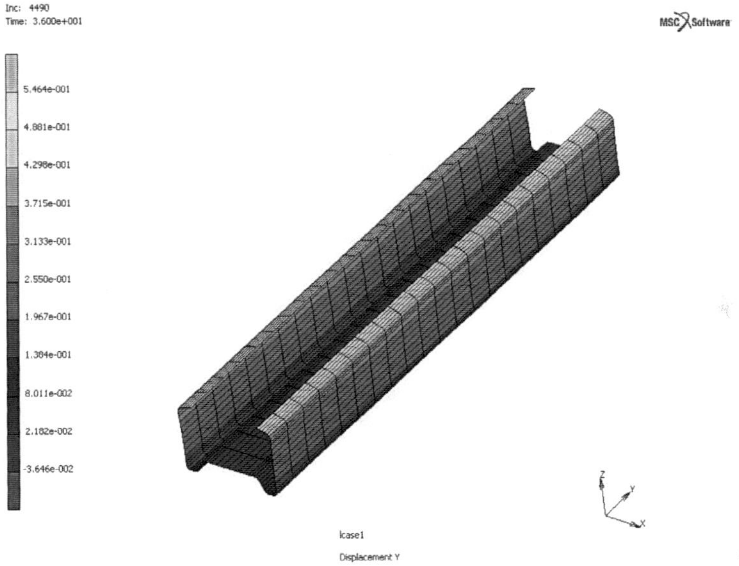

Figure 6. The geometry of final product

Figure 4 shows the final geometry of the product at the end of analysis. Based on the simulation the longitudinal membrane strain developed in the material is shown in Figure 5. As we may observe the longitudinal strain as the plates passes the forming stand gradually accumulates to the final residual values. It is readily seen that the fluctuations are in part due to elastic part of the strain. The sharp rises take place when the element came in proximity of forming stand, then it settles back. In the roll forming process of a U channel, web is the centre of a strip, while the flange is the part to be bent. The web first undergoes compression and then becomes tensile as it leaves the stand, while the flange first becomes tensile and then is compressed at the roll exit. This is the mechanism behind the residual strain in the finished product [7, 8]. In the current case the complexity of the cross section and numerous forming stands, cause more complex behaviour in the material.

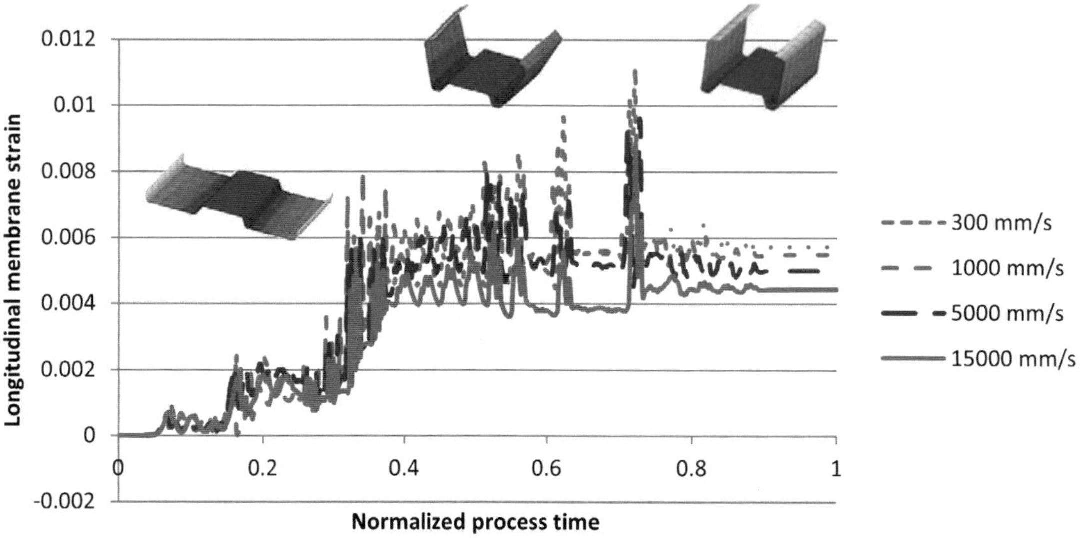

Figure 7. Longitudinal membrane strains at various process speeds

As for the comparison obviously Figure 5 is demonstrating the existence of a permanent deformation between 0.4 % to 0.6% in the final product depending on the speed level. The lower the speed is the more residual plastic strain observed in the model. This is in agreement with the result of [1,2,3] in a sense that lower strain rate constitute lower yield point, and consequently higher strain rate delays the material yield. Figure 6 demonstrate the rotation along lateral direction. The result shows that changes in the process speed only have a slight influence on the spring back behaviour of the product.

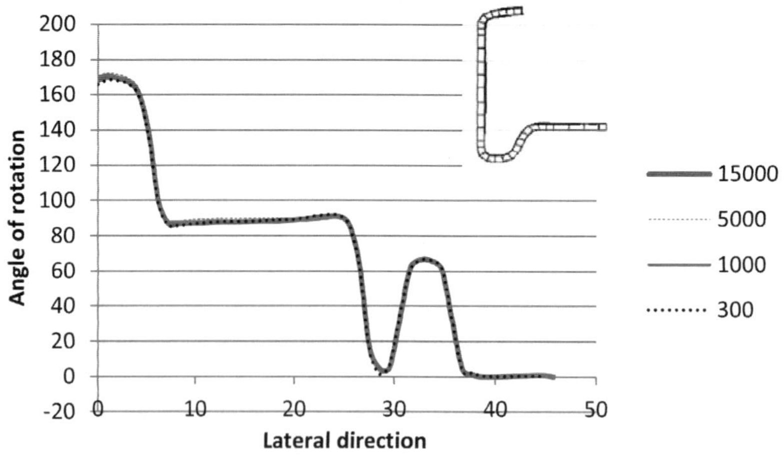

Figure 8. Angle of rotation along lateral direction

5. Conclusion

This study is focussed on a strain-rate effect in longitudinal strain of the CRF process of steel sheets. The strain-rate dependency of a plastic behaviour observed in most metals can affect the finished product's quality as well as process parameters hence the influence of the forming speed on the strain distribution should be explored. By incorporating Johnson–Cook constitutive model, the longitudinal strains induced in the roll forming operation has been investigated. In a 3D finite element analysis of complete set of forming stations with varying process speed, the strain rate impact on longitudinal peak strains was investigated. The results highlight the influence of speed in longitudinal strains and subsequent impact that can be observed in the geometry of the product. The impact of process speed on spring back was observed to be insignificant.

6. Acknowledgement

The authors are grateful to Hadley Group, for the technical support and data.

References

[1] Lindgren M, (2007 *J. Mater. Process. Technol.* **186** 77-81
[2] Han Z W, Liu C, Lu W P, Ren L Q and Tong J 2005 *J. Mater. Process. Technol.*, **159** 383-388
[3] Zeng G, Lai X M, Yu Z Q and Lin Z Q 2009 *J. Iron and Steel Research, Int.* **16** 32-37
[4] Bui Q V and Ponthot J P 2008 *J. Mater. Process. Technol.* **202** 275-282
[5] Wiebenga J H, Weiss M, Rolfe B and Boogaard A H 2013 *J. Mater. Process. Technol.* **213** 978-986
[6] Li S H, Zeng G, Ma Y F, Guo Y J and Lai X M 2009 *Thin-Walled Structures*, **47** 505-513
[7] Jeong S H, Lee S H, Kim G H, Seo H J and Kim T H 2008 *J. Mater. Process. Technol.* **201** 118-122
[8] Tehrani M S, Hartley P, Naeini H M and Khademizadeh H 2006 *Thin-Walled Structure*, **44** 184-196
[9] Paralikas J, Salonitis K and Chryssolouris G 2010 *Int. J. Adv. Manuf. Technol.* **47** 1041-1052
[10] Zeng G, Li S H, Yu Z Q and Lai X M 2009 *Materials & Design*, **30** 1930-1938
[11] Johnson G R and Cook W H, *Proc. of the 7th Int. Symp. on Ballistics,* 1983, 541-547
[12] Sheu J J 2004 *AIP Conference Proceedings* **712** 452-457
[13] *MSC.Marc,* Volume B: Element Library, Version 2012
[14] Vedantam K, Bajaj D, Brar N S and Hill S 2006 Johnson - Cook Strength Models for Mild and DP 590 Steels *Shock Compression of Condensed Matter - 2005: Proc. Conf. of Condensed Matter. AIP Conference Proceedings* **845**, 775-778

AUTHOR INDEX

AUTHOR INDEX